RESTORATION OF AQUATIC AND TERRESTRIAL SYSTEMS

Proceedings of a Special Water Quality Session Dealing with the Restoration of Acidified Waters in conjunction with the Annual Meeting of the North American Fisheries Society held in Toronto, Ontario, Canada, 12–15 September 1988.

Guest Editors

ROBERT W. BROCKSEN
Living Lakes, Inc., Washington, D.C., U.S.A.

and

JOE WISNIEWSKI
Wisniewski and Associates, Inc., Virginia, U.S.A.

Reprinted from Water, Air, and Soil Pollution
Vol 41, Nos. 1–4 (1988)

KLUWER ACADEMIC PUBLISHERS
DORDRECHT / BOSTON / LONDON

ISBN 0–7923–0111–0

Published by Kluwer Academic Publishers,
P.O. Box 17, 3300 AA Dordrecht, The Netherlands.

Kluwer Academic Publishers incorporates
the publishing programmes of
D. Reidel, Martinus Nijhoff, Dr W. Junk and MTP Press.

Sold and distributed in the U.S.A. and Canada
by Kluwer Academic Publishers,
101 Philip Drive, Norwell, MA 02061, U.S.A.

In all other countries, sold and distributed
by Kluwer Academic Publishers Group,
P.O. Box 322, 3300 AH Dordrecht, The Netherlands.

Printed in The Netherlands

WATER, AIR, AND SOIL POLLUTION / *Vol. 41 Nos. 1-4 September/October 1988*

SPECIAL ISSUE

RESTORATION OF AQUATIC AND TERRESTRIAL SYSTEMS

Proceedings of a Special Water Quality Session Dealing with the Restoration of Acidified Waters in conjunction with the Annual Meeting of the North American Fisheries Society held in Toronto, Canada, 12–15 September 1988.

Guest Editors

ROBERT W. BROCKSEN,

Living Lakes, Inc., Washington, D.C., U.S.A.

and

JOE WISNIEWSKI,

Wisniewski and Associates, Inc., Virginia, U.S.A.

Preface VII
Acknowledgements VIII
Foreword IX

SECTION I – PROGRAM OVERVIEW

P. Nyberg and E. Thørneløf / Operational Liming of Surface Waters in Sweden 3

A. Hindar and B. O. Rosseland / Liming Acidic Waters in Norway: National Policy and Research and Development 17

D. J. A. Brown, G. D. Howells, T. R. K. Dalziel, and B. R. Stewart / Loch Fleet – A Research Watershed Liming Project 25

D. B. Porcella / An Update on the Lake Acidification Mitigation Project (LAMP) 43

R. K. Schreiber / Cooperative Federal-State Liming Research on Surface Waters Impacted by Acidic Deposition 53

D. P. Dodge, G. M. Booth, L. A. Richman, W. Keller, and F. D. Tomassini / An Overview of Lake Neutralization Experiments in Ontario 75

R. W. Brocksen and P. W. Emler, Jr. / Living Lakes: An Aquatic Liming and Fish Restoration Demonstration Program 85

R. F. Huettl / 'New Type' Forest Declines and Restabilization, Revitalization Strategies: A Programmatic Focus 95

SECTION II – LAKES

J. M. Gunn, M. J. McMurtry, J. M. Casselman, W. Keller, and M. J. Powell / Changes in the Fish Community of a Limed Lake near Sudbury, Ontario: Effects of Chemical Neutralization of Reduced Atmospheric Deposition of Acids? 113

T. B. Adams and R. W. Brocksen / Dose-Response Relationships for the Addition of Limestone to Lakes and Ponds in the Northeastern United States 137

B. O. Rosseland and A. Hindar / Liming of Lakes, Rivers and Catchments in Norway 165

H. Sverdrup and P. Warfvinge / Lake Liming in Different Types of Acid Lakes Using Various Types of Calcite Powder and Methods 189

P. A. Bukaveckas / Effects of Lake Liming on Phytoplankton Growth in Acidic Adirondack Lakes 223

S. P. Gloss, C. L. Schofield, and R. E. Sherman / An Evaluation of New York State Liming Data and the Application of Models from Scandanavian Lakes to Adirondack Lakes 241

M. D. Marcus / Differences in Pre- and Post-Treatment Water Qualities for Twenty Limed Lakes 279

P. L. Riely and D. B. Rockland / Evaluation of Liming Operations through Benefit-Cost Analysis 293

SECTION III – STREAMS

T. I. Ivahnenko, J. J. Rento, and H. W. Rauch / Effects of Liming on Water Quality of Two Streams in West Virginia 331

A. Janicki and H. S. Greening / The Effects of Stream Liming on Water Chemistry and Anadromous Yellow Perch Spawning Success in Two Maryland Coastal Plain Streams 359

D. E. Arnold, W. D. Skinner, and D. E. Spotts / Evaluation of Three Experimental Low-Technology Approaches to Acid Mitigation in Headwater Streams 385

SECTION IV – WATERSHEDS

D. J. A. Brown / The Loch Fleet and Other Catchment Liming Programs 409

T. R. K. Dalziel, M. V. Proctor, and A. Dickson / Hydrochemical Budget Calculations for Parts of the Loch Fleet Catchment before and after Watershed Liming 417

J. E. Davis / The Use of Simulation Models as Resource Management Tools for Restoring Acidified Waters 435

R. W. Brocksen, H. W. Zoettl, D. B. Porcella, R. F. Huettl, K-H. Feger, and J. Wisniewski / Experimental Liming of Watersheds: an International Cooperative Effort between the United States and West Germany 455

References 473
Index of Authors 495
Subject Index 499
Announcement 502

Founded in 1971, with an international editorial board of 49 scientists, *Water, Air, and Soil Pollution* is an interdisciplinary journal for all the physical and biological processes affecting our flora, air, water, and solid earth in relation to environmental pollution. Because of its wide scope, the topics are diverse and include all aspects of sources, transport, deposition, accumulation, disposition, and mitigation of acid precipitation, atmospheric pollution, heavy metals, water pollution, waste water, wastes, pesticides, soil pollution, sewage sediments, etc.; effects of pollutants on human beings, vegetation, fish, aquatic species, microorganisms, animals, etc.; ecological implications of pollution; and pollution models.

Water, Air, and Soil Pollution is published monthly (double issues).
Subscription prices, per volume: Institutions $ 140.00, Individuals $ 57.00.
Second-class postage paid at New York, N.Y. USPS No. 913–880.
U.S. Mailing Agent: Expediters of the Printed Word Ltd., 515 Madison Avenue (Suite 917), New York, NY 10022.
Published by Kluwer Academic Publishers, Spuiboulevard 50, P.O. Box 17, 3300 AA Dordrecht, Holland, and 101 Philip Drive, Norwell, MA 02061, U.S.A.
Postmaster: please send all address corrections to: Kluwer Academic Publishers Group, c/o Expediters of the Printed Word Ltd., 515 Madison Avenue (Suite 917), New York, NY 10022, U.S.A.

PREFACE

The proceedings of a special technical session dealing with the addition of Ca based materials to waters and soils for the purpose of restoring or enhancing fisheries in acidified waters are contained in this volume. The session was part of the North American Fisheries Society's annual meeting sponsored by the Water Quality Section of the Society. While the concept of liming is not new, the extent of the recent research, demonstrations and applications reported at the conference is enlightening. Six countries are represented by government, state and private sector participants.

The scope of activities reported in the following 23 papers range from the simple to the very complex addition of liming materials to ponds, lakes, streams and watersheds. This volume of presentations is broad and impressive as are the methods used in the actual application of the neutralizing materials. Those programs that are demonstration application oriented are deploying techniques that are either new or modifications of existing technology.

The scope of the science reported in this volume is equally impressive. Subjects covered range from an Operational Program to the evaluation of the experimental low-technology applications to headwater streams. The apparent benefits to fish, as a result of neutralization, is reported as is the approach to calculating the estimated cost/benefit of applying an application of limestone to surface waters. Evaluation of changes in trophic levels containing biological organisms other than fish are reported, as well as the changes in complex hydrological budget calculations for parts of a catchment before and after watershed liming.

Also contained in this publication are the complex and often conflicting perceptions as to the reasons for applying liming materials to surface waters, catchments and watersheds. To the resource manager charged with the programmatic task of maintaining or enhancing his resource, be they fish or trees, liming is but one of many resource management tools that make scientific and economic sense. To others whose charge is maintaining an equitable balance between the social, political, economic and environmental desires of society, any simple resource management tool that hints of panacea will be viewed with suspicion.

To the reader who only scans the introductions contained in the papers included in this volume, it will quickly become apparent that the authors collectively do not represent liming as a panacea. Several of the papers present strict criteria for lakes and streams. These criteria are based upon years of collective research that indicate the greatest possibility for the successful treatment of a water body, and at the same time preclude selections of waters where liming may exacerbate a problem such as elevated levels of heavy metals. The successful demonstration of any technique or technology necessarily includes an extensive evaluation of both successes and failures, and by doing so, a basis is formed for reasonably selecting bodies of water that should be treated with neutralizing materials. If a body of water is to be treated, decisions on the best technique to be used can be made. With critical evaluations of success and failure, reasonable predictions for achieving desired results is possible. Many of the papers in this volume reflect the programmatic approach described above.

This volume can be used as an excellent example for students and practitioners of resource planning and management. Contained in this volume is documentation of the awareness of an environmental program; that is a discussion is undertaken of the acidity levels in surface waters and soils that impact the ability of biological organisms to reproduce and grow in a desired and healthy manner. The response to that problem was the re-evaluation of resource management practices using neutralizing agents, known otherwise as liming. From this evaluation, scientists and resource managers have defined areas of needed study and development and set about achieving answers as is well-reflected in the collective papers presented here.

The continuation of certain areas of study may produce results that present society with even better and more cost-effective approaches to neutralization of soils and waters. The watershed application research efforts described in this volume are a key example. If, as is hypothesized, the application of calcium based materials to watershed soils can result in a positive effect to the vegetation and a significantly longer lasting neutralization of receiving waters, and if liming in conjunction with fertilization can benefit forest productivity without any significant effects on the ecological balance of these systems, then we may be approaching a new era in environmental management and mitigation.

Dr. Robert W. Brocksen
Executive Director
Living Lakes, Inc.
and
Adjunct Professor
University of Wyoming

Mr. Joe Wisniewski
President
Wisniewski and Associates, Inc.

ACKNOWLEDGMENTS

The editors would like to acknowledge the input from all of the authors who contributed papers for this book. We also acknowledge the continual excellent guidance and advice of Gerrit Kiers and Billy McCormac. Special acknowledgment is given to James Davis for his contribution in co-chairing the conference from which these proceedings emanated. Overall appreciation is extended to Ms. Jennifer Smith who performed the many administrative and clercial duties associated with coordinating the papers for this publication as well as to Ms. Celeste Barnes for her assistance. Our sincere gratitude is extended to the many sponsors and members who fund and support Living Lakes, Inc., the number of participants being too lengthy to list here.

FOREWORD

The publication of this symposium proceedings, covering the liming of surface waters and watersheds to mitigate the effects of acidification, is timely and informative, and it could even be somewhat controversial. It is timely because the extensive experience with liming in Europe and North America has now matured sufficiently to warrant this summary. And the symposium and proceedings volume have been presented in time to fill a very real need for an evaluation of mitigation methods in the U.S. National Acid Precipitation Assessment Program's final report.

It is particularly informative because of the breadth of experience and perspective presented. Experience with liming programs gained over the past decade in six different countries has been summarized by key researchers, resource managers and administrators. Many different liming technologies have been covered, including very useful information on hydrology, chemistry, biology and economics.

But this symposium and proceedings could be controversial as well, because some might argue that efforts to evaluate surface water and watershed liming simply confuse the current discussions about controls for acidic emissions. That view, however, could only be founded on a very limited political perspective and would ignore the probable need for mitigative strategies no matter what legislative or regulatory actions are taken. The symposium participants and sponsors should be congratulated not only for their efforts but also for their demonstration that responsible scientific inquiry should not be hampered by controversy or by limited perspective.

Dr. Harold L. Bergman
Professor of Zoology and Physiology
The University of Wyoming

SECTION I: PROGRAM OVERVIEWS

Programmatic overviews describing the scope of liming/fertilization are presented in the following eight papers. Six countries are represented and the programs described vary in funding from a modest $200,000 U.S. per year to an annual budget of $7,000,000 U.S. The extent and intensity of the applications, demonstrations and research presented are as different as is the funding for these programs. In this section of overviews the reader will be exposed to most of the current research and technology associated with the application of calcitic materials for the purpose of neutralization of soils and water bodies.

OPERATIONAL LIMING OF SURFACE WATERS IN SWEDEN

Per Nyberg
Eva Thørneløf

ABSTRACT

From 1977 to 1982, liming of acidified waters was performed for a trial period using governmental subsidies. Based on the experience gained, an operational liming program was started in 1982 by the National Environmental Protection Board of Sweden. All waters with pH of less than 6 and/or an alkalinity of less than 0.05 meq L^{-1} have been eligible for subsidies. Individual water owners as well as organizations and federal agencies may apply for subsidies. Normally 85% of the costs are covered by government grants. Each county administration draws up a 5 yr liming plan and applies yearly for subsidies on a county basis. Since 1977, SEK (Swedish Kroner) 373 million have been spent on liming operations and SEK 33 million on administration and follow-up studies. A yearly budget of about SEK 110 to 130 million has been proposed for 1988 to 1991. The liming agent used has been powdered $CaCO_3$ with a particle size normally of 0 to 0.2 mm. At present, about 4,000 lakes have been treated. The results of the treatments are normally evaluated by analyzing water samples taken twice a year. These are analyzed for pH, alkalinity, conductivity, color, Ca and Mg. In addition, biological surveys are carried out in selected lakes and streams.

1. INTRODUCTION

A survey of the water quality of 6,908 Swedish lakes was conducted during the winter of 1985. It was calculated that about 16,000 lakes out of the 85,000 lakes in the country were acidified by airborne pollutants (Bernes and Thörnelöf, 1988). In addition, approximately 90,000 km of streams and rivers were found to be affected by acidification (Dickson, 1985). The survey also showed that some 7,000 lakes were acidic for other reasons, mostly due to the influx of humic water (Monitor, 1986).

Acid deposition has been known to seriously affect biota of surface waters in Scandinavia. The first effects were noticed in a Norwegian salmon hatchery at the beginning of this century (Leivestad et al., 1976). In Sweden, the problems had become severe beginning in the 1970's and the public began to be aware of the threat. Water owners and fishermen

Water, Air, and Soil Pollution **41** (1988) 3–16.

found that what had been earlier healthy populations of crayfish (*Astacus astacus*), roach (*Rutilus rutilus*), brown trout (*Salmo trutta*), Altantic salmon (*Salmo salar*) etc. were now declining or had already disappeared. This evidence combined with a scientific basis for the effects of acid deposition (Almer, 1972; Almer et al., 1978) influenced the Ministry of Agriculture in 1976 to commence a trial period of liming selected acidified surface waters using a system of government subsidies. Trials were run during 1977 to 1982 by the National Board of Fisheries, and some 1,100 lakes were treated. The results from the trial period showed that successful treatments, treatments resulting in improved water quality lasting over a sufficient period, had positive effects on the ecosystem. In addition, no adverse effects were observed (Fiskeristyrelsen and Statens Naturvårdsverk, 1981; Nyberg, 1984).

Based on these facts, in 1982 the Government decided to take further steps and to start liming on an operational basis. Technical and administrative results from the Swedish liming program, including the trial period as well as the operational phase, have been presented earlier (Bengtsson et al., 1980, Dickson, 1985; Fraser et al., 1985; Lessmark, 1987; Lessmark and Thörnelöf, 1986; Nyberg et al., 1986a).

The liming program is a component of the Swedish *Action Plan* against air pollution. The first Action Plan, covering a 3 yr period was accepted by Parliament in 1985 (National Environmental Protection Board, 1985) and the second Action Plan was accepted in the spring of 1988 (Acid Magazine, 1988a,b). The plans are focused on reducing the emissions of primarily S, but also NO_x, hydrocarbons and ammonia. In addition to the liming program, the Action Plan includes other activities to reduce the negative effects of acidification. For example, special grants are available for fishery management in limed waters that have been restored from a chemical point of view.

Lakes with low pH values have also been shown to have increased Hg concentrations in fish tissue (Bjorklund et al., 1984; Håkansson et al., 1988). Calculations indicate that about 10% of the lakes in Sweden should be black listed due to high concentrations of Hg in Northern pike (Håkansson et al., 1988). As another part of the Action Plan, an extensive applied research project is being carried out in a total of 96 lakes, to study the possibility of reducing the accumulation of Hg by different liming techniques, fertilization, intensive fishing and Se treatment (Häkansson, 1986a,b).

Acidification has also affected the groundwater, the soils and the forests in Sweden. Extensive research and monitoring, as well as experimental measures including liming, are also being carried out in the groundwater and the terrestrial ecosystem.

2. RESULTS AND DISCUSSIONS

2.1 The Aim and Organization of Liming Measures

Adverse effects due to acidification may be expected in surface waters of pH less than 6 (e.g., Engblom and Lingdell, 1983; Appelberg, 1986; Degerman, 1987). These effects are mostly due to high concentrations of toxic metals, mainly Al, but Fe and Mn may also be toxic (Anderson and Nyberg, 1984). The purpose of the liming measures in Sweden is, from a chemical point of view, to raise the pH and alkalinity to above 6 and 0.1 meq L^{-1}, respectively and, from a biological point of view, to detoxify the waters so that the naturally occurring flora and fauna can persist or recolonize the limed water.

Since 1982, the National Environmental Protection Board (NEPB) has had the major responsibility for liming activities in Swedish surface waters. This shift from the National Board of Fisheries, which lead during the trial period, was due to the fact that liming is necessary for natural resource conservation in general and not only for fisheries. The work is, however, run in cooperation with the National Board of Fisheries (NBF).

Application for liming subsidies can be made by municipalities, associations or single water owners. Application is made to the respective county administrative boards. Each board has a working group on liming measures with representatives of water owners, angling associations and the county fishery consultant. The county boards apply for grants for the whole county to the government. All counties have a running 5 yr liming plan which is updated every year based on the follow-up results, lake surveys and biological inventories. The NEPB, in cooperation with the NBF, acts as referee from a national point of view and proposes to the Government how the available grants should be allocated and shared between the different countries. The projects may concern single and very small lakes to large and whole catchments, including seepage and drainage lakes, streams and rivers.

It is possible to obtain government grants for liming acidified lakes with an alkalinity of less than 0.05 meq L^{-1} during summer and/or winter, i.e., hydrologically stable periods, and for acidified running waters which have an alkalinity below this concentration anytime during a year. The grants cover 85% of the costs for planning, purchasing of lime, transport and liming.

Some projects which have a high value from a fishery or resource conservation point of view -- salmonid-spawning streams and also projects which are especially difficult to finance may receive 100% subsidies.

All liming operations within each project -- methods, liming materials used, areas and rates of applications, costs and effects, i.e., results of analyses of water samples and other follow-up studies -- have to be reported yearly to a central database at the NEPB. The NEPB normally evaluate the data once a year and provide the county administrative boards, as well as single individuals or international organizations, with compilations of the data on request.

2.2 Characteristics of the Lakes Chosen for Liming

The median treated lake has been relatively small surface area (22 ha) and has a low maximum depth (9 m) (Lessmark and Thörnelöf, 1986). Most of the lakes also have retention times of less than 1 yr. Sixty treated lakes, however, have retention times of more than 5 yr (**Figure 1**). The median pH value of the lakes limed after the trial period is 5.7 before treatment. Alkalinity and conductivity are low (0.02 meq L^{-1} and 5.0 mS m^{-1}, respectively), and the lake water contains humic substances (water color 70 mg Pt L^{-1}) (data from the NEPB). These are median values and the limed lakes include extremely transparent waters (color < 5 mgPt L^{-1}), as well as very brown ones (<500 mgPt L^{-1}). All lakes are oligotrophic and normally inhabited by 3 to 6 species of fish before treatment.

2.3 Liming Material and Rates of Application

During the trial period, a number of liming agents used to neutralize acidity were tested. For different reasons, e.g., high costs, too high a content of organic material, metals or other

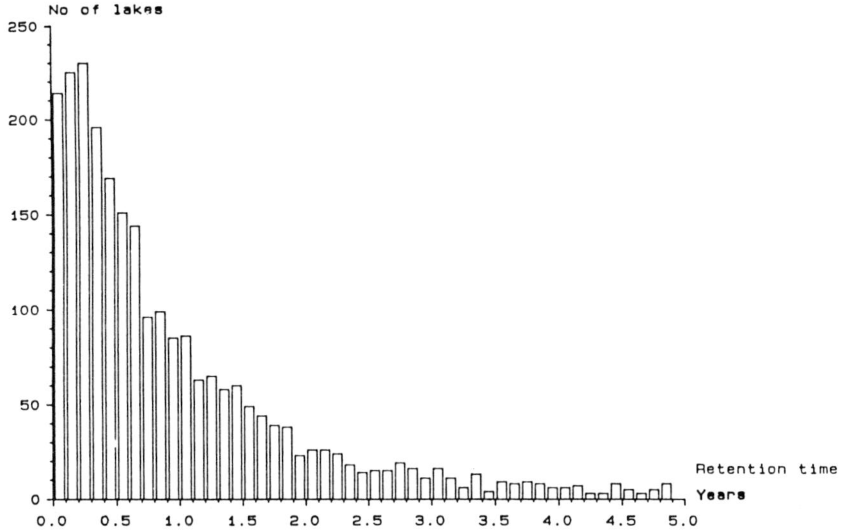

Figure 1: Theoretical retention times of the limed Swedish lakes (n = 2520).

(Sixty lakes with a retention time > 5.0 yr are not included in the figure.)

contaminants or too high solubility, practically all agents, except $CaCO_3$, are now excluded from use. About 80% of the lime used in Sweden is finely ground $CaCO_3$ with a particle size of less than 0.2 mm. The balance is almost exclusively a slightly coarser product (0.0 to 0.5 mm), while all other limestone products are of minor importance (**Figure** 2).

Pure limestone is a very clean product with low concentrations of metals. However, limits have been set for the highest acceptable concentrations of metals and other contaminants in all liming agents used on surface waters (Dickson, 1985). The recommended

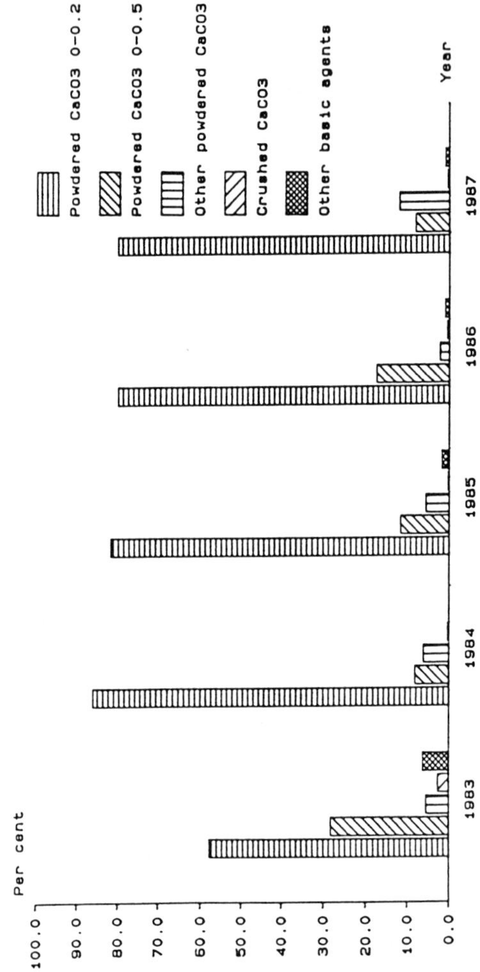

Figure 2. The relative distribution of the different lime agents used in Sweden.

product is finely ground $CaCO_3$ (0.0 to 0.2 mm) with a CaO content of about 50%. The dosage recommended varies between 9 g m^{-3} in extremely acidic lakes (pH < 4.5) with short turnover times (0.3 yr) to 10 g/m^3 in less acid lakes with long retention times (**Table I**). High dosage, however, should be avoided since a pH above nine may be deleterious to the fauna (Nyberg et al., 1986a). In lakes that require high dosages, it is recommended to apply the lime several times instead of giving high single dosages. In addition, lakes with a very short retention time are better treated by a combination of lake and seepage liming, and if necessary, treatment of inlet waters. When treating running waters, dosages of 10 to 30 g m^{-3} of even finer ground (0.0 to 0.1 or 0.0 to 0.2 mm) $CaCO_3$ are suggested, depending on the pH level.

Table I

Recommended dosages of $CaCO_3$ (g m^{-3}) (50% CaO) for the liming of lakes with different pH, alkalinity and retention time.

pH before treatment	Alkalinity before treatment (meq L^{-1})	Retention Time (yr) <0.5	0.5–0.9	1-3	>3
		Dosage (g m^{-3})			
5.6 - 6.0	<0.05	30	20	15	10
5.1 - 5.5	0	45	30	20	15
4.6 - 5.0	0	60	40	30	20
4.3 - 4.6	0	75	50	40	25
4.0 - 4.2	0	90	60	50	30

The aim of the liming measures is that no treated body of water should have a pH and alkalinity of less than 6.0 and 0.05 meq L^{-1}, respectively. The duration of a treatment is generally 2 to 3 times the retention time. The majority of the lakes have fast turnover rates which means that they have to be treated repeatedly. The dosages vary between 10 to 30 g m^{-3} depending on the retention time. The actual time for reliming, so far, has been based on the results of chemical samples. Different kinds of models for the calculation of the effects of a treatment, the duration time for reliming, etc. are available or are under development. They have not yet been used, however, to any great extent in the operational liming, but should be valuable tools in the future, especially for the planning and follow-up of whole catchment treatments.

Liming on suitable wetlands -- seepage areas -- will be done more frequently in the future (see below). Knowledge about necessary dosage levels and treatment areas required to achieve sufficient and lasting effects in the surface waters adjacent to the wetlands, however, is less developed than for direct treatment in lakes and streams. The present recommendations are that 0.3 to 1.0 t ha^{-1} of limestone powder of drainage area should be used in the northern and southern part of the country, respectively. The dosage is varied according to the amount of water drainage through the limed areas. For good effects in running waters, it is recommended that 3 to 50 t ha^{-1} be distributed on seepage areas.

2.4 Area and Methods of Application

Since 1983, most of the liming operations have been centered directly on lake surfaces, while a lesser amount has been applied directly into running waters. Liming on suitable land areas -- seepage areas -- was negligible in 1983, but has slowly increased (**Figure 3**). This method is expensive compared to other methods, especially when calculating the cost per ton of dissolved and leached calcite. The method, however, is favorable for other reasons, especially since the leaching of potentially toxic metals from the soil will decrease, and there is no risk for mechanical failure as with automatic dosers. Liming on seepage areas does bring about changes in the terrestrial flora and, hence, the planning of such treatments has to be done in close contact with the county administrative board to avoid treating areas of special interest.

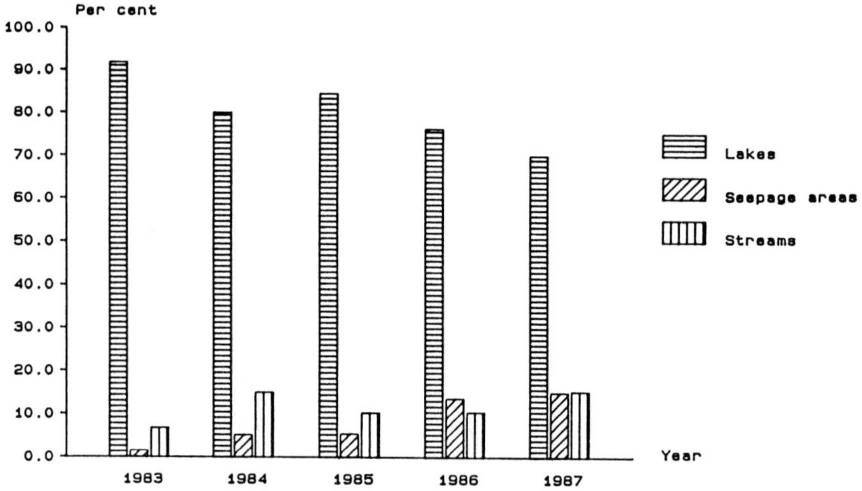

Figure 3: The relative distribution between different target areas treated with lime.

The fact that most liming has been carried out on lake surfaces is also reflected by what methods have been developed for application. More than half of the calcite has been spread by different kinds of floating equipment, mostly pontoon boats. The limestone powder is generally mixed with lake water and pumped out. About 20% $CaCO_3$ has been applied by different types of vehicles, generally trucks and tractors. These vehicles use compressed air and the lime is blown dry over the target areas. Smaller amounts of calcite have been distributed using automatic dosers, diversion wells and helicopters (**Figure 4**).

2.5 Extent of Liming and Costs

During the trial period, 1976 to 1982, about SEK 71 million (SEK 1= US$0.17) were spent on the treatment of some 1,100 lakes. Three thousand lakes have been treated since then. The

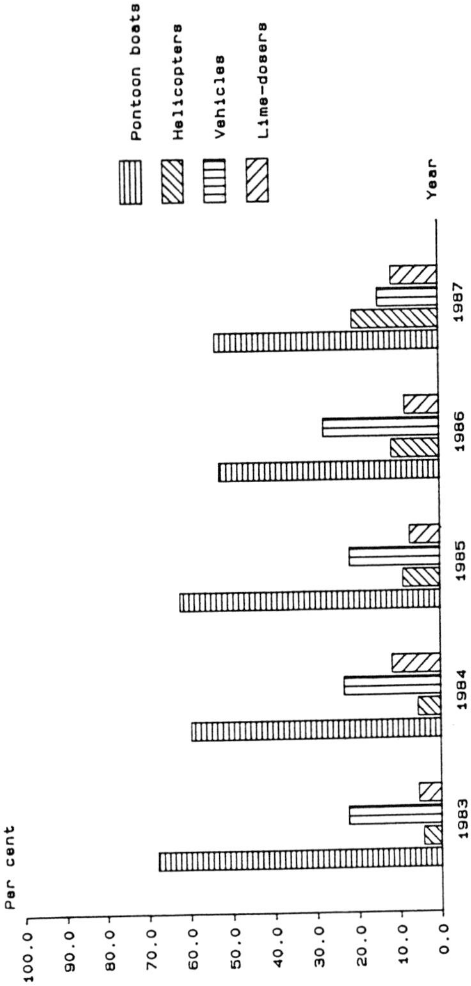

Figure 4. The relative distribution of different methods for lime application expressed as percent of the total yearly amount of lime applied.

yearly cost is currently about SEK 100 million. Approximately 75% of the total acidified lake area in the country (3,800 km^2) had received grants by 1986. The amount of $CaCO_3$ used for the liming of surface waters was 120,000 t in 1985 and will probably be slightly less than 200,000 t in 1988.

The total cost for a liming project is composed of three parts:

-Costs for the $CaCO_3$;
-Costs for the transportation of the lime; and
-Costs for applying the material.

In an ordinary lake project, these costs are about equal, and the total cost in the central and southern part of the country per ton of $CaCO_3$ is SEK 300 to 450 if the lime is applied by vehicles or floating equipment. In the northern part, liming is about twice as expensive due to higher transporation costs. The corresponding cost when using a helicopter is SEK 650-850. The costs for distributing the calcite into running waters using automatic dosers vary and depend on how much lime is distributed yearly in relation to the investments, e.g., the price of the doser, construction, electricity, costs for re-filling and maintenance (Tideström, 1984). In larger streams and rivers with continuous dosing throughout the year, the costs are comparable to other methods of application, while in smaller streams continuous dosing is an expensive method.

2.6 Follow-Up Studies, Monitoring and Applied Research

Within the area of surface water liming, there are different studies being carried out with three categories reflecting the level of ambition:

1. Regional follow-up water quality studies of the limed waters. This is performed by the county administrative boards and consists of water samples taken twice a year and analysis of pH, alkalinity, conductivity, water color and Ca. In addition, the boards also have performed extensive chemical and biological surveys.

2. Follow-up studies and monitoring performed by the NEPB and NBF (Institute of Freshwater Research). The following studies may be mentioned:

- Monitoring of effects of liming on phyto- and zooplankton (Ekström and Hörnström, 1986);

- Monitoring of effects of liming on fish and crayfish in lakes and streams. About 120 lakes and 30 streams are being tested (Nyberg et al., 1986 a,b);

- Research on metals and metal species in water and organisms (Andersson and Borg, 1988; Borg, 1986);

- Monitoring of the status of acidification of running waters by studying indicator organisms (Engblom and Lingdell, 1983); and

- Chemical and biological monitoring of unlimed reference lakes (Ahl, 1985).

3. Applied research conducted by different universities and institutes, e.g. the multi-disciplinary whole ecosystem research at the acidified and limed Lake Gårdsjön (Anderson and Olsson, 1985).

3. CONCLUSIONS FROM PRESENT EXPERIENCES

The water quality in acidified lakes has successfully been improved by liming as follows:

- Generally positive effects in aquatic ecosystems in cases where the pH has been above six after liming (Alenäs, 1986; Nyberg et al., 1986a,b; and Raddum et al., 1984);

- Decreased toxic effects of target metals and decreased metal concentrations in fish tissue (Andersson and Borg, 1987). A temporary risk for increased accumulation of metals in benthic invertebrates is, however, evident (Andersson and Borg, 1988; Raddum et al., 1984);

- Re-introductions of decimated fish populations, e.g., Arctic char (Salvelinus alpinus), have been successful following lime treatments (Nyberg et al., 1986b; Nyberg, 1987); and

- Adverse effects of liming have only been documented in two cases after the application of a relatively highly soluble liming agent (Ca - Mg silicate) resulting in pH above 9 (Nyberg et al., 1986a).

The liming of running waters offers greater problems due to extremely fast flushing rates and great variations in water quantity and quality during the year. The effects on biota, however, are generally positive as seen from the following:

- In catchments where upstream lakes have been treated, sometimes in combination with other liming methods, a successive recovery of benthic invertebrates (Nilsson and Johansson, 1985) and fish (Nyberg et al., 1986a) has been observed in the running waters;

- Continuous dosers that are functioning properly have positively affected the benthic invertebrates and fish; damage has occurred, however, immediately downstream of faulty dosers (Engblom and Lingdell, 1985; Lessmark et al., 1986); and

- In projects where the whole catchment area, including lakes, streams and seepage areas, has been treated, the results are mostly encouraging. One such project is the River Hogvadsan, which is one of the most important spawning rivers for Atlantic salmon on the Swedish west coast. The annual spawning run, which comprised about 200 fish during the 1950's and 1960's, declined drastically in the 1970's. Liming was started in 1978 and the spawning run, recorded in a trap, has increased to levels never recorded before (ca., 1600 in 1986) (**Figure 5**) Edman et al., 1988).

4. THE FUTURE

The only long-term way to permanently cure acidified waters, of course, is to reduce the emissions and the deposition of acidifying substances. In 1985 to 1986, a Nordic group of experts established what are known as *critical loads* for various air pollutants. These are the highest loads which will not result in long-term harmful effects in the most sensitive ecological systems (Acid Magazine, 1987). To come down to these loading levels in the southwestern part of Sweden, the 1980 levels of S deposition must be reduced by 75%, that is from the present 15 to 20 kg ha^{-1} to a maximum of 3 to 5 kg ha^{-1} (Acid Magazine, 1988a,b). In Sweden, the emissions of S were reduced by 60% during 1980 to 1985 (Statens Energiverk and Statens Naturvardsverk, 1987). Estimates show that nearly 4,500 lakes, with an area of 1,500 km^2, should return to normal if S deposition is reduced by 25% (Bernes and Thörnelöf, 1988). Swedish emissions currently contribute about 10% of the S and NO_x in the south of

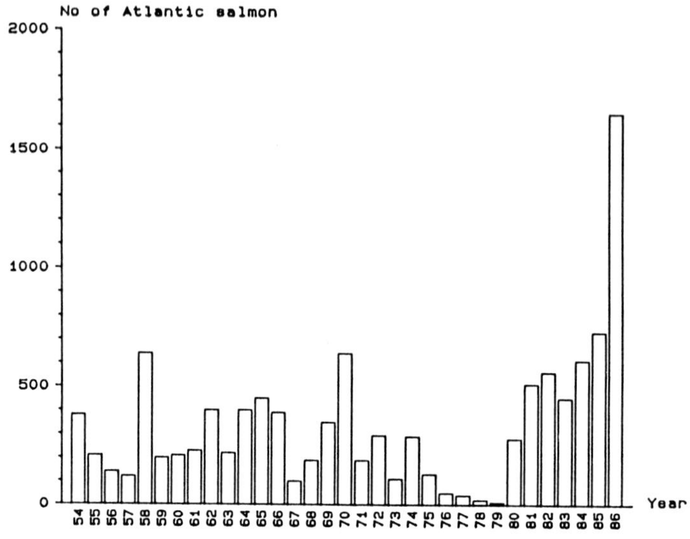

Figure 5. Catch of spawning migratory salmon in River HJogvadsan 1954-1986 (after Edman et al., 1986).

the country. So even with the most far-reaching countermeasures within the country, Sweden has to rely on European emissions also declining.

A fairly moderate deposition cutback should suffice to restore watercourses in areas where the soils still have a good neutralizing capacity. The cutbacks, however, will have to be enormous if lakes in areas of thin, impoverished soils are to be made healthy again. A 75% decrease in S deposition would gradually reduce the number of acidified lakes to somewhere around 3,500 (Bernes and Thörnelöf, 1988). Such a decrease, unfortunately, is not realistic for the near future which means that Sweden will have to continue liming.

Over the next 3 yr, it has been proposed by the NEPB that SEK 110, 120 and 130 x 10^6, respectively, should be spent on liming measures. In addition, SEK 14, 20 and 18 x 10^6 yr^{-1} should be spent during this period on fishery management, central follow-up studies, nationwide lake surveys and research.

LIMING ACIDIC WATERS IN NORWAY: NATIONAL POLICY AND RESEARCH AND DEVELOPMENT

Atle Hindar
Bjørn Olav Rosseland

ABSTRACT

Since 1983, the Norwegian environmental authorities have given financial support to operational liming and research following recommendation given by the Norwegian Liming Project (1979-1984). Liming all acid waters in Norway would require an annual expenditure exceeding NOK (Norwegian Kroner) 300 x 10^6. The funding level for 1988 is NOK 14 x 10^6 and will probably increase to NOK 30 mill. by 1990, including NOK 1 to 3 mill. for research and development. Priority is given to lakes and rivers whose fishing is open to the public, and to save or restore valuable fish populations. Local support in the form of voluntary labor is a condition for financial aid. Two salmon rivers are currently included in the program. The Norwegian Institute for Water Research (NIVA) plays an important role in liming research and development in Norway. The aims of this research are twofold: to document the chemical and biological response to neutralization and liming, and to improve liming strategy to obtain cost efficient liming activities.

1. INTRODUCTION

Acidification of lakes and rivers is among the most serious environmental problems of our time. In Norway, dramatic decreases in fish populations have been documented during the last decades, especially in southernmost Norway (Leivestad et al., 1976, Sevaldrud et al., 1980).

By 1983, 71% of the brown trout (*Salmo trutta* L.) populations and 43% of the perch (*Perca fluviatilis* L.) populations in the counties of Aust- and Vest-Agder had been lost (Sevaldrud and Skogheim, 1986), and the problems continue to expand to many parts of Scandinavia (Rosseland et al., 1985).

Liming as a remedy to counteract the effects of acid water was introduced in Norwegian hatcheries as early as the 1920's (Sunde, 1926; Dahl, 1927). Acid water was deacidified by use of limestone and shellsand. Shellsand and limestone gravel were also mixed with spawning substrate to improve the survival of fish eggs and alevins. Such simple efforts are

Water, Air, and Soil Pollution **41** (1988) 17–24.

still in use alone or as part of a larger liming strategy. The early attempts to neutralize acidic waters in Norway have been reviewed by Rosseland and Skogheim (1984).

Lime deacidifies acid water, changes the chemical properties and, thus, detoxifies Al and increases Ca. Calcium reduces the effects of toxic Al (Leivestad et al., 1980; Muniz and Leivestad, 1980; Brown, 1983).

The Norwegian Liming Project (1979-1984) intended to document "to what extent chemical countermeasures - primarily liming - are a remedy to reduce the harmful effects of acidification in lakes and rivers." Attention was paid to biological consequences, technical approaches and cost. At the termination of the project, no major objections or restrictions for liming were put forward. A practical handbook of liming acid water was prepared together with scientific documentation and a final technical report (Kalkingsprosjektet, 1985; Liming Project, 1987).

In 1983, funding was given for the first time to operational liming from Norwegian state authorities following recommendations given by the Liming Project. In 1988, the support amounts to NOK 14×10^6 (US\$ 2×10^6) of which about NOK 1×10^6 will be used for research and development.

The Norwegian Institute for Water Research (NIVA) has a long record of research activities related to acidification and liming. NIVA played a central role in the interdisciplinary research program: "Acid Precipitation - Effects on Forest and Fish" (The SNSF-Project, 1972-1980). This expertise was also used in the Norwegian Liming Project. Some of the projects still continue and new projects have been started.

In this paper we summarize events and policy making which led to the buildup of a national program for operational liming in Norway. We focus on the research program that will be prepared in 1988.

2. ADMINISTRATION OF THE LIMING ACTIVITY

From the start, in 1983, the Directorate for Nature Management (DN), formerly "Directorate for Wildlife and Freshwater fish" is responsible for the liming activities in Norway, both the operational liming and the research activities.

At the termination of the Liming Project, the directorate established a liming group in the city of Arendal. The group was to make the results from liming research available for the operational program and to make use of Norway's liming expertise. The group will probably be located in the research department of the directorate and will be moved to a new research foundation or to NIVA in 1988.

Today most of the decision-making regarding financial support is delegated to the fishery officers of the affected counties. Each county receives support from DN according to the area affected, number of inhabitants and local interest. The county authorities then distribute the financial support to projects that fulfill certain criteria.

Local fishermen associations, groups of land owners, and also some municipalities can apply for funds for liming activities. Funds are also available for stocking programs when reintroduction of fish populations are necessary. Local enthusiasm decides which lakes and rivers will be limed.

A committee consisting of state and county authorities, representatives of fishermen and land owner associations and research institutes advise DN on national liming policies.

3. NATIONAL POLICY

From the start, funding of liming operations follows certain criteria. An overall policy is: a) to improve the sport fishing possibilities in acidified areas, and b) to take care of resident fish populations. Guiding criteria for receiving support for local liming projects are:

- Fishing license is to be available for the public;
- The locality has qualities that make it attractive for sport fishing;
- A local interest is required and a plan for liming operations in the area should have been worked out;
- As few practical limitations as possible for spreading or dosing of lime is desired;

- A cost-benefit evaluation is recommended; and
- Water chemistry data is needed.

These criteria for receiving support assure reasonable use of money. If the applications for money are on the same level as the total funding, however, the possibility exists that dubious projects will be funded. Some projects have been funded in spite of inadequate information on limestone requirements, liming strategies etc.

In Sweden, support from the state authorities has increased very fast compared to Norwegian funding (**Figure 1**). The Norwegian policy has followed the recommendations from

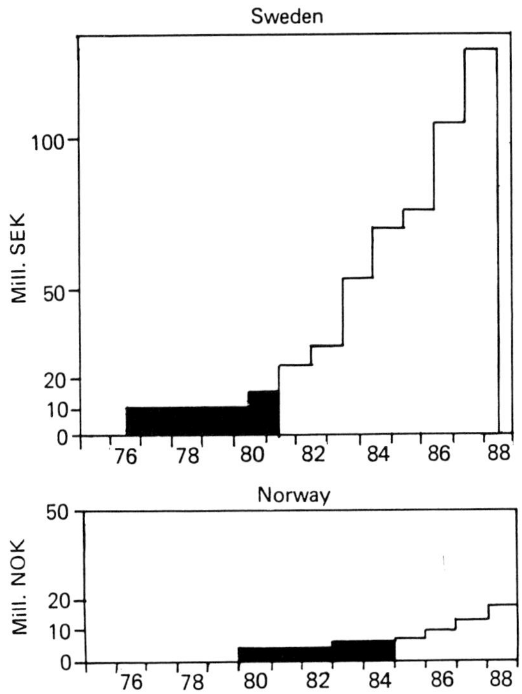

Figure 1: Annual funding of liming activities by state authorities in Sweden and Norway. Filled areas represent research programs.

the steering committee of the Norwegian Liming Project who recommended a gradual increase in the support.

The result of the Norwegian policy is that the administration of the program is gradually built up according to the increase in the funding. Routines for documentation of water quality prior to and following liming are adequate and research and development yields information regarding liming strategies and their effects. The lack of an operational system of national statistics makes it impossible, however, to evaluate the program in detail.

The question arises as to what extent the program has decreased the rate of loss of fish populations or, at least, stabilized a negative trend.

The national policy results in a wide distribution of a limited amount of money supporting several liming operations. Most of the money (about 40 to 50%) is given to the two most-affected counties in southern Norway (Aust- and Vest-Agder). When as much as 60% of the trout populations are wiped out, most of the liming is carried out in lakes and rivers that have no fish populations left. Others may have fish populations, but as in the barren lakes an additional stocking program is necessary in most cases. Stocking with trout populations from unaffected regions or from central hatcheries is common.

The limited fundings often lead to unsatisfactory limitations to the liming projects. Such effects might be that most spawning sites in running waters are not limed due to high costs. Stocking with "new" populations or even new species in limed lakes at regular intervals is often the result. Distribution of money to localities that are not necessarily optimal from a biological point of view, but that might have a low budget for liming is common.

The result is that locally adapted fish populations may still be wiped out although the possibilities of sport fishing in affected areas are improved.

Standard liming operations of surface water in the more than 33,000 km^2 acidified area in Norway would require annual expenditures exceeding NOK 300 x 10^6 (US$ 44 x 10^6). Elevated costs associated with lakes and rivers with difficult access are not considered in this estimate.

A liming program has been worked out at the Directorate for Nature Management (DN, 1987). The program recommends that support from state authorities be increased to NOK 31 x 10^6 (US$ 4.5 x 10^6) in 1990. The program states that NOK 30 to 40 x 10^6 will be adequate to deacidify catchments corresponding to 10% of the acidified area of 1980. An area corresponding to 20% is recommended to be treated in 1997 at a cost of NOK 70 to 80 x 10^6. The sums are all in 1986 money.

The Ministry of the Environment is not obligated to the program, and neither is the Government. A broad debate has not yet arisen among politicians concerning liming policy. Pressure from the public will probably decide how close to the recommendations the yearly support will be.

4. RESEARCH AND DEVELOPMENT

Limestone is widely used in agriculture, production of cement and as a bleaching agent in the paper industry. The possibilities of use in acid lakes, rivers and in hatcheries have been known for many years, but the major research activities concerning liming have been within the last 10 to 15 years. The national liming projects of Sweden, Norway, Canada and the U.S.A. have introduced liming of acid water as a branch of the acidification research.

Conclusions drawn by the Norwegian Liming Project form the basis of the liming program in Norway (Liming Project, 1987). The operational program was not initiated until the major conclusions were ready. Since then, research, development and monitoring activities have been aimed at producing adequate information to increase the quality of and the support to the liming program. Although many fundamental processes are involved, liming research is applied science. The money is allocated to research projects that are supposed to supply answers of importance to the program.

Research conducted by the Liming Project focused on the following:

- Dissolution properties of different neutralizing agents in laboratory and in whole scale liming projects;
- Ecosystem effects (phytoplankton, zooplankton, bottom fauna and fish);
- Physiological effects on different fish species;

- Technical solutions to lake liming and liming of running water; and
- Liming strategies that include the above mentioned elements.

Since the termination of the project in 1984, research and development have been at a minimum. Chemical monitoring of some lakes and rivers dominates. An exception is the Hovvatn project and the salmon river liming of the River Vikedalselva and River Audna. Systematic and qualified test fishing to document liming effects also suffers from lack of financial support.

At present NIVA has a leading position within liming research and development in Norway. The liming projects run in 1988 that involve NIVA are:

- Lake Store Hovvatn, limed in 1981 and relimed in 1987;
- The whole catchment liming at Tjønnstrond, limed in 1983;
- The River Vikedalselva, limed from 1987;
- The catchment Vegår, including a 6 km^2 lake and a salmon river. The Lake was limed in 1985, 1987 and a tributary from 1987; and
- Two smaller lakes, one limed with dolmite powder from 1985, the other with different calcite powders from 1981.

New projects that are funded from 1988 include:

- Liming of discharge areas as a measure in remote areas or in connection to lakes with short hydraulic retention time;
- Systematic testing of different neutralizing agents for application on spawning sites and brooks; and
- Optimization of a stream liming doser at the river Vikedalselva. Dosing is governed by a feedback system that involves a pH sensor about 700 m downstream from the doser.

The ongoing projects at NIVA fit well into a national research program that will be available in 1988. Three major research fields will have priority within the national plan:

- Further testing of different limestone products and application techniques to optimize the lime quantity and the cost efficiency. Water quality targets will be central. Simulation models will be developed on the basis of research in the field;

- Liming strategies that improve the water quality of spawning sites in running water. This research will aim at protecting existing populations of spawning brook trout. A system that ensures the use of local populations in stocking programs will probably be developed; and

- The study of the re-establishment of stocked salmon populations.

The research program is a result of the awareness of liming as a countermeasure in acid lakes and rivers for many decades ahead. Research and development can ensure the best use of money if other conditions are adequate within the national liming program.

ACKNOWLEDGMENT

We greatly acknowledge the valuable comments on the manuscript given by Dr. Richard F. Wright.

LOCH FLEET - A RESEARCH WATERSHED LIMING PROJECT

D.J.A. Brown
G.D. Howells
T.R.K. Dalziel
B.R. Stewart

ABSTRACT

A research watershed liming project is being carried out at Loch Fleet in the Galloway District of southwest Scotland. This acid loch originally supported a brown trout fishery, but the number of fish caught by anglers declined during the 1950's and the fishery eventually disappeared about 20 yr ago. After 2 yr of pre-treatment data collection, the first set of land lime applications took place in the spring of 1986, and the resulting surface water chemical changes are very encouraging. Increases in pH and Ca concentration and decreases in Al (especially the toxic labile monomeric fraction) concentrations have been observed in the streams and waters draining the treated subcatchments in the loch itself and in the loch outlet. Also, the normal trend for pH and Ca concentration to fall during periods of high flow in the main feeder stream has been reversed. In the spring of 1987, the loch was restocked with brown trout. Initial indications are that there has been very good survival of these fish, and that they managed to spawn during the 1987/88 winter. The chemical and biological changes brought about by the land treatments are described in detail.

1. INTRODUCTION

It is now clear that *susceptible* soils on hard geologies throughout western Europe have been accumulating acidity (and S) for many decades, possibly for 150 yr. This acidification has been exacerbated by rapid afforestation in some instances, and it is now accepted that acid deposition over the same time period has made a significant contribution to this increase in soil acidity, and the concomitant acidification of surface water.

The recovery of soil is crucial to the restoration of surface water quality acceptable to fish. This is because most of the water in streams or lakes will have first passed through at least one layer of soil where it can dissolve mineral elements. If the rain is acid and the soil has become acidified, the surface water may be too high in acidity, too high in toxic dissolved

Water, Air, and Soil Pollution **41** (1988) 25–41.

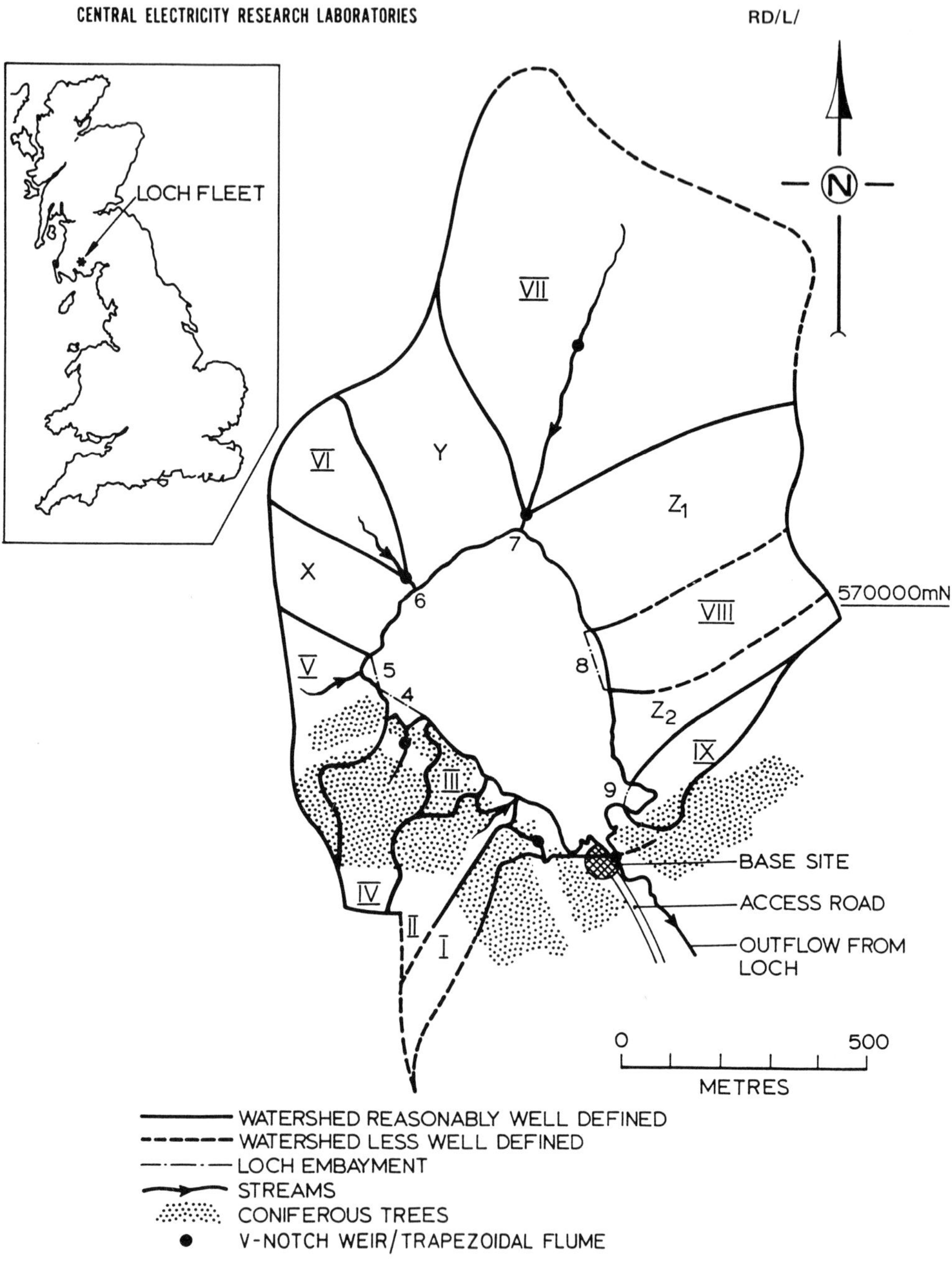

Figure 1. Loch Fleet Catchment showing subcatchments (sectors).

Al and too low in dissolved Ca to support fish life. Reductions in acid deposition are a necessary condition for reducing such levels of acidity and Al, but this will also be accompanied, in the short term at least, by the lowering of levels of essential base cations in the water (Wright et al., 1988).

The naturally slow time scale for replenishment of lost soil Ca will limit the rate of reattainment of acceptable water conditions for fish in affected areas. If rapid restoration of soil and, hence, of acidified surface waters is an overriding concern in any particular local situation, then it is possible this can only be achieved in the short term by local restoration of the basic elements removed from the soil over a century and a half of acid deposition and by commercial forestry. The Loch Fleet Project was designed to test the feasibility of such local restoration.

2. SITE DESCRIPTION, HISTORY AND BACKGROUND DATA

A full description of the site's history -- background rain, soil and surface water chemistry and biology -- is given by (Howells 1986).

The main conclusions are summarized below:

- Loch Fleet (**Figure 1**) is a small (17 ha) upland (340 m) acid loch at a remote site in southwest Scotland. It is distant from significant pollutant sources, and unlikely to be affected by land use changes during the course of the project. The underlying geology is granitic.

- The vegetation of the area, in general, has altered over past centuries from deciduous woodland to birch scrub, heath and montane species and, most recently, man-made coniferous forest.

  In the 17th century, sheep monoculture was introduced with regular muirburn to maintain a stable heath vegetation. Since 1945, large areas of the region have been densely planted with coniferous plantations by the Forestry Commission. Approximately 10% of the Loch Fleet catchment area was afforested in 1961 with sitka spruce and lodgepole pine. The remaining 90% of the 111 ha catchment is moorland in character dominated by heather (*Calluna vulgaris*) and flying bent (*Molinia caerulea*).

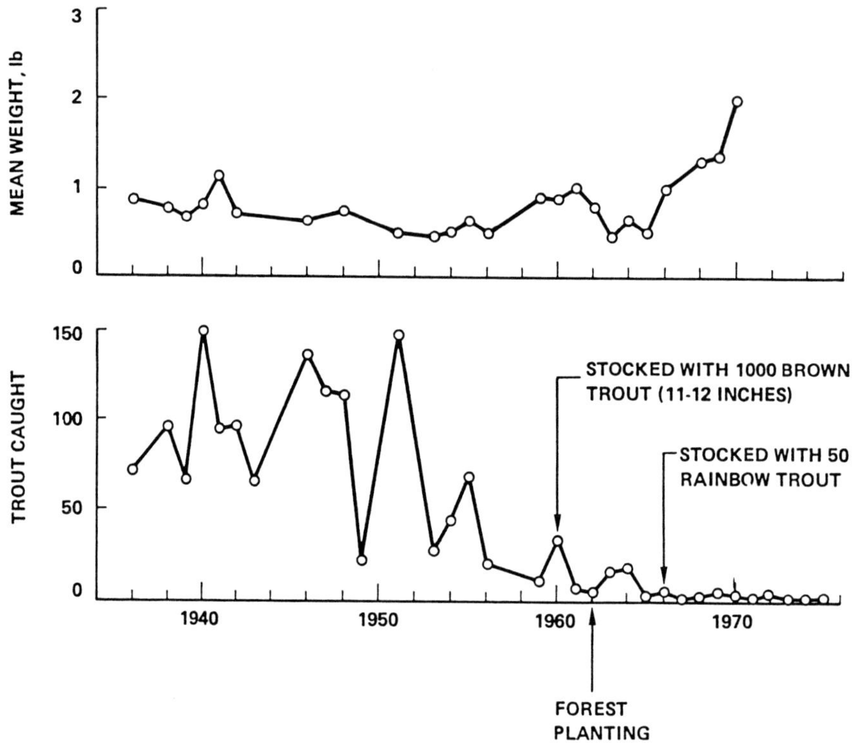

Figure 2. Loch Fleet: brown trout fishery since 1935.

- Rainfall is approximately 2200 mm yr^{-1}, predominantly (80%) from the west and southwest. Sodium chloride and other marine derived ions are the major solutes present. Mean weighted rain pH is 4.84. Deposition of acidity is estimated to be about 0.1 g $H^{+}m^{-2}$ yr^{-1}.

- The soils of the catchment, which have been surveyed by the Macaulay Institute, are shallow and almost entirely (92%) organic peats, peaty gleys, and peaty rankers. The soil pH (in $CaCl_2$) ranges from pH 2.5 to pH 4.6. It is most acid in the forested southern section of the catchment. Exchangeable bases are low and base saturation ranges from 3 to 19%.

- The loch water (mean weighted pre-treatment pH 4.4) is more acid than bulk deposition. The additional acidity is mobilized by overland and surface horizon flows (90 to 98% of total) in this (unusually) saturated catchment.

- Loch water composition is dominated by drainage from moorland areas. Drainage from the forested section of the catchment is marginally more acid, but tends to be more variable in quantity and composition. Loch turnover is estimated at 173 days.

- At one time Loch Fleet was recognized as "a connoisseur's trout fishery." In the 1930's and 1940's there was an annual catch of approximately 100 fish with an average weight of approximately 350 g, however, this declined sharply in the post-1950's (**Figure** 2).

  This population was self-sustaining since continuous recruitment via upstream migration is prevented by a waterfall approximately 7 km downstream. The loch was stocked on two occasions as shown, but from the mid-1960's, only a very few fish were caught, the last catch being in 1975.

- Electrofishing and netting surveys in 1984 and 1985 to determine the presence or absence of fish in the loch and outlet stream showed there were no fish in the loch. However, eels (*Anguilla anguilla L.*) were found immediately downstream of the loch outlet. Both eels and brown trout were present below the waterfall. During this time, the loch water had a pH range of 4.0 to 4.8, a Ca concentration of about 1 mg L^{-1} and a total Al concentration of about 200 *u*g L^{-1}. Survival tests using this water showed it to be toxic to trout.

- Other pre-treatment biota were characteristic of upland oligotrophic lakes where only 103 invertebrate taxa were present, of which one-third were beetles. Macrophytes are sparse along the loch margin and dominated by *Juncus*, *Isoetes* and *Lobelia*; *Mougeotia* is present as an epiphyte. Some *Sphagnum* is present as a benthic species in limited areas of the loch.

- Lake sediment cores, investigated by University College London, indicate a major change in the catchment at the time of pre-afforestation ploughing in 1961/62, but diatom remnants suggest that acidification did not occur significantly until about 1970. Prior to this, the pH was estimated to be about 6.0. Thus, although the loss of fish can be explained in terms of present water quality, their decline was evident at least a decade prior to acidification of the main body of the loch as indicated by diatom analysis. It is possible that episodes of adverse chemistry in the main spawning stream were preventing reproduction for some time prior to the main body of the loch becoming acidified.

3. PROJECT OBJECTIVES AND PLANS

The underlying assumption is that the decline and loss of the brown trout population in Loch Fleet have been due to a deterioration in water quality to a point at which it can no longer support fish. The aim of the project is to improve the water quality so as to allow reintroduction of trout to the loch.

The principal objectives are:

- to demonstrate that the water chemistry of the loch can be brought into a range suitable for trout by one or more of several treatments to the catchment or water, and
- to demonstrate the suitability of the water for a self-sustaining trout population, once suitable chemical conditions have been achieved in portions of the loch.

As a result of the work, it is expected that practical land, lake or stream management procedures can be developed and will be applicable elsewhere.

From the beginning of the project (1984), it was envisioned that it would fall into four phases: an initial phase (1984-86) of collecting and collating existing and baseline information during which embayments of the lake would be curtained off to provide *macrosamplers* to catch the drainage water; a second phase (1986) when sectoral treatments would be initiated; a third phase (1987) when additional treatment would be applied and water quality tested further; and a final phase (1988 to 1989) when, if necessary, additional parts of the catchment may be treated, a trout population restocked and its re-establishment otherwise investigated.

The results of the first phase (Phase 1) have been reported by Howells (1986), and the long-term (2 yr) surface water chemical effects of the first set of treatments that were applied in the spring of 1986 (**Phase** 2) will be described in this paper. The initial chemical effects of the second set of treatments (**Phase 3**) will be briefly referred to and the biological effects, including the results of the fish restocking exercise, will also be described.

4. TREATMENTS APPLIED (1986) AND THEIR EFFECTS

A variety of base materials were analyzed and Ca yield tests were performed on their effects on the percolates from soil cores taken from near the Loch Fleet catchment by the Environmental Advisory Unit of Liverpool University. This led to the choice of carboniferous limestone dust, with a nominal particle size distribution 100% less than 62 *u*m obtained from ICI. Buxton and the lysimeter studies by EAU, suggest applications between 10 and 30 t ha^{-1} would be effective. Thus, in April 1986, the following treatments were applied: Sector IV, limestone dust as a slurry between the trees at 20 t ha^{-1}; Sector VI, limestone dust as a powder at 30 t ha^{-1}; Sector VI, limestone dust as a powder at 30 t ha^{-1}; Sector VII, limestone dust as a powder at a range of 10 t ha^{-1} onto the wetland boggy areas only. (The specific areas where the limestone would best be applied which represented c. 25% of the area of this subcatchment were decided after consultation with Dr. P. Nyberg and Mr. H. Hultberg from Sweden.) Thus, a total of 350 t was applied manually to restrict contamination of adjacent sectors.

The pH and Ca concentrations in the main inlet stream (Sector VII), the water draining Sectors IV and VI, and the loch outlet between March 1986 and January 1988 are shown in Figures 3 to 6, respectively. The concentrations of three Al fractions, as defined by Driscoll et al. (1980), are also shown.

4.1 Sector VII - Main Inlet Stream - Bog Liming - (Figure 3)

Prior to treatment, the Ca concentration in the inlet stream was around 1 mg L^{-1}. The pH fluctuated between 4.1 and 5.0. The total Al concentration ranged from 120 to 290 *u*g L^{-1} with low pH and high Al concentrations being associated with periods of high flow. In

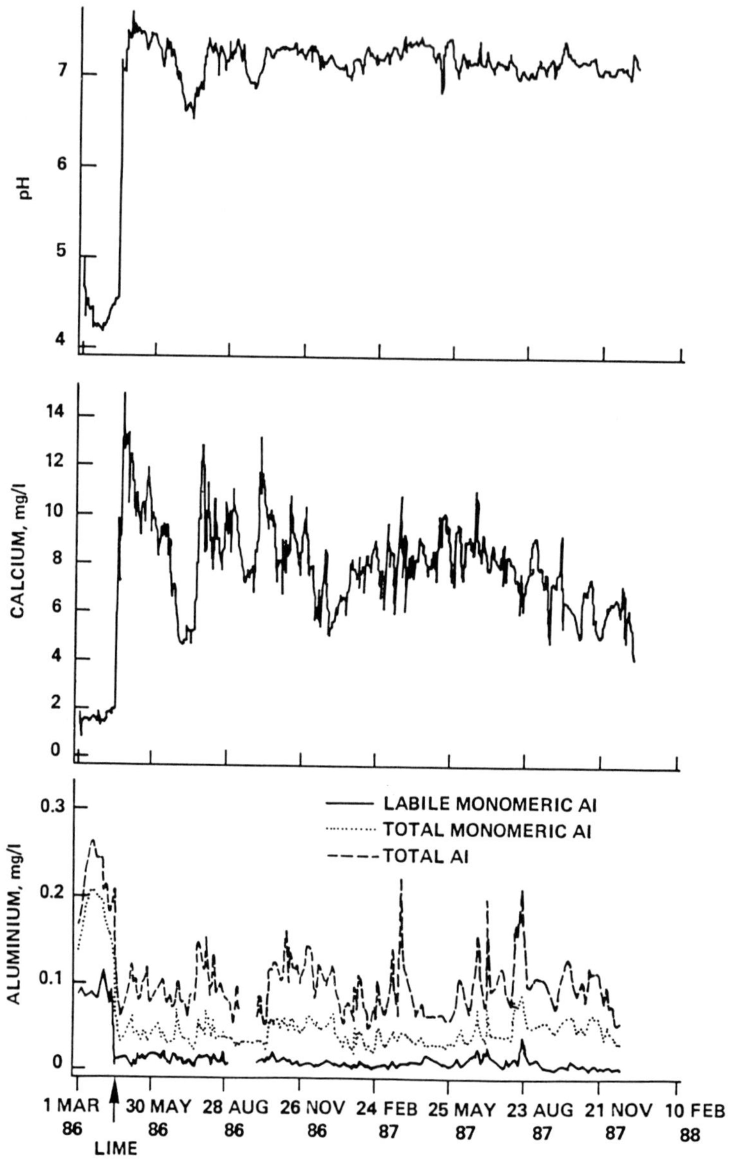

Figure 3. Water chemistry in the Altiwhat Stream (Sector VII) showing when limestone (lime) was applied to the sector.

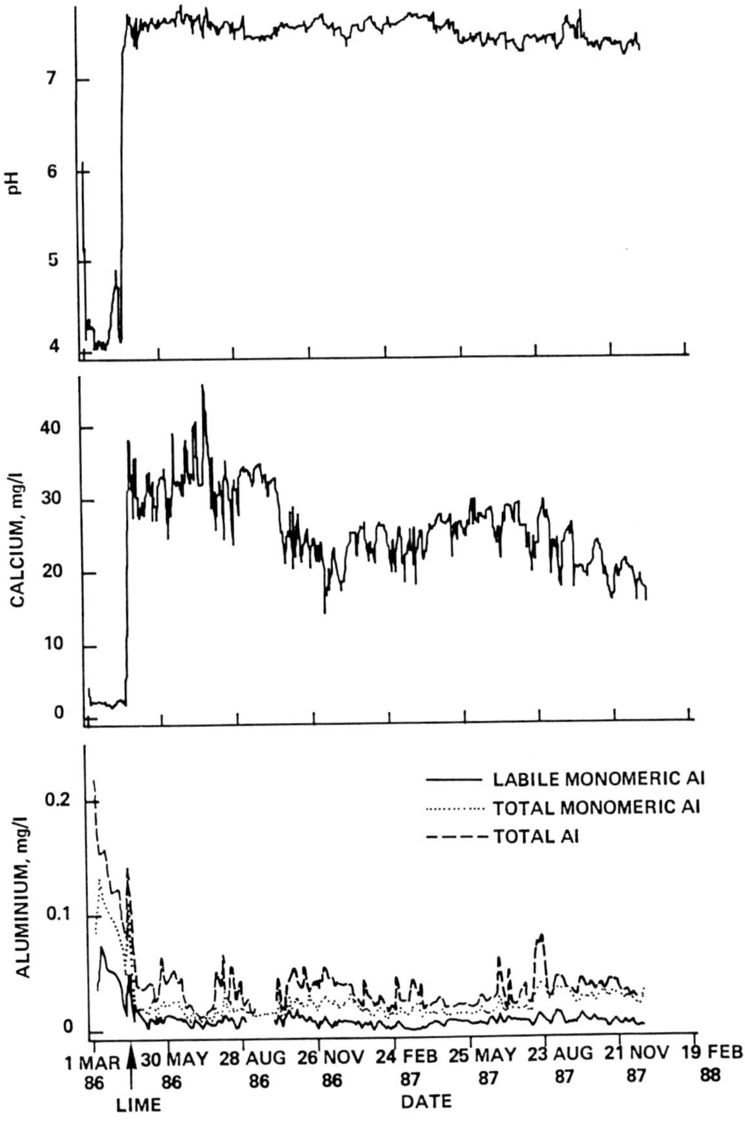

Figure 4: Water chemistry of drainage from Sector IV (Forested) showing when limestone (lime) was applied to the sector.

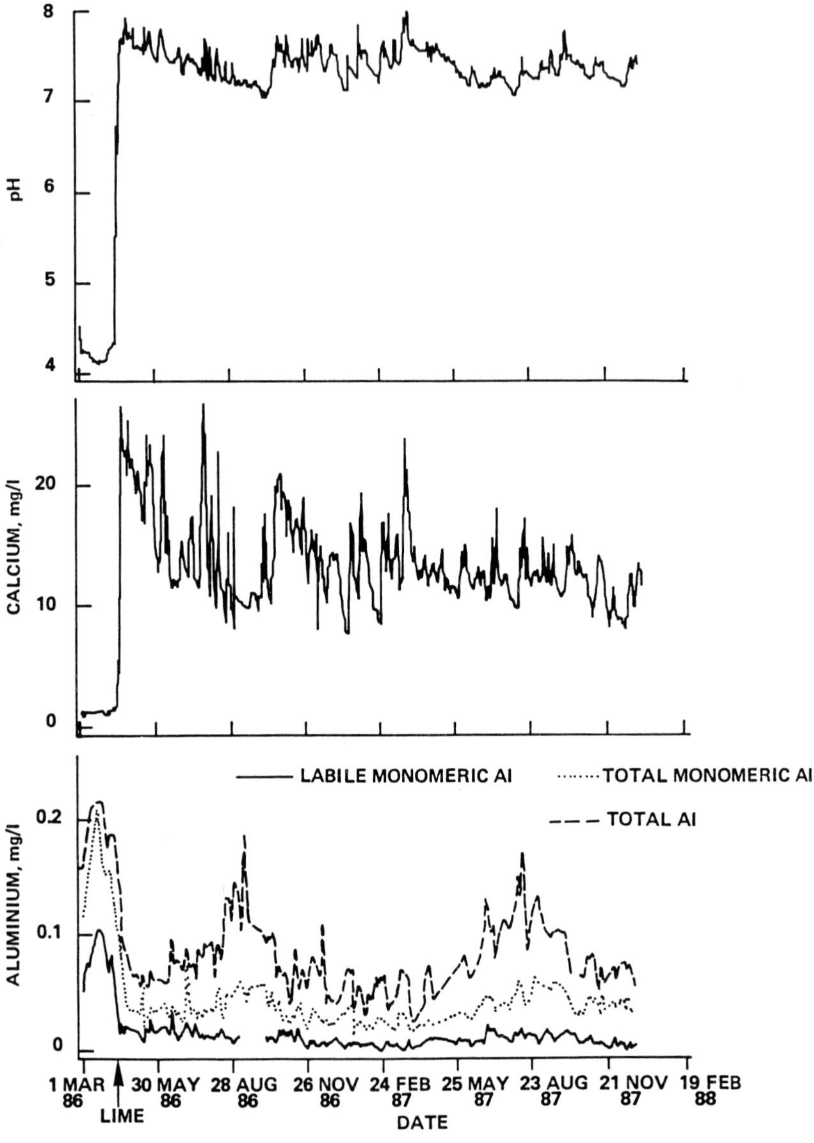

Figure 5: Water chemistry of drainage from Sector VI (Moorland) showing when limestone (lime) was applied to the sector.

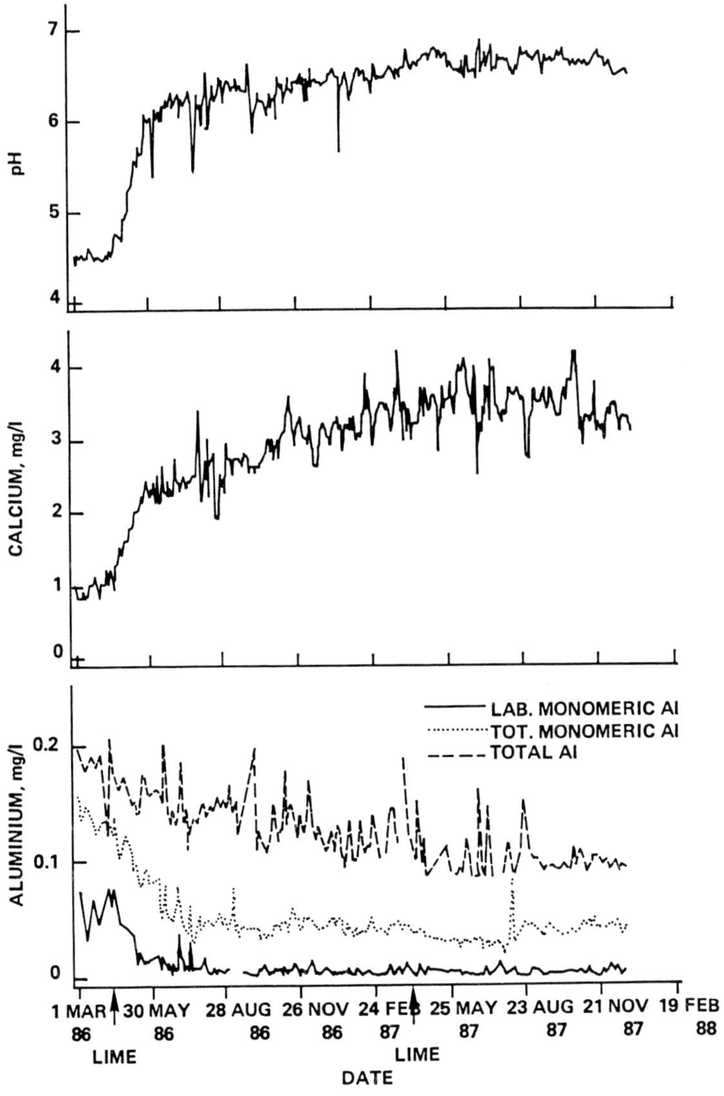

Figure 6: Water chemistry at the loch outlet showing when limestone (lime) treatments were applied to catchment sectors in 1986 and 1987.

general, the concentration of labile monomeric Al (the fraction considered to be the most toxic to fish) was around 100 $ug\ L^{-1}$.

Immediately following treatment, the pH rose to above 7.0, the Ca concentration increased to over 10 mg L^{-1} and, within a week the total Al concentration was less than 100 ug^{-1}. During subsequent dry spells in July and September 1984, the previously observed pre-treatment trend of higher pH's being found during periods of low flow was reserved. There was a tendency for the pH to drop, but only to a minimum of 6.5, and for the Ca concentration to also fall when the flow was low. However, as soon as the flow increased, both pH and Ca concentrations increased again. The concentration of total Al also increased following this first increased flow, but the fractionation results show that most of this increase was in the polymeric fraction and that the concentration of labile monomeric Al has remained consistently below 20 $ug\ L^{-1}$ since the treatment.

4.2 Section IV - Forest Drainage - Slurry Liming - (Figure 4)

The flow from this sector is much less than from Sector VII, and before treatment the Ca concentration was higher at approximately 2 mg L^{-1}. The pH of the water draining this sector fluctuated widely with pre-treatment values around 4.0 when flows were greater than 1 L s^{-1}, and when flows were below this, the pH rose over a period of 1 mo (e.g., February 1986) to around 6.0. The Ca concentration also rose during this period and the total Al, in general, at concentrations of 150 to 200 $ug\ L^{-1}$ (i.e., lower than from Sector VII), fell to values around 50 $ug\ L^{-1}$.

Since treatment, the chemistry has been more stable (although this may be due partly to lower flows). The pH has been around 7.5 and Ca concentrations around 30 mg L^{-1}, and total Al concentrations have been maintained lower than in Sector VII (frequently less than 50 $ug\ L^{-1}$) with a labile monomeric concentration of around 10 $ug\ L^{-1}$.

4.3 Sector VI - Moorland Drainage - Powder Liming (Figure 5)

Flows in this sector are slightly greater than in Sector IV, but are still much less than in Sector VII. The pattern of response, however, is similar to that observed in Sector VII with the pre-treatment relationship between pH and flow being reversed after treatment. The pH and Ca concentrations have consistently been above 7.0 and 10 mg L^{-1}, respectively, in the water draining this catchment since treatment. The response has been similar to that already described for Sector VII.

4.4 Loch Outlet - (Figure 6)

As expected, the outlet, which represents an integrated sample of catchment runoff, pre- and post-treatment chemistry, has been less variable than at the tributary sampling sites, and the response to the treatments less immediate. The pH rose fom 4.5 to 6.3 over a period of about 2.5 mo, and the Ca concentration increased from 1 to 3 mg L^{-1} over the same period. The decrease in total Al was less dramatic with a decline in total concentrations from just below 200 *ug* L^{-1} to values typically between 100 to 150 *ug* L^{-1}. However, the fraction present as labile monomeric form has dropped from about 50 to 60 *ug* L^{-1} to around 10 *ug* L^{-1}.

5. TREATMENTS APPLIED (1987) AND THEIR EFFECTS

The treatments tested in the spring of 1987 included the same limestone dust used in 1986, but at lower application rates (5 t ha^{-1}) with and without fertilizer application. The application of limestone pellets (produced by British Coal in a process involving the use of bentonite) at a rate of 7.5 t ha^{-1} and a muirburn of the open moorland. In general, the first year's post-treatment results indicate that the lower application rate of powder is not so successful at achieving either the degree of neutralization in the forested sector or the effective length of treatment in the moorland sector as were the higher rates previously applied. The initial results, from the only slightly higher rate of application of pellets, however, are very encouraging in that the pH and Ca concentration of the runoff increased from 4.2 to 6.0 and from 0.5 to 5.0 mg L^{-1}, respectively, over a short period, and then continued to rise up to 6.5 and 6.0 mg L^{-1} over the next 8 mo. Thus, aggregation of the limestone into pellets of this nature may well be a way of ensuring somewhat slower release of the neutralizing capacity of the limestone. The muirburn was not very successful. Due to the saturated nature of the peat, the muirburn did not penetrate to any depth and did not significantly alter the quality of the water draining the sector. The reference to Figure 6 suggests that, in general, the much smaller quantities of lime applied in the second set of treatments (c 90 t compared with 350 t applied in 1986) did not have a significant effect on the water quality of the loch at its outlet.

6. BIOLOGICAL EFFECTS

6.1 Terrestrial Effects

At the start of the project, a survey was undertaken of the existing vegetation within the catchment. This survey was repeated, although on a less frequent sampling grid, 18 mo after the first set of treatments was applied. In addition, forest assessment involving foliar analyses in treated and untreated sections of the catchment has been undertaken and conditions in the boggy wetland areas have been studied intensively for post-liming.

The general vegetation resurvey suggests that liming at 30 t ha^{-1} may favor a move towards dominance of *Molinia* over *Calluna*, and the foliar analysis showed that liming significantly decreased the Al concentration of conifer needles. Although just out of the 95% confidence level, the effect of liming was to increase foliar nitrogen concentrations.

The wetland areas of the catchment are characterized by the occurrence of *Sphagnum*, and since this calcifuge is without roots and its tissues are directly exposed to the surrounding water, it was to be expected that the various *Sphagnum* species in the catchment would be adversely affected by lime applications. The extent to which this happened was studied intensely in an area of 2.5 ha in the upper part of Sector VII, about one-third of which was covered by *Sphagnum*, principally *S. papillosum*. A year after treatment, the *Sphagnum* was bleached and presumably dead, although apparently dead stems are known to regenerate. Where *S. papillosum* was protected (e.g., by being upstream of applied lime, or raised on a hummock), it was undamaged. *Sphagnum* species tolerant of lime treatment and found alive were *S. inundatum*, *S. recurvum*, and *S. capillifolium*. The potential damage over time, the possibility of recolonization by tolerant species of *Sphagnum*, and the risk of erosion in damaged areas will continue to be kept under surveillance.

6.2 Aquatic Effects

The aquatic macrophytes in the loch have been sampled on a yearly basis since the start of the project. Prior to liming, *Littorella uniflora* and *Obelia dortmanna* dominated sandy littoral areas, particularly along the western shore. *Isoetes lacustris* and *Sphagnum spp.* were widely distributed in deeper water and *S. auriculatum* was the dominant macrophyte in water 2 to 5 m deep.

Eighteen months after catchment liming, more than 90% of *Sphagnum* samples showed significant signs of die-back. A new macrophyte species, *Potamogeton polygonifolius*, has become established in shallow water. There is evidence of an increase in *Isoetes echinospora* while *Juncus bulbosus* and *Utricularia intermedia* are significantly more abundant.

The aquatic macroinvertebrates have been sampled regularly since the beginning of the project from the littoral and deeper parts of the loch and from the inlet and outlet streams. In general, in the season following liming, the density of most invertebrate groups was lower.

In the case of shore samples, these have gradually returned to what would appear to be preliming levels, but the deep water numbers have remained low. Both abundance and diversity in the two streams increased post-treatment. Mayfly nymphs of the family Beatidae were found for the first time in the summer of 1987 in the outlet stream.

One particular group of invertebrates that has been studied in detail is the water beetles. In the absence of fish at the beginning of the project, these were the top aquatic carnivores in the system. Since the introduction of fish (see later), some interesting changes in species dominance have occurred, which appear to be related to their camouflage status, and hence, their ability to avoid predation by fish.

As previously mentioned, survival studies using trout ova and yoke sac fry confirmed the suspicion that chemical conditions in the loch and its afferent streams prior to liming were acutely toxic to these life history stages (Turnpenny et al., 1988). Following liming, a program of fishery investigations was put into effect to test the suitability of the improved water quality for a self-sustaining brown trout population. The program followed the same pattern as for the pre-intervention studies, namely that the viability of captive trout representing successive stages of the life cycle was tested in its relevant stream-bed (eggs and fry), stream surface water (fingerlings) and lake (yearling-plus) habitats. Once viability had been demonstrated for each of these life history stages during at least one full annual cycle, approximately 300 free-living fish, from three separate but local sources, were introduced into the loch in May 1987. Throughout the summer of 1987, the presence of free-surviving fish was obvious by the amount of rising observed, and several of the larger fish were caught in spawning condition by traps in the inlet and outlet streams in November. No redds were identified in the stream beds over the winter, but the recently observed trout fry in both inlet and outlet streams confirms that successful reproduction occurred.

7. CONCLUSIONS

The chemical and biological results from the project to date have been very encouraging. Model calculations are in hand to answer questions about how long the treatment will last so that relative costs and efficiencies of the different treatments can be assessed. The return to a self-sustaining fishery, which results to date suggest is possible, remains to be monitored. Also, it is recognized that the hydrology of the Loch Fleet catchment, with a majority of the water entering the streams and loch via surface and subsurface flow, may render it especially suitable for treatment of this type. Nevertheless, it is perhaps not too early to conclude that, if fishery restoration to an acid lake is an overriding short term objective, then catchment liming appears to be a suitable method of achieving such an objective.

ACKNOWLEDGMENTS

This paper is published with the permission of the Project sponsors, the Central Electricity Generating Board, the South of Scotland Electricity Board, the North of Scotland Hydro Electric Board and British Coal for which we are grateful. We are also indebted to the following people for the supply of data/information used in this paper. Miss K. Paterson and members of Technical Services, SSEB for chemical analyses and data handling, Mrs. M.V. Proctor and Dr. A. Dickson of CERL for data processing, Mr. G. Hudson and staff of the Macaulay Land Use Research Institute for vegetation survey, Miss E. Wilson of CERL for foliar analyses, Professor R.S. Clymo of Queen Mary College, London for the *sphagnum* survey, Dr. P. Raven of University College, London for the Aquatic Macrophyte Survey, Dr. R.J. Aston and staff at CERL Freshwater Biology Unit for macroinvertebrate data, Dr. G.N. Foster of the Balfour Browne Club for information on the water beetles, and to Dr. A.W.H. Turnpenny and the staff at the CERL Marine Biology Unit for the studies on fish.

Finally, we thank the Swedish State Power Board for funding the participation in the management of the project of Mr. H. Hultberg, Dr. P. Nyberg and Mr. I. Andersson.

AN UPDATE ON THE LAKE ACIDIFICATION MITIGATION PROJECT (LAMP)

D.B. Porcella

ABSTRACT

Limestone addition is a management tactic to mitigate acidic conditions in lakes. The Lake Acidification Mitigation Project is a long-term integrated project to study the ecological effects of liming and develop the technical information necessary for an effective liming program. Three Adirondack lakes have been limed: Cranberry Pond, a small, 60-day hydraulic residence time fishless lake; Woods Lake, a larger fishless lake with longer residence time (214 days); and Little Simon Pond, a large lake (160 ha) containing populations of brook trout, lake trout and other fishes, with a residence time of 450 days. Woods Lake has been limed twice and the stocked brook trout fishery has been maintained; Cranberry has been limed once and then allowed to reacidify with loss of the stocked fishery. Little Simon Pond showed no ill effects of liming on the extant populations. In general, short term evaluation of ecological effects has shown no major deleterious effects of liming. Methods for predicting reacidification are extremely accurate and use of different limestone particle-size fractions is recommended to achieve longer term in-lake neutralization.

1. INTRODUCTION

The Lake Acidification Mitigation Project (LAMP) was organized to address issues raised as a result of Swedish and other liming experiences. The major objective of LAMP is to learn how aquatic ecosystems respond to chemical perturbation from the addition of calcite (liming) and subsequent reacidification. Reacidification following base application is of particular interest because it represents an important experimental tool to assess the response of lakes to acidification processes.

Lakes represent a continuum of physical, chemical and biological characteristics that vary in time and space. A range of these characteristics exemplified in three study lakes was examined by measuring chemical changes over time and space, relating the chemistry to structural and functional responses of lake communities, and then identifying the most

Water, Air, and Soil Pollution **41** (1988) 43–51.

significant control variables. Specific objectives included the evaluation of changes in stocked and unstocked fish populations and other biota in relation to liming and reacidification, the risk of metal toxicity to fish during reacidification after repeated liming, and whether direct lake liming protects biota against episodic or littoral zone acidic events.

The Adirondack Mountains have high acidic deposition, contain a number of low ionic strength and low alkalinity lakes, and have a relatively good baseline of data (Schofield et al., 1985). Based on accessibility, permission to lime, ability to perform hydrologic measurements and chemical and biological characteristics, three clear-water drainage lakes were selected: Cranberry Pond, Woods Lake, and Little Simon Pond. The first two lakes were fishless prior to liming, whereas Little Simon Pond contained *Salvelinus namaycush* (lake trout) which we concluded were under stress based on observations that only older year classes occurred in the population. The Little Simon Pond fish community also included *Salvelinus fontinalis* (brook trout) and other fish species. The major differences in the three lakes are associated with the mean hydraulic residence time: 63, 214 and 450 days, respectively. Woods Lake is the only one of the three with a substantial pre-LAMP record of study (Goldstein et al., 1985).

2. METHODS AND MATERIALS

LAMP was organized in late 1983 with field sampling beginning in January 1984. Seventeen months later (May 30-31, 1985) a helicopter limed the two short residence time lakes, Cranberry and Woods, with a water slurry (71% solids) of fine particle (median diameter - 2 μm) calcite (98% calcite) and 0.15% sodium polyacrylate dispersant to aid suspension (Scheffe et al., 1986). 2Slurry was delivered to the lakes at 6.5 t hr^{-1}; about 7 t of dry calcite was added to Cranberry Pond and 23 t to Woods Lake. On a volumetric basis, the liming increased nominal water concentrations of calcite to 34 mg L^{-1} in Cranberry and 28 mg L^{-1} in Woods. The helicopter sprayed slurry on the lake surface in proportion to volume with added calcite of 0.1 t ha^{-1} of lake area to achieve uniform concentrations in the water column and sediment treatment from sedimented calcite particles. However, settling was less than expected because rapid dissolution of the small particles minimized sedimentation. Four days after chemical perturbation of the two lakes, yearling and fingerling brook trout were introduced to both lakes.

Cranberry was allowed to reacidify. The reacidification occurred by the end of November 1985. Much of the pH decrease was due to washout of ANC by dilute storm waters with pH greater than 5.

Woods Lake was relimed in September 1986 when ANC had decreased to about 20 $u$eq L^{-1}. In the second liming of Woods Lake, 35 t of calcite were added, producing nominal concentrations of 44 mg L^{-1} of calcite. The first liming of Little Simon Pond occurred in August 1986 when 86 t of calcite were added to attain nominal concentrations of 13 mg L^{-1} of calcite. In both the 1986 limings, several size-ranges of powdered calcite were used so that coarser material would settle to the bottom sediments, hypothesizing that this calcite would slowly dissolve to prolong the treatment duration. Initial results indicate marked increases in treatment duration, illustrating the importance of appropriate design, especially calcite particle size. These latter experiments have not been completed. As a result, the remainder of this presentation summarizes the results of the liming of Woods and Cranberry in May 1985.

3. RESULTS

The spatial and temporal distribution of water entering and exiting the lake along with the chemical composition of those flows were expected to have the greatest effect on in-lake chemical concentrations and mass balances. Staubitz and Zarriello (1989) showed that runoff in the two lakes was highly variable with high peak flows and poorly sustained low flows. Diffuse seepage from shallow groundwater responded rapidly to seasonal recharge and entered the lakes along the lake shore. Error analysis indicated the hydrologic balance was accurate within 10% on an annual basis.

Within the first 24 hr of base application, both lakes responded like systems closed to atmospheric CO_2 because the dissolution of very small calcite particles (median diameter 2 um) exceeded the rate of atmospheric CO_2 invasion (Fordham and Driscoll, 1989). Rapid dissolution of calcite coupled with very low concentrations of DIC prior to treatment resulted in pH increases in the upper mixed waters from 4.9 to 9.4 in Woods and from 4.6 to 9.1 in

Cranberry Pond. Solutions were at saturation with respect to $CaCO_3$ solubility. Stoichiometric increases in pH, ANC, DIC, and dissolved Ca^{++} were observed as:

$$H^+ + CO_2 + H_2O + 2CaCO_3 = 2Ca^{++} + 3HCO_3^-. \qquad (1)$$

Over the next four weeks, the upper mixed waters equilibrated with atmospheric CO_2, facilitating dissolution of suspended particles of calcite, and releasing additional ANC and Ca^{++}. By the end of the four weeks, calcite dissolution efficiency attained 86% in Woods and 79% in Cranberry, according to budgets based on Ca^{++}. In effect, all of the added calcite dissolved in the two lakes by the end of the summer.

Similarly, Driscoll et al. (1989a) noted marked changes in the concentrations of the trace metals, Al, Mn, and Zn following liming. There was a shift in the speciation of Al from acid soluble to labile (inorganic) monomeric forms. The shift in concentration was not sufficient to increase toxicity. Over the summer period following manipulation, pH values decreased as the mixed layer equilibrated with atmospheric CO_2. Within 1 to 2 mo of liming, the concentrations of the three metals were greatly reduced, typically by an order of magnitude. Because thermal stratification limited penetration of base into the 5% of lake volume contained in the hypolimnion, trace metal concentrations were considerably higher there than in the mixed layer.

The mechanisms for immobilization of trace metals following base treatment are not entirely clear. Chemical equilibrium calculations suggest that base addition causes pH increases and over-saturation conditions for readily-forming Al and Mn minerals. Throughout the study, waters were under-saturated with respect to the solubility of Zn-containing minerals. Therefore, retention of Al and Mn in the lakes may have been due to direct precipitation while Zn retention probably occurred by absorption or co-precipitation. Acidic Adirondack lakes are generally net sinks for Al and conservative with respect to transport of Mn and Zn (Driscoll et al., 1989a). Liming greatly increased retention of all three metals.

For the long-term chemical response to liming, Driscoll et al. (1989b) noted that during the fall season following treatment, completely mixed water columns and a typically elevated runoff sped reacidification, especially in Cranberry Pond. When pH values decreased below

5.5 in Cranberry, transport of elevated concentrations of inorganic Al through the lake was evident, though the concentrations were not higher than those prior to liming. Annual ANC budgets suggest that little $CaCO_3$ penetrated to the sediments, limiting prolonged release of ANC from sediment dissolution. Hydrolysis of Al, due to the elevated lake pH after liming, served to consume ANC. Also, there is evidence of exchange of water column Ca^{++} with H^+ in sediments after treatment followed by release during reacidification. However, these processes did not significantly accelerate or attenuate the reacidification rate. The rate of reacidification could largely be explained by the flushing of ANC by hydrologic inputs.

Young et al. (1989) concluded that observed calcite dissolution exceeded predicted values (Scheffe et al., 1986) significantly because of 1) thermocline inhibition of calcite settling, 2) hydrodynamic dispersion, and 3) atmospheric CO_2 influx. Enhancement factors for CO_2 exchange were 3.4 to 5.9. Immediately after treatment, sediment pH gradients changed from positive to negative downward. Amounts of total Ca in surficial sediments increased in parallel with decreases in sediment base neutralizing capacity (BNC). This result suggested that ion exchange was important for sediment neutralization. In Woods Lake during the subsequent 15 mo of reacidification, only partial return to preliming conditions occurred, typical of active exchange mechanisms in diffusion limited systems. It is apparent that sediment interactions become quantitatively important errors if neglected when calculating mass balances of limed systems.

The physical and chemical variables in the LAMP study fit easily into a modeling framework. Using a lumped parameter dynamic model (ALaRM), DePinto et al. (1989) evaluated the importance of inflow, sediment interactions with water, seasonal effects and other variables that affect the duration of treatment. ALaRM was confirmed by its accurate simulation of the observed reacidification profiles of Woods Lake and Cranberry Pond, predicting the near-0 ANC of Cranberry within 6 mo, and ANC concentrations of about 20 $\mu eq\ L^{-1}$ for Woods at 15 mo just prior to reliming. ANC, pH and Ca are the state variables followed in these profiles. Results showed that these profiles were controlled primarily by hydrologic flushing and secondarily by sediment-water ANC transfer. It is apparent that a prediction tool must consider these variables to provide a reasonable estimate of reacidification.

Clearly, a study of ecological effects requires both detailed physical, chemical and biological measurements and the integration of these measurements into some cohesive whole. Although LAMP does not apply a formal modeling structure for the biotic components, investigators have related food web variables to physical and chemical variables to evaluate the ecological effects of liming.

Bukaveckas (1989) discussed phytoplankton dynamics during liming and reacidification in Cranberry and Woods as affected by light, temperature and chemical variables. Corresponding changes of productivity and biomass observed in the mixed and deeper layers of the water column suggest that the effect of liming on phytoplankton growth rates and loss rates was not restricted to the maximum depth of penetration of calcite particles. The neutralized (mixed) and acidic layers exhibited similar increases in phytoplankton abundance, but changes in species composition were limited to the mixed zone. Physiological capacities for using CO_2 vs bicarbonate forms of DIC appeared to play an important role in species composition, but other factors such as nutrients seemed to regulate abundance. Post-treatment increases in phytoplankton abundance were attributed to 1) increased nutrient availability following Al precipitation, and 2) reduced zooplankton grazing pressure.

Without analysis by an ecological model, the zooplankton responses to liming and species shifts of food organisms and predators cannot easily be deciphered. Schaffner (1989) attributed zooplankton responses to phytoplankton changes and fish stocking effects that could cause changes in both zooplankton production and loss terms. Most rotifer taxa were greatly reduced in numbers within a week following base addition. The micro-crustacean community also responded to liming. *Diaptomus minutus* was dominant in both lakes, but declined after liming. Longer-term population responses appeared to be related to the dynamics of individual populations and changes in predation pressure. This caused some species to become abundant again in Cranberry Pond after reacidification while remaining rare in Woods Lake. Four additional crustacea became prominent in the lakes after liming: the cladocerans *Bosmina longirostris* and *Daphnia catawba* in Cranberry Pond, and *D. catawba* and the cyclopoid copepod *Cyclops scutifer* in Woods Lake.

Evans (1989) characterized the macroinvertebrate fauna which formed the initial base of the food supply for the stocked trout prior to liming. Within 2 to 3 mo after liming and trout introductions, all limnetic insect populations (notonectids, corixids, dytiscids, and *Chasborus americanus*) were near or below the detection limit of 0.02 individuals m^{-3}. Available evidence strongly indicated that brook trout predation was the major factor reducing limnetic insect populations.

The combination of zooplankton and macroinvertebrate data provided the initial conditions for exploring food chain limitations on trout production. After fingerling (0^+) and yearling (1^+) brook trout, *Salvelinus fontinalis*, were reintroduced into the two previously acidic and fishless lakes (4 days after liming), they were monitored by semi-annual trapnetting and continuous operation of outlet traps to measure emigration. Stomach samples were collected during the ice-free season. Gloss et al. (1989) concluded that growth and condition of both ages of fish were comparable to those of circumneutral lakes, while water quality remained suitable. During reacidification, Cranberry Pond fish growth and survival declined rapidly. About 55 percent of the 0^+ aged and 25% of the 1^+ aged fish emigrated. Some limited survival of fish within the lake occurred but, practically speaking, in-lake mortality accounted for the remainder. Annual survival in Woods Lake fish was somewhat lower than other limed or circumneutral Adirondack lakes, averaging 35% for age 0^+ and 25% for age 1^+ fish. Occurrence of false annuli on scales during summer suggested that growing conditions were not optimal, although annual growth was similar to other waters. Initially, large conspicuous invertebrates made up most of the caloric intake by fish over 200 mm total length. However, these taxa were replaced in importance by zooplankton late in the first summer and during the second year in Woods Lake.

Reproduction was insignificant in Woods Lake, occurring only on manmade spawning areas, both calcitic and granitic substrates. Reproduction appeared limited by low velocity groundwater flows in the lake and high acidity in tributary streams. The emergence period for fry from the manmade substrates peaked in mid-May, well after episodic snowmelt inflows produced toxic conditions in littoral areas.

Schofield et al. (1989) used a bioenergetics model based on diet analysis to compare actual vs estimated production to test whether liming is a suitable management technique for maintaining adequate recruitment and growth. They showed that the yearling and fingerling brook trout stocked in Woods and Cranberry after liming produced 10 to 14 kg ha^{-1} yr^{-1} at average standing crops of 8 to 10 kg ha^{-1} during the first year after stocking. These values were similar to levels observed in other limed and circumneutural lakes which collectively support a limited range of biomass (10 to 20 kg ha^{-1}) and production (7 to 16 kg ha^{-1} yr^{-1}) compared to more fertile waters. The estimated level of food consumption required to maintain initial production rates observed in the two limed lakes was 80 to 100 kg ha^{-1} yr^{-1}. This level of predation was sufficient to substantially alter the size, structure and composition of the invertebrate prey community within a few months of stocking. However, mean production efficiency remained at levels of 20 to 25% and the development of adverse water quality conditions in the first year after liming appeared to be of greater significance than food limitation in regulating production and biomass of these stocked brook trout populations. By 5 mo after liming, reacidification in Cranberry Pond resulted in rapid declines in biomass and production as a result of both increased mortality and reduced growth.

Although these analyses of responses to liming do not answer all of the possible questions, the results help to define continuing research. A key question concerns whether model predictions of bulk water chemistry can be related to the responses of biota that utilize littoral habitats or other areas subject to episodic acidic inputs. Likewise, the effects of changes in prey density caused by the interaction of liming and predation by the reintroduced fishery may cause changes in the fishery that could not be predicted from these short term measurements. LAMP is continuing to measure long-term ecological responses as part of its ongoing research. Future research includes a watershed liming experiment which will address some of the above questions, along with measurement of the effects of soil liming on base saturation of soil ion exchange and vegetation responses to liming.

4. CONCLUSIONS

1. No deleterious effects of liming have been observed during LAMP. The observed changes in community structure are desirable and result from water quality changes and stocking of trout.

2. Estimates of fish community productivity of limed lakes do not vary significantly from otherwise circumneutral lakes.

3. Reacidification of limed lakes can be predicted precisely if hydrologic and in-lake processes are known. The major uncertainty in forecasting arises from hydrologic variation.

4. Liming enhances the removal of metals from the water column.

5. Significant differences in the effectiveness of liming result from different application techniques and limestone preparations.

ACKNOWLEDGMENTS

The Lake Acidification Mitigation Project (LAMP) was funded by the Electric Power Research Institute and the Empire State Electric Energy Research Corporation. Liming of Little Simon Pond and reliming of Woods Lake was performed by Living Lakes, Inc., Robert W. Brocksen, Executive Director. This paper was prepared from material supplied by the principal investigators: Carl Schofield, Steven Gloss, Joe DePinto, Charles Driscoll, Thomas Young, Paul Bukaveckas and Ward Staubitz. I gratefully acknowledge their help and that of our advisors on this project: Richard D'Auteuil, William Dickson, Steven Gherini, Arnold Gnilka, John Gunn, Terry Haines, Arne Henriksen, Hans Hultberg, William Keller, Ken Mills, Per Nyberg, Kent Schreiber and Harald Sverdrup.

COOPERATIVE FEDERAL-STATE LIMING RESEARCH ON SURFACE WATERS IMPACTED BY ACIDIC DEPOSITION

R. Kent Schreiber

ABSTRACT

In the eastern and north-central United States, lakes and streams with low acid neutralizing capacity are at risk from acidity. Resource management agencies are interested in developing mitigation strategies that protect or restore fisheries in these waters. Addition of limestone (calcium carbonate) to improve water quality and prevent episodic depressions of pH during precipitation events and spring runoff is one mitigation technique being used. The ecological changes that accompany such treatment of streams and lakes are being investigated in a cooperative program between the U.S. Fish and Wildlife Service and individual states. Streams in Massachusetts, West Virginia and Tennessee, and a lake in Minnesota are included in this 5-yr research program. Intensive monitoring during pre- and post-liming tracks a suite of physical, chemical and biological parameters that influence the re-establishment or maintenance of healthy fisheries. Supporting studies on liming being conducted at Adirondack lakes in New York focus on fisheries management. A model on the influence of liming on light attenuation and thermal stratification is also being developed. Management guidelines are to be generated from the program results.

1. INTRODUCTION

Acidic deposition has affected a diversity of aquatic environments, including some surface waters in the northeastern United States that have been damaged by the loss of alkalinity and increased availability of toxic metals (e.g., Haines and Akielaszek, 1983; Arnold et al., 1985; Haines et al., 1986). Aquatic resources in other areas such as the Southeast and Upper Midwest are also potentially at risk; some are already demonstrating symptoms of acidic stress (e.g., Wiener and Eilers, 1987).

Water, Air, and Soil Pollution **41** (1988) 53–73.

After the establishment of the National Acidic Precipitation Assessment Program (NAPAP) in 1980, a 10-yr federal interagency research effort was initiated to better understand the effects of acid deposition on the environment. Part of this program also addresses "ways to remedy or otherwise ameliorate the harmful effects which may result from acidic precipitation" (Interagency Task Force on Acid Precipitation, 1982). Previous research and field operations, particularly in Sweden, have shown that liming or the addition of $CaCO_3$ to acid-damaged or acid-stressed surface waters can increase alkalinity or acid neutralizing capacity (ANC) (Fraser and Britt, 1982). The ecological and long-term consequences of liming, however, have not been adequately studied. A group of international scientists convened in 1982 to identify high priority research on liming (International Liming Workshop, 1982). The results of this workshop provided the basic guidance for the present five-year federal Acid Precipitation Mitigation Program (APMP) managed by the U.S. Fish and Wildlife Service (Schreiber and Rago, 1984; Villella and Schreiber, 1986; Schreiber and Britt, 1987).

A pilot mitigation research project on 10 small acidic lakes in the Adirondack Mountains of New York examined the response of stocked brook trout (*Salvelinus fontinalis*) to agricultural limestone treatment (Schofield et al., 1986). Total limestone dissolution efficiencies for the lakes ranged from 17 to 59%; initial dissolution efficiency was best predicted by the dose rate (Gloss et al., 1988). When pH was 6.5 or greater in the treated lakes, trout survival was 80%, but declined with re-acidification. Procedures for APMP were developed in part on the basis of the experience gained in these preliminary studies. Additional investigations continuing in the Adirondacks include an assessment of how the spatial distribution, abundance, and quality of prey items affect the survival of brook trout in early life stages (emergent fry to fingerling) in treated lakes. Another study is investigating the effects of liming on thermal stratification and light attenuation.

Other programs also contributed important insight on field and laboratory procedures and research design of the APMP. The National Surface Water Survey (NSWS), conducted by the U.S. Environmental Protection Agency, was designed to document the chemical and biological status of lakes and streams in regions of the United States that were potentially sensitive to acidic deposition. It was also used to select regionally representative surface waters to quantify future changes in aquatic resources through a long-term monitoring program (Brakke et al., 1985). Another major study, the Lake Acidification Mitigation Program (LAMP), sponsored by the Electric Power Research Institute, deals more specifically with liming research. Detailed information is being obtained on the chemical, physical, and biological effects of liming and reacidification of acid lakes, focusing on three Adirondack lakes (Porcella, 1987). Likewise, mitigation research in Canada has focused basically on rehabilitation of acid lakes (Booth et al., 1986; Watt, 1986). Recently, in a new operational liming program initiated by private business, the treating and monitoring of lakes and streams in the U.S. have begun (Brocksen, 1988). Information from these programs will be valuable in extrapolating the results of APMP to broader geographic use.

The primary objective of APMP is the analysis and evaluation of the ecological responses of surface waters treated by liming. Although the focus of the research is on re-establishing or protecting the fisheries resources, the program incorporates a comprehensive research design that examines the physical, chemical and biological conditions that influence the welfare of the entire aquatic ecosystem (Schreiber and Rago, 1984; Brown and Goodyear, 1987). An integral part of the APMP is the cooperative agreements and co-funding (85% federal, 15% state); the responsibility for the research is shared by the U.S. Fish and Wildlife Service and the cooperating state agency. The program is intended to produce guidance documents for resource managers electing to rehabilitate or protect surface waters affected by acidity. This paper describes the procedural features in the research design that are required to specifically evaluate the ecological effects of liming.

2. METHODS AND MATERIALS

2.1 Site Selection

Specific programmatic, hydrological and research criteria were used to select sites in various geographic regions since the results of the research are intended to address management needs in a variety of aquatic ecosystems (Brown and Goodyear, 1987). Stream sites were selected in the Northeastern, Mid-Atlantic and Southeastern regions (MA, WV, and TN, respectively), and a lake was chosen in Minnesota (**Figure 1**). Each site represents surface waters now

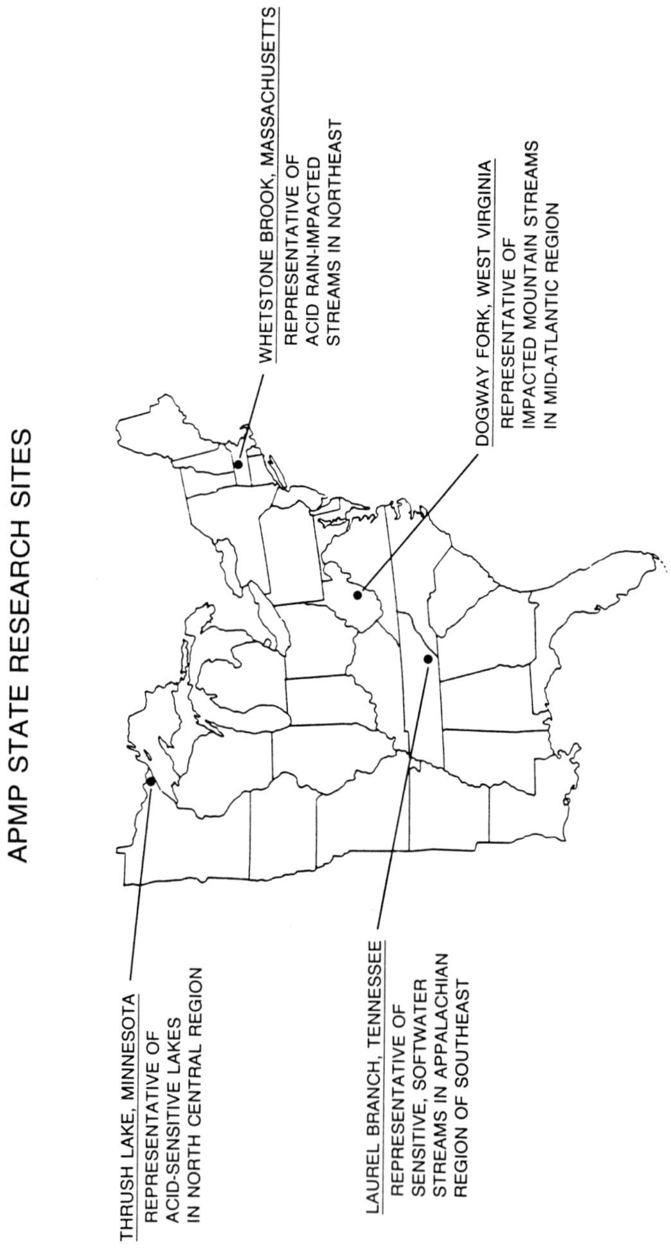

Figure 1. Location of research sites in the APMP.

experiencing acid stress in that region and is subject to an annual average precipitation pH of less than 5.6. In addition, the pH of the surface water is less than 6.0, either frequently, or for 2 or more weeks, or during critical periods with respect to the sensitive life stages of important biota.

In the New England region the site chosen was Whetstone Brook in Franklin County, Massachusetts. Whetstone Brook is a tributary of the Millers River, the drainage in the state most vulnerable to acidification. The brook presently has a low buffering capacity and is subject to spring depressions of pH 4.6 to 4.8. The acid neutralizing capacity can be less than 0.0 *u*eq L^{-1}. The present fish community includes brook trout, brown trout (*Salmon* trutta), longnose dace (*Rhinichthys cataractae*), brown bullhead (*Ictalurus nebulosus*), and Atlantic salmon (*Salmo salar*, occasional transients). Historically, it was one of the best trout streams in the Millers River drainage. The trout population now, however, lacks some year classes and is in poor condition. The forage fish are also reduced in numbers with some species having been lost (Simmons and Halliwell, 1986).

In the Mid-Atlantic region, Dogway Fork in West Virginia is in Pocohontas and Webster counties in the Monongahela National Forest. It is a tributary of the Cranberry River, an important trout fishing area whose pH has decreased significantly since 1969 and now exceeds 6.0 only during low flow. Stocking is restricted in some areas because of the acidic conditions. No fish now exist in the main stem of Dogway Fork; only a few brook trout occur in a small tributary. Dogway Fork has the poorest water quality of any Cranberry River tributary, with a mean pH of 4.5 and a mean ANC of less than 0.0 *u*eq L^{-1} reported in 1987 (Zurbuch et al., 1988).

In the Southeast, Laurel Branch, a second order Appalachian stream, is in the Tellico Wildlife Management Area of the Cherokee National Forest, Monroe County, Tennessee. It is a tributary of the North River in the Tellico River drainage and is a soft water stream subject to ANC and pH depressions during episodic events. The present fish community is diverse and includes rainbow trout (*Salmo gairdneri*), brook trout, brown trout, blacknose dace (*Rhinichthys atratulus*), northern creek chub (*Semotilus atromaculatus*), and northern hog sucker (*Hypentelium nigraicans*). This stream periodically undergoes ecological stress from acidity (Weaver, 1988). For example, an instream bioassay in April 1980, when pH was 4.8 and the ANC less than 0.0 *u*eq L^{-1}, demonstrated severe stress to rainbow trout. The treatment strategy for this stream focuses on protecting the existing biological resources. To provide additional support for detecting changes in physical, chemical and biological variables caused by liming, the investgators have included a control stream, South Fork Citico Creek, in the Tennessee project. This site will not be limed and will permit paired stream sampling. Comparison of data between these streams in contiguous watersheds will increase the confidence in the projections made for Laurel Branch. Studies on this stream will also specifically investigate the ways in which macroinvertebrate availability influences rainbow trout production in the lime-mitigated vs non-treated waters.

In the North Central region the study site is Thrush Lake in Minnesota. It is located in the Superior National Forest in Cook County and is part of the Lake Superior drainage via the Cascade River. Thrush Lake, 6.6 ha, is the highest elevation lake in Minnesota, located on a ridge that receives the greatest regional precipitation; its mean annual pH is 4.6 and its sulfate loading rate is high. The area also has the deepest snowpack in the state. The present fish population in the lake includes brook trout, lake trout (*Salvelinus namaycush*), fathead minnow (*Pimephales promelas*), and pearl dace (*Semotilus margarita*). Recent stockings of lake trout have been unsuccessful. The lake is subject to ANC and pH depressions during episodic events, principally snowmelt (Wright et al., 1988).

These study sites represent a broad range of the aquatic systems affected by acidity. The sites have been characterized in terms of habitat features, water quality, and biota before lime treatment. After treatment, the ecological response of the system will be monitored to determine significant changes. In addition to the results of the individual sites, intersite comparisons should provide new understanding of aquatic system responses to liming.

3. RESEARCH METHODS AND PROTOCOLS

To ensure comparability among the projects, APMP has standardized sampling locations and frequency, analytical protocols, quality control and data handling procedures (Brown and Goodyear, 1987). Sampling stations are established at the sites to monitor physical, chemical, and biological parameters in the pretreatment, transition and post-treatment phases of the program (**Figures 2 and 3**). Field and laboratory experience gained in the program will assist in developing future research and management strategies. Current components of the research design are outlined in the following sections.

3.1 Biological Parameters

The APMP biological parameters (**Table I**) emphasize the monitoring of fisheries resources. The fisheries protocols emphasize the types of data to be collected, allowing investigators in the states to select the methods most effective for their geographic regions and for the species present at the study site. The species composition of the fish community is determined during each spring and fall and recruitment success is monitored in the fall. Populations are estimated annually in streams and twice in lakes, before and after liming. Age and growth are determined twice in both lakes and streams -- once before liming and again at the end of the project.

Three *in situ* bioassays are also scheduled. In the lake system, bioassays are conducted once during the pretreatment phase, once during the transition phase after liming, and again during the post-treatment phase. In-stream bioassays are completed once during pretreatment, once during operational liming, and once at the end of the project to determine affects of doser shutdown. The study design for streams uses bioassays conducted in the segment upstream from the doser as a "control." Both the lake and stream designs include bioassays at two other control sites -- one at the site of test stock origin (i.e., hatchery) and another, if possible, in a nearby water body not affected by acid deposition. A field, rather than laboratory, examination of the fish for signs of acid or toxic stress is emphasized. Because decomposition begins rapidly after death, detailed examination of gills later in the laboratory is not sufficient. Specimens are preserved in formalin for later histological examination, if necessary.

The macroinvertebrate community is monitored to detect any short-term system responses to liming and to help explain any observed changes in the fish community. Qualitative sampling is conducted quarterly during ice-free periods in both lakes and streams to determine species presence and relative abundance. The zooplankton community in lakes may

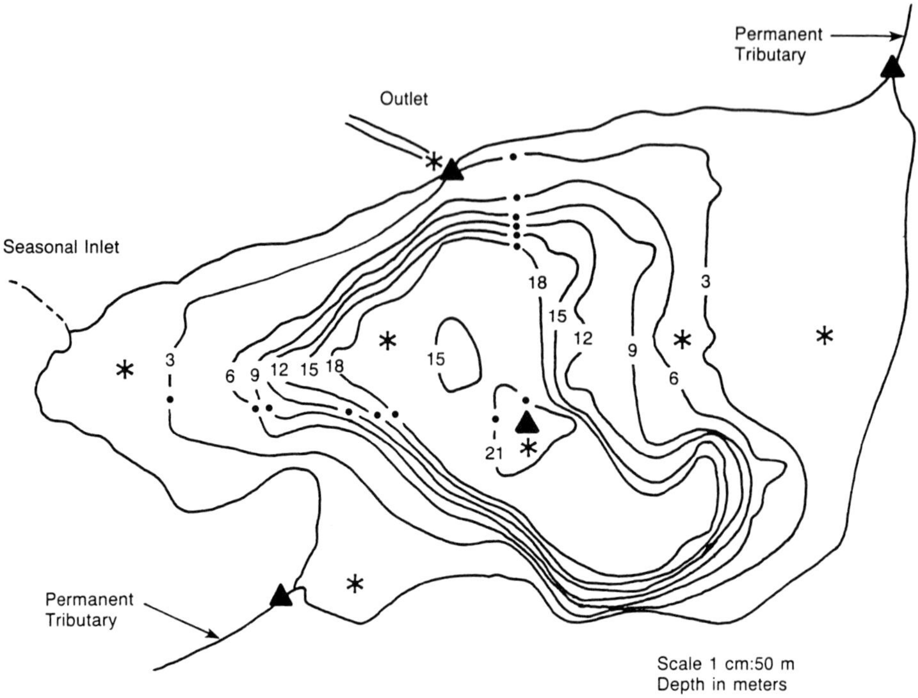

Pre- and Post-Treatment Monitoring

▲ Physical and Chemical Water Quality Sampling Stations (Surface and Water Column)

• Sediment Sampling Stations

Transition Monitoring

* Physical and Chemical Water Quality Sampling Stations (Surface and Water Column)

Figure 2. Map of hypothetical lake illustrating physical and chemical water quality and sediment sampling locations.

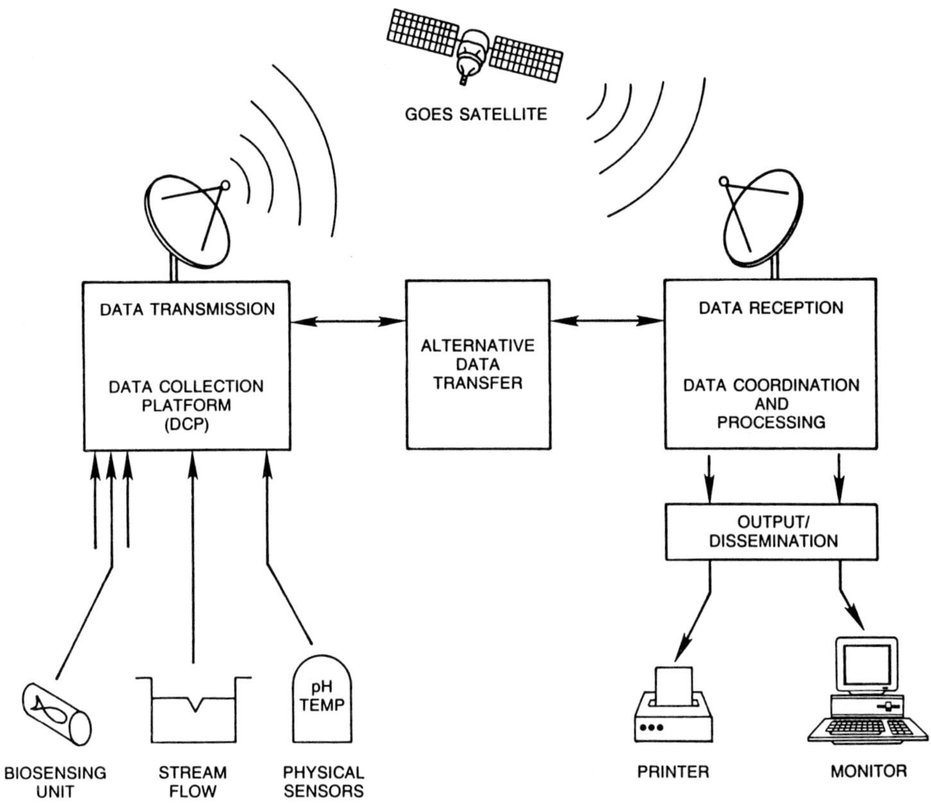

Figure 3. Schematic of APMP monitoring stations.

Table I

APMP biological parameters and their frequency of measurement.

	Frequency	
Parameter	**Lake**	**Stream**
Fish		
Community characterization	Semi-annually	Semi-annually
Recruitment success	Annually	Annually
Population estimate	One pretreatment, one at end of project	Annually
Age and growth	One pretreatment, one at end of project	One pretreatment, one at end of project
In-situ bioassay	One pretreatment, one transition, one post treatment	One pretreatment, one operational, one during doser shutdown at end of project
Microinvertebrates		
Qualitative	Quarterly	Quarterly
Quantitative	———	Quarterly
Zooplankton (quantitative)	Monthly	———
Chlorophyll	Monthly (growing season)	Monthly (growing season)
Macrophyte distribution	Annually	———

demonstrate short-term responses to liming and may explain changes in the fish community. Zooplankton sampling is, therefore, conducted monthly during ice-free periods. Samples are collected with a conical plankton net.

In the stream systems, macroinvertebrates are sampled quantitatively and performed quarterly to allow more powerful statistical analysis of any changes in the community structure. For quantitative sampling, the Surber square-foot bottom sampler is used. In the relatively remote sites in this study, this sampler has three advantages over the use of multi-plate artificial substrates: 1) it eliminates the trip to the field to pre-expose the artificial substrates and the extra trip to either install or collect the samplers, 2) it eliminates the risk of losing samples by either vandalism or high flows, and 3) it allows more consistent sampling in streams with highly variable flows.

Chlorophyll _a_ concentrations in water samples are measured monthly during the growing season to detect gross changes in phytoplankton. Initially, chlorophyll is being analyzed in both lakes and streams. Preliminary data, however, indicate low chlorophyll concentrations in water samples from the lotic systems, and stream preriphyton sampling is being considered as a better estimate of primary productivity.

Macrophyte distribution is monitored annually in the lake system and mapped in combination with habitat features, such as substrate, that are important to fish species. Habitat features in the stream systems, including aquatic vegetation, are also mapped and updated annually.

In cooperation with the Tennessee Valley Authority (TVA) and Tennessee Technological University, the Laurel Branch site has been equipped with two satellite relay remote sensing platforms that record data on biomonitoring studies, pH, temperature, specific conductance and stream flow (Morgan _et al_., 1986). This set-up is unique in APMP and provides the ability to view the conditions of Laurel Branch and the fish response at any time. One platform is in the control segment above the doser site and another is at the downstream end of the study area. The platforms contain minimonitors that collect and transmit stream data on a predetermined cycle. *In situ* biomonitoring chambers record fish behavior (**Figure** 4). Automatic Isco water samplers are activated during periods of elevated fish stress and the samples are analyzed for metals suspected to be at or near toxic levels.

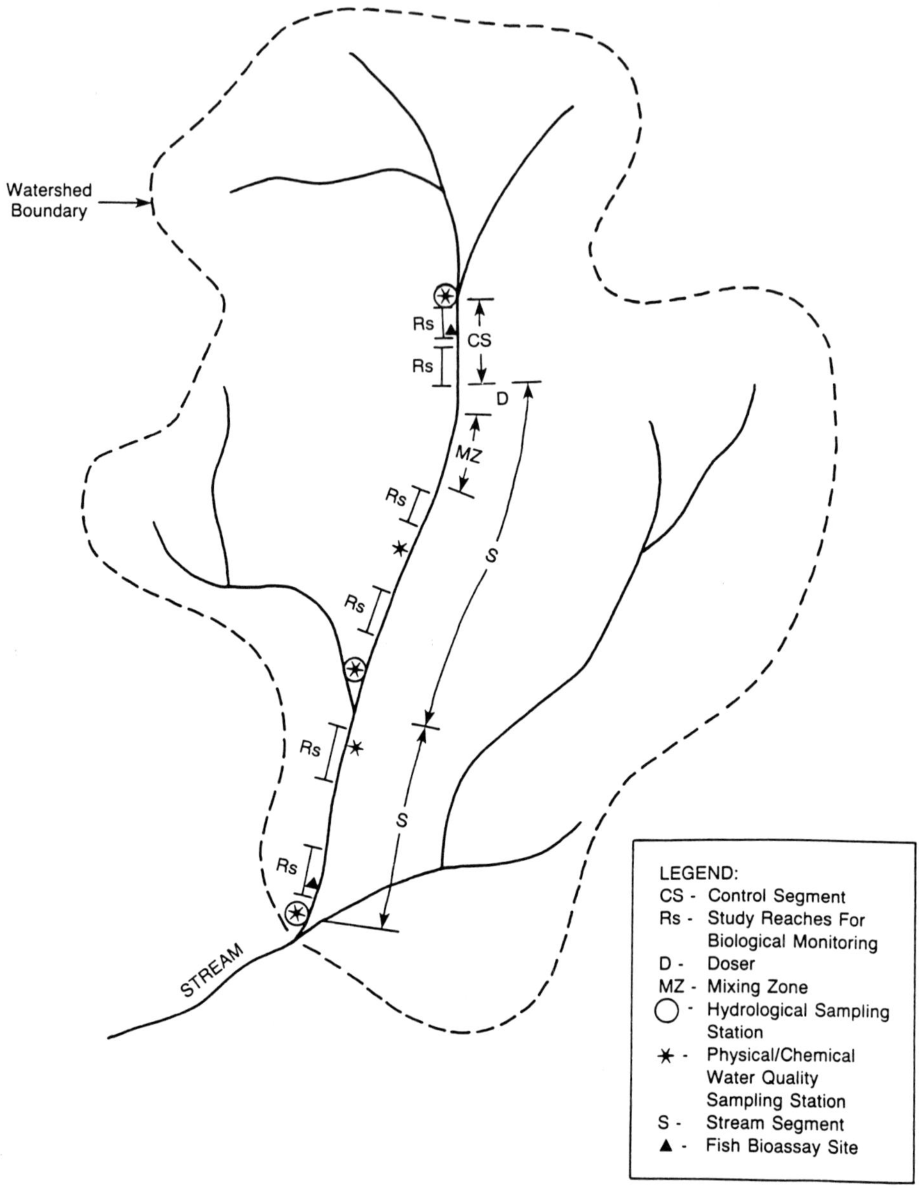

Figure 4. Remote water quality monitoring station employed on APMP Laurel Branch study, Cherokee National Forest, Tellico Wildlife Management District, Tennessee.

3.2 Water Quality Parameters

Although the major thrust of APMP is directed toward the biological effects of liming -- especially fisheries resources -- a concerted effort is being made to monitor the chemical dynamics of the systems under investigation.

A total of 24 physical and chemical water quality variables is monitored in APMP (**Table II**). These variables were selected because of their significance to the biological community and for calculating mass balance budgets needed to evaluate chemical response to liming. Also, they are important in quality control (i.e., calculating ionic balances). Total N is used as an indicator of primary and secondary production potential, and the N/P ratios are used to indicate limiting nutrients. Fluoride is monitored because of its complexation potential with Al and its importance in determining ionic balances. Total Ca is used to describe the Ca dynamics of the limed system.

3.2.1 Sample collection. The APMP requires monitoring of numerous chemical parameters during and after liming and the potential presence of undissolved $CaCO_3$ in the water samples could create problems. Because of the nature of this research program and the great importance of several parameters for data interpretation, the following special precautions are being taken in sample collection:

1. Samples for several critical parameters (ANC, pH, dissolved Ca and monomeric Al) are collected directly into syringes using a peristaltic pump equipped with inert, flexible tubing. This arrangement ensures closed-headspace samples and eliminates the possibility of erroneous data due to interaction of the samples with air.

2. Lime particles in a water sample may continue to dissolve in the syringe and cause chemical changes, thus affecting the values for ANC, pH, and dissolved Ca. These samples are therefore filtered by in-line filtration before they enter the syringe. Because pH and ANC are routinely determined on filtered samples, unfiltered syringe samples are also collected prior to liming to determined the effects of filtering, if any, on these parameters. This is done throughout the program for stream samples. Lake samples are filtered only during transition, because this short period is the only time undissolved lime should be present.

Table II

APMP water quality parameters and corresponding methods for measurement/analysis.

Parameter	Method
Acid neutralizing capacity (ANC)	Titrimetric, two end-point
Aluminum, total dissolved	Atomic absorption (furnace)
Aluminum,totalmonomeric	Atomic absorption (furnace)
Ammonium,dissolved	Colorimetric, automated (phenate)
Calcium, dissolved	Atomic absorption (direct aspiration)
Calcium, total	Atomic absorption (direct aspiration)
Chloride, dissolved	Ion choromatography
Fluoride, dissolved	Potentiometric, ion-selective electrode
Inorganic carbon, dissolved (DIC)	Oxidation (IR detection)
Magnesium,dissolved	Atomic absorption (direct aspiration)
Nitrogen, total	Coloroimetric, automated (phenate)
Nitrate, dissolved	Ion chromatography
Organic carbon, dissolved (DOC)	Oxidation (IR detection)
Oxygen, dissolved	Membrane electrode
pH	Potentiometric
Phosphorus, total	Colorimetric, automated (absorbic acid)
Potassium, dissolved	Atomic absorption (direct aspiration)
Silica, dissolved reactive (streams only)	Colorimetric (heteropoly blue)
Sodium, dissolved	Atomic absorption (direct aspiration)
Specific conductance	Wheatstone bridge
Sulfate, dissolved	Ion chromatography
Temperature	Combination probe/thermometer
Transparency (lakes only)	Secchi disk
Turbidity	Field turbidimeter

3. To reduce the length of time needed to collect samples and the possibility of contamination, water for the remaining parameters is collected in a cubitainer, which must be processed in the laboratory within 24 hr of collection.

4. For quality control on field samples, two sets of duplicates are collected. For the stream projects, one duplicate must be in the control segment, the other in a treated segment (Figure 3).

3.2.2 Frequency of Sampling. In lakes, water samples are collected and analyzed monthly. There are, however, some exceptions:

1. Total N and total P are measured quarterly because only a gross estimate of change in productivity is necessary.

2. Turbidity is measured only during transition because this is the period when it could change significantly. Secchi depth readings are sufficient during other periods.

3. During the transition period, beginning the week before liming, most parameters are measured every other day and then weekly thereafter until the chemistry stabilizes.

4. During high inflow periods (i.e., spring runoff), most parameters are measured weekly.

In stream systems, water samples are collected and measured 12 to 20 times per year on a flow-weighted basis (i.e., determined by the annual stream hydrograph). In addition, the telemetry platforms enable more frequent monitoring in Tennessee. Minimonitors on these platforms collect data on pH, specific conductance, temperature, and dissolved O_2, which are reviewed daily. The real-time data quickly indicate occurrence of episodic events and possible doser malfunction.

3.2.3 Sample Analysis. For most parameters, analytical protocols used by the NSWS (Hillman et al., 1986) have been adopted. Some modifications have been made:

1. Because ANC is a critical parameter and may change if undissolved lime is in the sample, a 24-hour maximum holding time for the samples has been established for APMP. To meet this timing with the large number of samples that need to be analyzed in this short time, ANC determinations are made using the fixed end-point titration method for low alkalinity waters (U.S. EPA, 1983).

This method, commonly referred to as the two end-point method, has been shown to give values consistent with the more time-consuming Gran plot method (Omernik and Griffith, 1986). In addition to allowing the laboratory personnel to complete all the desired analyses within the required time-frame, it also reduces the variability among the different laboratories in technique and computation that is sometimes associated with the Gran plot method.

2. Because of its importance and ease of analysis, several measurements of pH are taken at each sampling site. A field measurement gives a value indicative of the pH affecting the biota at that time and is critical because of the potential for continued dissolution of lime in the sample water. Once the samples are delivered to the laboratory, a closed-headspace pH is determined according to the survey protocol (Hillman et al., 1986). This value, obtained by directly injecting the sample into a closed system to prevent atmospheric exposure, gives an indication of the pH of the water at the actual time of collection and serves as a quality control check on the field value. A third pH is determined just before the titration for each ANC determination. Finally, an air-equilibrated pH value is determined on a separate sample to approximate the degree of CO_2 saturation.

3. The methyl isobutyl ketone extraction for monomeric Al is conducted in the laboratory rather than the field, primarily because of safety reasons (the chemical is flammable and an irritant).

4. Because lime particles may be suspended in the water and may agglomerate with time, turbidity is measured in the field.

5. Each state has a backup laboratory to ensure sample analysis within specific holding times, should the primary laboratory have problems. It is anticipated that the backup laboratory will be used rarely, if at all, but it must still meet the operating criteria imposed on the primary laboratory (e.g., audits and maintaining APMP specifications). Because this laboratory will be used only infrequently and for relatively few samples, the decision was made to use one state's primary laboratory as the backup for another state.

3.3 Sediment Parameters

The APMP measures three sediment parameters: pH, total Ca and porewater Ca. Sediment samples are collected quarterly in the stream systems and semi-annually in the lake. The collection protocol specifies a coring device because of the requirement for analysis of pore water. The analysis for exchangeable H^+ is not performed because of the difficulty of the analysis and questionable usefulness to this program.

3.4 Hydrologic Parameters

APMP hydrologic parameters include precipitation, evapotranspiration and snow storage data obtained from the nearest meterological station (**Table III**). Information on precipitation chemistry is normally taken from the nearest station in the National Atmospheric Deposition Program or the National Trends Network. However, because of the regional variation in precipitation chemistry in Minnesota, its monitoring has been upgraded to include a precipitation chemistry monitor in the Thrush Lake watershed. Hydrological records at the stream sites, which include upstream flow, tributary inflow, and discharge, are particularly important in establishing dosing rates.

4. DATA MANAGEMENT

To accomplish the goals of APMP, an error-free data base is required for analyses and interpretation. Errors can occur in data recording, sample chain of custody, machine coding and verification, data filing and storage, and data transmission. Therefore, sound data management is an integral part of the total quality assurance and quality control program (**Figure 5**). Prototype data forms are provided to the investigators and data flow is controlled and closely monitored from field collection through final storage (Brown and Goodyear, 1987).

5. MITIGATION TREATMENT

To aid in evaluating and comparing options for application strategies in APMP, a microcomputer decision assistance program called "DeAcid" was developed (Fraser et al., 1985). DeAcid factors three items into the selection process -- technical feasibility, effectiveness in meeting water quality objectives, and cost. Based on site chemistry, it conducts a number of technical assessments to determine the amount of neutralizing agent required to meet the target pH and alkalinity.

Table III

APMP hydrologic parameters and their frequency of measurement.

Parameter	Lake	Stream
Precipitation	Monthly	Monthly
Snow Storage	Monthly	Monthly
Evapotranspiration	Monthly	Monthly
Lake Level	Monthly; weekly during transition	—
Discharge	Continuous (outlet)	Continuous
	—	Flow-weighted (upstream)
	—	Flow-weighted (tributary inflow)

APMP Data Management Scheme

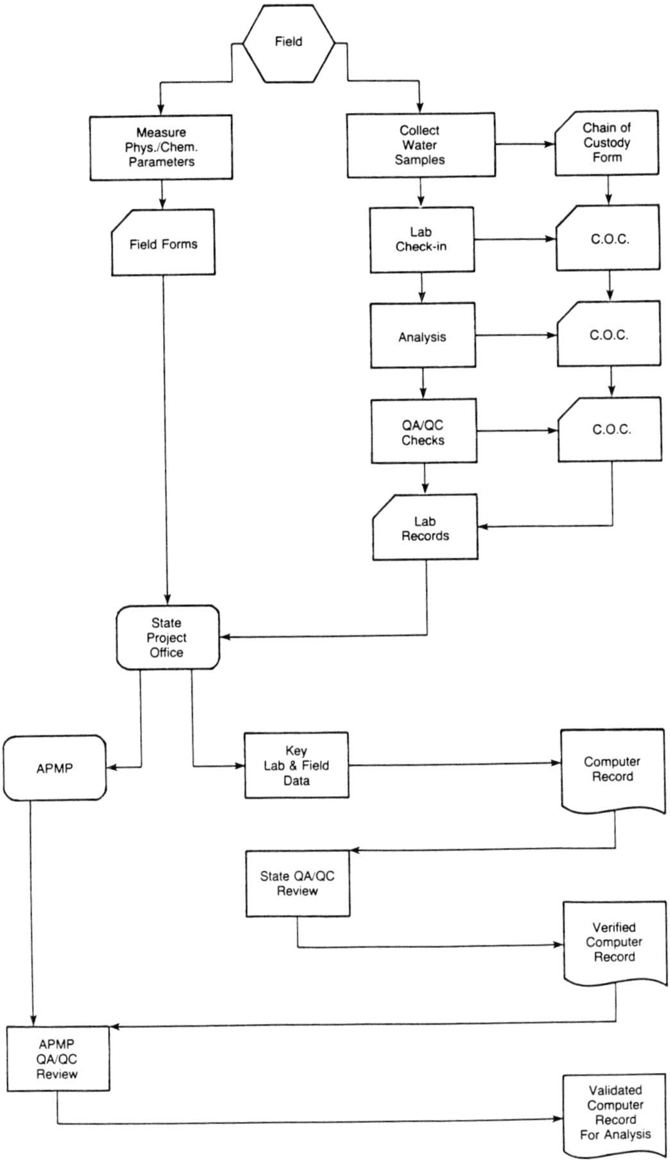

Figure 5. The data managment process. Schematic of information flow in the APMP research program.

Liming at APMP sites, originally scheduled for fall 1986, was delayed until 1988 to provide an additional year of baseline data. For Thrush Lake, treatment was in May; a boat equipped with a slurry box was used for the application. The dose requirement, based on protective treatment, was approximately 5 t of limestone, averaging 2 *u*m in particle size.

Treatment of streams is more complicated because the application must be continuous for at least the periods of flow when pH and alkalinity are depressed (Fraser et al., 1985). The limited availability of equipment, the limited access to remote sites, the need for continuous application, and the relatively high costs compared to lake liming offer particular challenges. In addition, stream sites require somewhat extensive construction work for doser installation, and these effects on baseline data must be assessed.

In West Virginia, rotary drums have been used to treat several small streams (Zurbuch, 1984). This application method appears suitable for selected streams and is the method used for Dogway Fork. The streams in Massachusetts and Tennessee will be treated with limestone dosers powered by mechanical energy produced by an instream turbine. In all streams a measured amount of slurry will be released into the stream depending on flow, pH and ANC. Doser operation at the stream sites will begin in late summer or early fall of 1988.

6. SUMMARY

To ensure healthy and productive commerical and sport fisheries, resource managers are interested in remedial measures that may be used to restore and protect aquatic ecosystems from acidity. With that goal, research on mitigation of surface waters impacted by acidity has been an active part of the NAPAP. The management and coordination of the federal mitigation research effort are provided by the U.S. Fish and Wildlife Service. In 1985 the Service designed and initiated a 5-yr research program, APMP, to evaluate the ecological responses to mitigative liming of surface waters. The APMP is being accomplished through cooperative research projects in Massachusetts (Whetstone Brook), West Virginia (Dogway Fork), Tennessee (Laurel Branch) and Minnesota (Thrush Lake). In each of the projects the physical, chemical and biological changes that accompany liming are being investigated by a standardized approach. The approach stresses uniformity among the projects and specifies sampling and analytical protocols and quality assurance procedures. The insight achieved in APMP will assist in determining the ecological feasibility of liming as an interim mitigative measure in combating the effects of acid deposition on aquatic resources.

ACKNOWLEDGMENTS

I thank my colleagues, Jim Brown, Rita Villella and Bill Hartman for their constructive comments and the session and journal reviewers for their technical review of the draft manuscript. Much of the information provided in this program overview is taken from a previously published article (Schreiber et al., 1988).

AN OVERVIEW OF LAKE NEUTRALIZATION EXPERIMENTS IN ONTARIO

D.P. Dodge
G.M. Booth
L.A. Richman
W. Keller
F.D. Tomassini

ABSTRACT

In 1981, an experimental neutralization program was initiated in Ontario to investigate the feasibility of using neutralization as an interim measure for the protection of acid-sensitive lakes and the restoration of acidified lakes. Aerial applications of powdered limestone ($CaCO_3$) to two lakes were used to increase whole-lake pH and alkalinity, and experiments were initiated to investigate the neutralization of lake trout (*Salvelinus namaycush*) spawning shoals with crushed limestone. Intensive studies are continuing to assess temporal patterns in water chemistry and to evaluate the long-term responses of zooplankton, phytoplankton, zoobenthos and fish communities.

1. BACKGROUND

In July 1981, the Ministers of the Environment and Natural Resources announced a 5 yr investigation of the feasibility of using neutralization to protect Ontario lakes that are in danger of acidification, and to rehabilitate lakes already acidified.

The mandate of the program was to provide Ontario's Natural Resource Managers with information regarding the feasibility of neutralization as a management technique while negotiations for abatement of acidic precipitation continued and control technology were developed. The study was in no way intended to divert effort or attention away from the necessity of emission abatement. Neutralization was and continues to be viewed as a temporary measure only. Neutralization alleviates the symptoms of the problem but does not eliminate the cause.

One of the initial objectives of this program was to test the feasibility of whole-lake liming to rehabilitate an acidified lake. Proof of successful rehabilitation would be the

Water, Air, and Soil Pollution **41** (1988) 75–83.

Table I

Pysical and chemical characteristics of study lakes. Chemical data are presented as a range of values measured from June 1982 to October 1983 (August 1983 for Bowland Lake). Some chemical characteristics for lakes that were not neutralized were measured in September 1986. Water samples were taken from the epilimnion at mid-lake.

	Bowland Lake	Trout Lake	Miskokway Lake	Fowke Lake	Laundrie Lake
Surface Area (ha)	108.0	290.0	237.6	245	375
Maximum Depth (m)	28.0	37.8	42.0	53	20.4
Mean Depth (m)	7.0	11.0	14.8	14	4.9
Retention Time (yr)	2.0	2.7	1.8		
pH	4.78-5.76	5.64-6.20	5.68-6.39	6.9-7.3	5.3-5.6
Alkalinity (mg L^{-1})	-0.89-0.47	0.78-1.23	0.80-1.15	14.87-25.23	-0.93-1.12
Total Al (μg L^{-1})	100-160	16-63	13-50	30-40	44-102
Conductivity (μs cm^{-1})	35-42	28-29	24-27	50-55	23-35
Color (true) (Hazen units)	1.5-7.4	7.0	16.6	9	13

restoration of a healthy, self-sustaining sport fishery. Because of the importance of lake trout *(Salvelinus namaycush)* as a sport fish in Ontario and the sensitivity of lake trout to acidification, the study concentrated initially on rehabilitating a lake trout fishery. While the emphasis was primarily on the sport fishery, a viable ecosystem was required to support such a fishery. Therefore, many other biological, chemical and physical aspects of the ecosystems were monitored for their responses to lake neutralization.

The second objective of this program was to test the feasibility of whole-lake neutralization to protect an acid-sensitive lake. Success was judged by the ability to restore effective acid-neutralizing capacity to the lake without inducing negative biological effects.

A further objective was to evaluate site-specific neutralization, such as the addition of limestone to natural spawning shoals, as a cost-effective means of protecting the acid-sensitive development stages of lake trout. This technique could be used in acid-sensitive lakes where whole-lake neutralization would be an expensive option, or to protect shallow, nearshore spawning areas of neutralized lakes if these were not adequately protected from spring-melt acid episodes by whole-lake treatments.

2. STUDY SITES

Physical and chemical characteristics of all lakes are given in Table I; Figure 1 shows the study area. The selected lakes all had low concentrations of metals related to smelter emissions, such as Cu and Ni, so that the effect of neutralization on the biota could be addressed without introducing metal toxicity as an additional variable.

One acidic lake, Bowland Lake, and two lakes sensitive to acidification, Trout and Miskokway Lakes, were selected for intensive studies.

Before neutralization, Bowland Lake had a self-sustaining yellow perch (*Perca flavescens*) population only. Although lake trout had been previously reported in Bowland Lake, none were captured in 1983. Trout Lake has a complex fish community of 17 species, including lake trout and smallmouth bass (*Micropterus dolomieui*). The sport fish community in Miskokway Lake was similar to Trout Lake (Table II).

A fourth lake, Fowke Lake, with a self-producing native lake trout population was chosen

Table II

Fish species collected in Trout Lake, 1982 to 1986.

Salvelinus namaycush	- Lake trout
Coregonus artedii	- Cisco
Umbra limi	- Central mudminnow
Catostomus commersoni	- White sucker
Phoxinus eos	- Northern red-belly dace
Notemigonus crysoleucas	- Golden shiner
Notropis cornutus	- Common shiner
Notropis heterolepis	- Blacknose shiner
Pimephales notatus	- Bluntnose minnow
Semotilus atromaculatus	- Creek chub
Ictalurus nebulosus	- Brown bullhead
Culeae inconstans	- Brook stickleback
Lepomis gibbosus	- Pumpkinseed
Micropterus dolomieui	- Smallmouth bass
Perca flavescens	- Yellow perch
Etheostoma exile	- Iowa darter

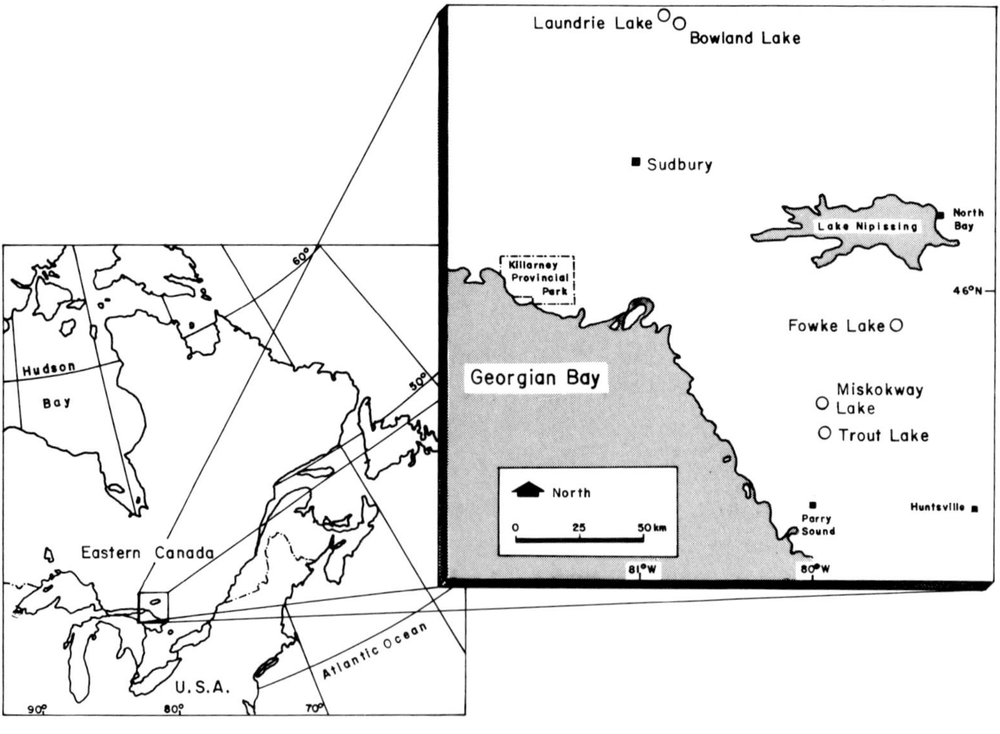

Figure 1: Location of study lakes.

as a non-acidic control lake for bioassay experiments. A moderately acidic lake, Laundrie Lake, directly downstream from Bowland Lake, was chosen for shoal liming experiments (site-specific neutralization). A mature lake trout population appeared to be spawning but no recruits had been captured in Laundrie Lake.

Field work began in the spring of 1982. Bowland Lake was neutralized in August 1983, and Trout Lake was neutralized in May 1984. Miskokway Lake was not neutralized and was eliminated from the study as a candidate lake for neutralization in 1984. Miskokway, however, was still tested for water chemistry and fish bioassays. Phase I, the first 5 yr of the study, was concluded in the fall of 1986 (Booth et al., 1987). A second phase of the study has since been initiated and will be completed in 1991.

3. PROGRAM ELEMENTS

3.1 Neutralization

Calcite (limestone ($CaCO_3$)) was chosen because it is safe to handle, inexpensive and does not cause "pH shock." For Trout and Bowland Lakes, delivery of calcite from nearby airports in Canso water bombers proved the most cost-effective.

A model developed in Sweden (Sverdrup, 1983) was used to calculate the amount of calcite required. This model accounts for lake pH, lake mean depth, particle size and dissolution efficiency of the neutralizing agent, and the base neutralizing capacity of the sediment.

Eighty-four t of dry, powdered limestone (particle size - mean diameter 9.1 *u*m) were dropped in August 1983 on Bowland Lake, from a Canso water bomber carrying a payload of approximately 2.4 t. The lake was stratified during neutralization. However, since lake trout were not planted until after the fall turnover, it was not critical that the whole lake be neutralized immediately. Chemical sampling during and after neutralization showed that 47% of the limestone dissolved (Molot et al., 1986). This dissolution was lower than expected and may have resulted from electrostatic forces in the dry powder causing clumping of calcite particles which dissolved less efficiently.

In an effort to boost the dissolution efficiency of the Trout Lake application, a finer particle size (mean diameter 5.4 μm) of calcite (147 t) was mixed in a 70% water slurry using a sodium polyacrylate dispersant (Busperse 39 from Buckman Laboratories). The slurry and dispersant enhanced dissolution efficiency (to 86%) without reducing the payload of calcite delivered by the Canso on each trip, making this technique more cost-effective than using dry powder. Trout Lake was neutralized in May 1984 before the lake was fully stratified in order to maximize the volume of the lake which received calcite (Molot et al., 1986).

Shoal liming on Laundrie Lake was done with three grades of calcite: 1/8", 3/4", and a mixture of these two plus 6" rock. Several replicates of each treatment were placed in 1 m^2 plots at five sites to test the survival of lake trout eggs in artificial incubators. Water chemistry was measured from within the incubators on alternate days during spring-melt.

At Miskokway Lake, one-half of a shoal which was known to be used by spawning lake trout, was neutralized to test whether lake trout would avoid the light colored calcite. Baskets were buried randomly in both neutralized and unneutralized sections to assess the distribution of naturally spawned eggs.

3.2 Water Chemistry

In Bowland and Trout Lakes, water samples were collected monthly before and after neutralization (ice-conditions permitting). Water samples from the deepest mid-lake basin were volume-weighted to provide an estimate of whole-lake conditions and concentration depth profiles were sampled for some parameters. The pH, alkalinity, Ca and Al were measured, in addition to nutrients, other cations and anions and trace metals.

In Bowland Lake, extensive chemical sampling was conducted over the period of addition of calcite to describe the immediate chemical responses of an acidic lake to neutralization.

An intensive 3-to-8 week program was implemented in Bowland and Trout Lakes during March and April, the period of spring-melt, in 1983, 1984 and 1985 to sample the quality of nearshore water. Transects running perpendicular to shore were sampled at 0, 1 and 2 m below the bottom of the ice. Additional sites located over possible lake trout spawning shoals were sampled at the surface, on the bottom and interstitially in the shoal rubble. The purpose of this sampling program was to determine whether whole-lake neutralization mitigated nearshore episodes of increased acidity and high Al.

3.3 Biological Sampling Programs

3.3.1 Bowland Lake. Surveys in Bowland Lake documented the status of the planktonic, macrophyte and benthic invertebrate communities, pre- and post-neutralization. Benthic algal communities were mapped and described during shoreline cruises made in the late summer from 1982 to 1986. A littoral zone survey assessed macrophyte distribution before neutralization in the summer of 1982 and again in 1985 after the lake was neutralized. The benthic invertebrate community in Bowland Lake was surveyed in 1982, 1984 and 1985. Diver searches, bottom sampling with dredges and cores, and colonization rates for artificial substrates were used to provide a quantitative assessment of the benthic community.

Phytoplankton and zooplankton are collected monthly at the mid-lake station using a 3 L closing water bottle and a Schindler-Patalas trap, respectively. Portions of samples from discrete depths are pooled to give a single volume-weighted sample.

Following neutralization, fingerling, yearling and 2 yr-old hatchery lake trout and adult lake trout from other lakes were stocked in Bowland Lake in late fall 1983 and again in 1985. The fish community in Bowland Lake was initially assessed in late summer and fall of 1982, and there was a detailed assessment throughout the open water period in 1983. From 1984 to 1987, the community was assessed in the spring and fall. A variety of standard gear was used (e.g., emergent lake trout traps, gill nets, minnow traps and trapnets). A representative sample of yearling yellow perch was tested each year for Hg concentration.

Bioassays of sensitive stages of lake trout were used to simulate responses that might be occurring in nature. Each fall, hatchery lake trout eggs are transported to the study lakes and the circumneutral reference lake (Fowke Lake) and placed in incubators at 1 m depths on shoals around the lakes. These eggs are allowed to develop through the winter until after ice-out in the spring. Periodic sub-sampling provided estimates of mortality at various stages of development and hence the time of year at which mortality of any eggs and fry occurs.

A second bioassay experiment conducted annually exposed juvenile hatchery-reared lake trout in cages to worst-case water conditions immediately under the ice during spring-melt and again shortly after ice-out. Mortality rates are used as an index of toxicity of nearshore water quality before and after neutralization.

3.3.2 Trout Lake. Phytoplankton and zooplankton are collected monthly as at Bowland Lake. Fish communities were assessed each spring with standard netting gear as at Bowland Lake and in the fall with additional effort using electrofishing boats for the collection of spawning lake trout.

Lake trout egg bioassays were conducted each winter in conjunction with the bioassays on Bowland and Fowke lakes. Caged lake trout fry were exposed under the ice each spring.

4. CONTINUING QUESTIONS

Although lake neutralization improved water quality, the technique provides only a temporary solution. Continuing acidic precipitation and runoff will cause the neutralized lake to return to its previous acidified state. Additional application of limestone to Bowland and Trout Lakes is not planned now. The effects of reacidification on the lakes remain to be determined.

There are uncertainties associated with productivity and biomass changes in lakes that have been neutralized. More research is needed to determine if the forage base in acidified lakes is suitable to sustain an introduced fishery. Better understanding of fish growth rates is also needed to distinguish between natural variability and changes induced by indirect or direct effects of the treatment. It is also still uncertain if invertebrate communities will return to original preacidified assemblages and remain healthy.

Overall, whole-lake neutralization is a promising technique for lake rehabilitation; however, more monitoring is needed to address unanswered questions about long term effects. Further trials are necessary to test the neutralization of lake trout spawning shoals on a management scale.

ACKNOWLEDGMENTS

The Experimental Lake Neutralization program was jointly funded by the Ontario Ministries of Natural Resources and the Environment and was part of the Acid Precipitation in Ontario Study (APIOS).

Program design and data interpretation are a joint effort involving a Technical Task Force, Steering Committee, and BAR Environmental. We would like to acknowledge the work of the members of those groups. Many of the individual projects within the program are performed through the private sector. A listing of all participants in Phase 1 of the program can be found in the Final Technical Report, 1987.

LIVING LAKES:
AN AQUATIC LIMING AND FISH RESTORATION DEMONSTRATION PROGRAM

Robert W. Brocksen
Paul W. Emler, Jr.

ABSTRACT

Living Lakes, Inc. (LLI), a not-for profit organization, sponsors an applied aquatic resources restoration demonstration program for acidified waters. In 1986 and 1987 LLI began evaluation of seven different liming technologies on 22 lakes and 10 streams in 6 states. Lakes and streams have been treated in the states of Massachusetts, Rhode Island, New York, Pennsylvania, Maryland and West Virginia. The Living Lakes program commits to ongoing water quality analysis for the lakes and streams treated for the life of the program and, if necessary, retreatment. Living Lakes has recently begun investigations into the treatment of partial or entire watersheds in conjunction with the Loch Fleet project in Scotland, investigations in the Black Forest in Germany, and studies in the Woods Lake watershed in the Adirondack region of the state of New York, co-sponsored by the Electric Power Research Institute.

1. INTRODUCTION

The potentially adverse effects of acidification on surface water quality and the resulting effects on fisheries have been known for many years. The use of neutralizing chemicals is a common management tool which has been used in American agricultural practices for centuries (Ruffin, 1832; Rorison, 1980). For several decades, similar liming techniques have been practiced on aquatic resources to promote fisheries in acidic waters (Fraser et al., 1982). The causes of this surface water acidification are issues of debate that are not likely to be resolved in the near future. However, regardless of the sources of the acidity, a need exists for resource management techniques to protect or maintain aquatic biota in affected waters. Even if acidic inputs decline substantially, such mitigation techniques can be used to help restore biological productivity more quickly than the years required for a water body to recover naturally.

Water, Air, and Soil Pollution **41** (1988) 85–93.

LLI was incorporated as a not-for profit corporation in December 1985 to design and demonstrate an aquatic liming and fisheries restoration demonstration program for acidified waters. The primary objective of LLI is to demonstrate the effectiveness of using lake and stream liming techniques to neutralize acidic waters. This program builds on the knowledge gained from liming research and resource management activities in Norway (Hindar and Rosseland, 1988), Sweden (Nyberg and Thornelof, 1988), Great Britain (Brown et al., 1988) and the United States (Porcella, 1988; Schreiber, 1988). The ultimate goal is the restoration and protection of ecologically, recreationally or economically important fisheries. Progress towards this goal includes the following components: 1) mitigative liming to neutralize acidified systems and restore fisheries; 2) maintenance liming to neutralize acid-stressed systems and maintain existing fisheries and 3) research to enhance the effectiveness or applicability of future liming operations. The overall objectives or measures of success are site-specific water quality targets that will provide a water quality capable of supporting the desired fisheries for the site.

This paper describes the criteria used to select sites for treatment, the methodologies used, the types of sample analysis, and presents changes in pH and ANC for 23 lakes treated in the program to date.

2. METHODS AND MATERIALS

Site characterization and selection are two of the most critical procedures of the field implementation operations. Sites are not restricted by type, ownership or geographical region of North America; however, a minimum set of environmental, biological and technological criteria (Living Lakes, Inc., 1987) has been established for selection of candidate lakes and streams. Generally, a lake or stream must: 1) have a suitable habitat to support fish populations; 2) have alkalinity or acidity indicative of systems that are biologically stressed or vulnerable to the adverse effects of acidification; 3) have acceptable metal concentrations; 4) not be affected bypoint source discharges and 5) not be naturally occurring acidic systems such as acid bogs, etc.

Water samples collected in the field are sent to contractor laboratories for analyses of a wide range of parameters (**Table I**). Samples collected just prior to and following treatment are analyzed for the full suite of parameters, while at other times during the year, a modified suite is analyzed. For specific sites where metals may be a concern, fish tissues are analyzed for Al, Cd, Hg, Pb and Zn. These analyses are conducted using methods described in USEPA (1983), Goulden (1970), Barnes (1975) and Driscoll (1984).

Treatment application strategies are determined by the site hydrological characteristics, the water quality targets, size and accessibility of the site and fisheries status. Dosage calculations from the DeAcid Computer Program (Warfvinge et al., 1984) generate data used to develop a site-specific treatment plan so that sites are treated with the correct volumetrically-based dose of calcite. Basic site information requirements that provide input to the treatment application and DeAcid Model are water chemistry, hydrology and sediments. The size and accessibility of the site determine the method of material delivery. For lakes, inaccessible sites are most efficiently treated using application by helicopter, accessible sites over 40 ha can be treated using a barge, while smaller sites may be treated by slurry box. A stream may be treated utilizing automated battery-powered, water-powered or gravity feed devices, diversion wells and fines. In treating lakes, a limestone slurry is applied, where as in streams, either a slurry mix, dry powder or limestone chips, depending on the application method, are used. Research has identified calcitic limestone, the neutralizing agent used by LLI, as the material of choice, although a variety of other substances is available for treating the acidity of surface waters. Calcium carbonate, the major component of calcitic limestone, is a natural component of lake and stream buffering systems. The reactivity of $CaCO_3$ is moderate which protects fish from pH stock.

3. RESULTS AND DISCUSSION

To date, over 150 water bodies have been reviewed as potential program candidate sites with 33 lakes and streams (**Table II**) accepted into the program. Both public and private sites have been treated in Massachusetts, New York, Pennsylvania, Rhode Island, Maryland and West Virginia. While most program sites are demonstration sites, one treated in 1987, Flat

Table I

Full and Reduced suite of analytes sampled in the Living Lakes, Inc. Program

Full Suite	Reduced Suite
pH	pH
Acid neutralizing cpacity	Acid neutralizing capacity
Conductivity	Conductivity
Ca	Ca
Al (total)	Al (total)
Al (monomeric)	Al (monomeric)
Total P	Total P
Total N	Total N
Cd	
Fe	
Mn	
Pb	
Zn	
Dissolved organic C	
Dissolved inorganic C	
Nitrate	
Sulfate	
Cl	
F	
K	
Mg	
Na	
Ammonium	
Silicate	

Table II

SITE NAME	DATE TREATED	LAKE SURFACE AREA - acres (ha) or STREAM TARGET ZONE LENGTH - miles (km)		SITE TYPE
Baker Pond, MA	6/23/86	29.6	(12.0)	Seepage Lake
Curlew Pond, MA	6/27/86	43.2	(17.5)	Seepage Lake
Gallows Pond, MA	Spring 1988	43	(17.4)	Seepage Lake
Laurel Lake, MA	10/20-21/87	50.9	(20.6)	Drainage Lake
Lawrence Pond, MA	6/25/86	137.8	(55.8)	Seepage Lake
Lout Pond, MA	6/27/86	12.6	(5.1)	Seepage Lake
North Pond, MA	9/15/86	24.7	(10.0)	Drainage Lake
Round Pond, MA	11/14-15/87	21	(9.3)	Seepage Lake
Schoolhouse Pond, MA	6/23/86	21.0	(8.5)	Seepage Lake
Shubael Pond, MA	11/21/86	54.8	(22.2)	Seepage Lake
Snake Pond, MA	6/25-26/86	83.0	(33.6)	Seepage Lake
South Pond, MA	10/19/87	25	(10.1)	Drainage Lake
Bacon Ridge Branch, MD	February-June Annually	1.0	(1.6)	Stream
Mattawoman Creek, MD	February -June Annually	8.1	(13.0)	Stream
Barnes Lake, NY	10/16/87	7.2	(2.9)	Seepage Lake
East Pond, NY	10/31-11/1/86	57.3	(23.2)	Drainage Lake

Table II (Cont'd.)

SITE NAME	DATE TREATED	LAKE SURFACE AREA - acres (ha) or STREAM TARGET ZONE LENGTH - miles (km)		SITE TYPE
Heavens Lake, NY	10/29/86	22.2	(9.0)	Drainage Lake
Lake Kanacto, NY	8/9/86	9.9	(4.0)	Seepage Lake
Lake Madeleine, NY	10/28-31/86	303.8	(123.0)	Drainage Lake
Lake Tejekna, NY	8/9/86	4.9	(2.0)	Seepage Lake
Little Simon Pond, NY	8/7-9/86	155.6	(63.0)	Drainage Lake
Woods Lake, NY	9/16-17/86	56.8	(23.0)	Drainage Lake
Gifford Run, PA		4.7	(7.5)	Stream
McNerney Run, PA		3.1	(5)	Stream
Powell Creek, PA		10.6	(17)	Stream
Stony Creek, PA	Continuous	14	(22.1)	Stream
White Deer Lake, PA	10/26/87	48.9	19.8	Drainage Lake
Wolf Run, PA		4.3	(7)	Stream
Wallum Lake, RI-MA	11/13-15/87	322	(130)	Drainage/ Seepage Lake
Crouch Run, WV	1985	0.7	(1.2)	Stream
Flat Top Lake, WV	7/13-7/20/87	300.1	(121.5)	Reservoir
Laurel Fork, WV	Feb-June Annually	4	(6.4)	Stream
Yokum Run, WV	12/3/86	0.7	(1.2)	Stream

Top Reservoir, is a research site in which one arm of the reservoir was treated and the second arm left untreated to determine the effects on water quality and the benthic population. Three sites, Round and Gallows Ponds in Massachusetts and one in Minnesota, are technology transfer sites. Living Lakes, Inc. supplied the technical assistance and equipment necessary for treatment and the lake owner's associations treated the ponds.

During 1986, 15 lakes in New York and Massachusetts were treated using helicopter application of limestone slurry (70% calcite by weight). In 1987, seven new lakes were treated utilizing a slurry box on a power boat or a barge with spray arms that disperses a limestone slurry. In the spring of 1988, one new lake in Massachusetts was treated using the slurry box technique. During 1985 and 1986, two streams in West Virginia were treated with limestone fines. Two streams in Maryland were treated in 1987 with automated slurry dosers. In Pennsylvania and West Virginia during 1988, streams are being treated using a variety of techniques including water-powdered mechanical dosers, water powered turbine dosers, water-powered column dosers and diversion wells (Arnold, 1988; Ivanhnenko, 1988).

Post-treatment monitoring data for the lakes treated in 1986 (Brocksen et al., 1987) and 1987 indicate that the water quality objectives were met for all but one site and illustrate the effectiveness of applying limestone to increase the ANC of surface waters. Table III presents summary data for pre- and post-treatment pH and ANC. Sites that do not have 1986 data were treated in 1987. Gallows was treated in 1988, and post-treatment data are not yet available. Additional data analyses is reported in Adams and Brocksen (1988). At ten of the treated sites, fish have been stocked in accordance with fisheries management plans developed in cooperation with state agencies or private owners. Continuing biological monitoring will provide information on the success of these stockings.

In both 1986 and 1987, two films were produced describing the LLI program. A short 13 minute film, "Upon the Waters...", was made in 1986 that discussed the field monitoring and helicopter liming applications in New York and Massachusetts. In 1987, the film was expanded to include new lake treatment techniques, stream treatment and the QA/QC program.

Table III

	Pre-treatment		Post-treatment Fall 1986		Post-treatment Fall 1987	
SITE	**pH**	**ANC**	**pH**	**ANC**	**pH**	**ANC**
Lawrence	6.02	8.7	7.08	199	7.05	130
Shubael	6.72	64	7.23	212	7.06	174
Schoolhouse	6.15	8.6	7.18	209	7.03	131
Lout	6.21	34	6.96	192	6.84	144
Curlew	5.99	17	7.24	209	7.00	144
Baker	6.27	25	7.22	221	7.08	153
Snake	6.46	26	7.04	141	6.71	91
North	6.60	66	7.11	317	7.39	328
Round	5.99	11	--	--	7.21	219
Wallum	6.01	34	--	--	7.03	176
South	6.14	16	--	--	7.37	331
Laurel	6.51	41	--	--	7.10[1]	216
Gallows	4.99	-5.6	--	--	--	-
Madeleine	6.25	31	6.82	160	6.79	139
Heavens	6.12	39	6.91	180	6.84	135
East	6.06	25	6.60	119	6.65	95
Kanacto	4.83	-7.0	6.81	166	6.41	123
Tejekna[2]	4.51	-26	6.70	120	4.87	4.
Little Simon[3]	5.51	3.7	7.03	187	6.4[4]	90-13
Woods[3]	5.97	16	7.26	303	6.5-7.0[4]	200-6
Barnes[2]	4.86	4.7	--	--	7.07	291
White Deer	7.34	236	--	--	7.38	494
Flat Top	6.83	114	--	--	7.02	219

[1]Unofficial data, spring 1988
[2]Values for epilimnion samples only
[3]Research sites
[4]Data provided by Lake Acidification Mitigation Program Septemb
1987

Table 1. Summary data of pH and ANC measurements for 23 lakes treated under the Living Lakes Program

4. SUMMARY

Field monitoring of current sites will continue during the existence of the program. New lake sites will be limed in 1988, while liming new sites past 1988 will depend on the extension of the program and the availability of new sites. During 1988, an international watershed project was initiated. This project is described by Brocksen et al. (1988) elsewhere in this volume of papers and constitutes an exciting and promising method for enhancing terrestrial and aquatic systems impacted by excess acidity.

ACKNOWLEDGMENTS

The authors gratefully acknowledge the help of Jennifer Smith in the typing and editing of this manuscript. While many people and organizations have contributed to the formation and functions of the LLI Program, none are more crucial to the organization than Leslie DeMarco, Executive Assistant at LLI.

"NEW TYPE" FOREST DECLINES AND RESTABILIZATION/ REVITALIZATION STRATEGIES

A PROGRAMMATIC FOCUS

R.F. Huettl

ABSTRACT

Annual terrestrial inventories carried out in West Germany since 1984 indicate that for the last 4 yr approximately 50% of the forests were damaged. The damages were and are commonly attributed to the adverse effects of acid rain and other air pollutants. As most decline types are associated with nutritional disturbances, it was found that the application of lime and/or fertilizers could reduce or even remove the damages. Therefore, the West German government has established a nationwide program by which up to 80% of the costs of fertilization in forests revealing "new type" forest damages are subsidized. In practice, a clear distinction is made between compensative and ameliorative fertilization methods. Compensative fertilization means the application of Mg containing lime (generally 3,000 $kg \cdot ha^{-1}$) to neutralize further acidic deposition in forest soils. The ameliorative fertilization is focused on the specific application of nutrients, particularly in stands with acute nutritional deficiencies. This is accomplished by means of readily soluble Mg and K fertilizers at application rates of 500 to 1,000 $kg \cdot ha^{-1}$. In cases of insufficient P supply or for the activation of microbial decomposer chains in organic top soil horizons, P fertilizers are utilized. For ameliorative fertilization, foliar and soil analyses are recommended as a sound basis for fertilizer selection. In many cases, particularly in the States of Bavaria and Baden-Wuerttemberg, Mg containing lime in combination with fast soluble Mg and/or K (P) fertilizers are applied. Between 1984 and 1987, approximately 400,000 ha of "new type" forest damage areas have been limed and/or fertilized in West Germany. As positive as the application of lime and/or fertilizers appear to be, there exists some potential risks and limitations of these practices with regard to restabilization and revitalization of declining forestt stands.

1. INTRODUCTION

According to the German forest damage inventory of 1987, the "new type" forest declines have for the first time decreased by approximately 2% (VWD, 1987). This slight improvement is

Water, Air, and Soil Pollution **41** (1988) 95–111.

probably related to more favorable climatic conditions during the recent vegetative periods. However, the forest damage inventory of 1987 is not completely comparable to the previous surveys (conducted since 1983). As in 1987, the number of observations was reduced substantially. In those states of the Federal Republic of Germany where the damage inventories were carried out by applying the same methods as during the previous year, the damaged area showed increases of about 10%. In any case, the recent forest damage report also indicates that more than 50% of the German forests show decline symptoms. Considering the fact that the applied inventory methods are based on a multi-causal parameter (i.e., foliage losses), the results cannot be used for scientific research focused on the determination of causal mechanisms. However, the survey data indicate without doubt a significant environmental problem, particularly for a densely populated and highly industrialized nation as West Germany, which places a high value on its forest resource.

2. SUCCESSFUL RESEARCH

Since the beginning of the 1980's, the potential causes of the "new type" forest damages were investigated with much financial effort and man-power. In West Germany alone, more than 300 research projects are being conducted related to this problem. Present research results indicate that the forest damages are obviously due to various causes which act together depending on specific site conditions, tree species and anthropogenic impacts (Huettl, 1988). So far, no conclusive ideas exist on how to evaluate the single causes. Nevertheless, an overwhelming majority of the scientific community involved in "new type" forest research believes that the recent declines are in some way related to the adverse effects of air pollution. It is also commonly agreed that many decline types in various tree species are associated with nutritional disturbances.

Based on deficiency symptoms as well as on foliage and soil analysis, it appears that the nutrient supply of forest ecosystems was poor at many sites over the last decades. Particularly for Mg and Ca, poor parent materials, the Mg content was frequently reduced below the threshold value for sufficient supply. For instance, the analysis of needles which reflect the nutrient supply in the soil for a series of Norway spruce and silver fir stands in the southern Black Forest, show a pronounced decrease of Mg content within the time period of 1975 to 1983 (Huettl, 1985).

A dramatic deterioration of K supply was also detected at specific sites. The historic comparison of needle analysis of about 20 Norway spruce stands on moraine sites in southwestern Germany revealed a reduction of K contents of more than 60% from 1961 to 1983.

Such acute nutritional disturbances are not only reported from German forests. In France, the Netherlands, Belgium, Austria, Switzerland, Sweden, Poland, Czechoslovakia, the German Democratic Republic, the Soviet Union, the U.S.A. and Canada have also reported similar findings in damaged forest ecosystems.

Furthermore, the relatively high H^+ deposition rates into forest ecosystems have resulted in increased leaching losses of essential cationic nutrients such as Ca, Mg and K from the rooted soil horizons. In this context, historic changes of the chemical soil status were produced and related to anthropogenic impacts resulting in accelerated soil acidification and nutrient impoverishment. Furthermore, under these soil conditions, the destruction of clay minerals leads to the mobilization of metal ions like Al which may impede uptake of cationic nutrients or may even cause direct root and mycorrhiza damage.

The direct impact of gaseous air pollutants such as 0_3, SO_2, NO_x and NH_x on the foliage, are well known. The effects of dry deposition of air pollutants such as HNO_3, H_2O and formaldehyde in the tissue of leaves and needles require more study. The combination of acid precipitation, particularly with acid fog, and dry deposition leads to increased leaching of nutrients such as K, Mg, Zn, Ca, Mn, and other tissue constituents like carbon hydrates from the foliage. Increased foliar leaching can occur even without the pre-damaging effects of air pollutants solely due to the influence of acidic precipitation (Bosch et al., 1986).

Increased N inputs into the forest ecosystems of western Europe have also contributed to the overall problem. Such emissions originate from the burning of fossil fuels, but also from manure produced in intensive livestock production. Elevated N input, particularly of NH_x, increases nutrient losses from the canopy as well as from the soil. Due to soil related processes, NH_x deposition contributes considerably to the process of soil acidification. At particular sites, the nutrient element turnover can be disrupted resulting in an imbalanced nutrient supply, and finally, in visible nutrient deficiency symptoms.

In addition, many forest sites have soils derived from extremely nutrient poor materials which contain only low supplies of Mg, Ca and occasionally, K. Typical sites are the German chains of central mountains such as the Black Forest, the Bavarian Forest, the Fichtel Mountains, the Harz Mountains, and to some degree, the Palatine Forest.

Forest management practices applied over the last 200 yr have also contributed regionally to the degradation of forest sites. Examples of practices contributing to this degradation are the establishment of off-site species, wrong stand composition in relation to tree species and tree age and type of utilization, e.g., forest pasturing lifestock, litter-racking, bark utilization, as well as elevated populations of wild game animals.

It will take many decades before forest ecosystems damaged by atmospheric deposition will be regenerated, even with recently initiated measures to reduce air pollution. For revitalization and restabilization of damaged forests and for protection of endangered ecosystems, it is necessary to apply suitable silvicultural practices. In this respect, one such important practice is the improvement of the chemical status of forest soils, particularly of the nutrient supply.

3. REVITALIZATION AND RESTABILIZATION STRATEGIES

From former fertilization and liming trials, it is known that the vitality of forest trees and stands can be improved over a longer time period even when atmospheric deposition has had an adverse effect. In numerous recent trials, tests were conducted to determine whether proper fertilization utilizing fast soluble fertilizers can mitigate or even reverse net type forest damages associated with nutritional disturbances. In the cases of Mg and K deficiencies, convincing results have been obtained (Huettl, 1988; Kaupenjohan, 1988). The same was found for Mg containing lime applications in forest ecosystems impacted by Mg deficiency. The revitalization effects produced by these recent fertilization and lime experiments have now been observed for 5 to 6 yr. Based on previous experiences, positive effects can be sustained over at least 10 to 20 yr. Furthermore, in cases of acute deficiencies, the application of fast soluble fertilizers resulted in a rapid revitalization of declining trees. Yellowed needles of younger conifers became green within only a few months after fertilization. For older stands, this revitalization takes about two vegetative periods. However, it was also found that once a critical degree of damage had been reached, no revitalization was possible.

The application of modest amounts of limestone on acidified soils generally leads to a pronounced improvement of the chemical soil status. With this technique, anthropogenically accelerated soil acidification can be mitigated and, under certain circumstances, can even be reduced.

4. PRACTICAL APPLICATION

The positive effects of fertilization and liming on the vitality of forest ecosystems was expected, at least to some degree, due to the experiences from former fertilization and liming trials. The rapid and sustained effect of these techniques was impressive.

Based on all these findings, the West German Government established a program in 1984 to subsidize up to 80% of the total costs (materials and application) of fertilization and liming in forests revealing "new type" forest damages.

In practice, a clear distinction is made between compensative and ameliorative fertilization methods. Compensative fertilization implies the application of Mg containing dolomitic lime (generally 3,000 kg ha^{-1}) to neutralize further acidic deposition in forest soils. Recently, this type of ecosystem restabilization technique has also been referred to as "soil protection liming."

Ameliorative fertilization focuses on the specific application of nutrients, particularly in stands with acute or latent nutrient deficiencies. Ameliorative fertilization is accomplished by means of readily available Mg and K fertilizers at application rates of 500 to 1,000 kg ha^{-1}. In cases of insufficient P supply or for the activation of microbial decomposer chains in organic top soil horizons, P fertilizers are utilized. For ameliorative fertilization, foliar and soil analyses are recommended as a sound basis for fertilizer selection.

In many cases, particularly in the states of Bavaria and Baden-Wuerttemberg, Mg containing dolomitic lime in combination with readily soluble Mg and/or K(P) fertilizers are applied. Between 1984 and 1987, approximately 400,000 ha of "new type" forest damage area have been limed and fertilized. It is estimated that about 150,000 to 200,000 ha will be treated in 1988.

5. POTENTIAL NEGATIVE ASPECTS OF FOREST LIMING AND FERTILIZATION

As positive as the results appear to be, the potential risks and limitations of liming and fertilization with regard to revitalization and restabilization of "new type" forest damages must be addressed.

In steep mountainous terrain, fertilizers and lime applied at times of heavy precipitation, particularly during the non-vegetative period, might be washed away with surface waters before they can penetrate the rooted substrate and thus be available for plank uptake. Very shallow soils or skeletal rich substrates might experience the same problem. Losses have to be expected when excessive amounts of fertilizers are applied.

Once a critical degree of damage has been reached, no revitalization by fertilization is possible. This finding indicates that it might be worthwhile to apply nutrients not only at times when visible symptoms are already occurring, i.e., in cases of acute deficiencies, but to use fertilization as a preventive tool. No positive fertilization response can be expected in cases where a strong negative influence of air pollutants causes direct acute damage. Examples of this type of impact are the SO_2 damaged forests in the Ore Mountains of Czechoslovakia or the former SO_2 damages to forest trees close to the Sudbury Smelter in Canada.

Heavy application of limestone may lead to enhanced mineralization of organic matter. This may result in various undesired ecological impacts. Recently, Wenzel and Ulrich (1988) have discussed potential risks and how to minimize these risks. According to these authors, compensative liming is aimed at reducing the risk of acid toxicity and its effects in forest trees and other vegetation. Compensative liming alone cannot lead to the deacidification of the soil, to a replenishment of nutrient supply, and to the establishment of a stable soil status. Large scale liming practices in ecosystems call for a sound evaluation of potential risks (Wenzel and Ulrich, 1988; Ulrich, 1986).

5.1 Increased N Mineralization, N Displacement in the Soil and Increased Nitrate Concentration in Seepage and Ground Waters

Increased N cycling in forest ecosystems due to liming and its potential negative aspects on the soil (N losses) and the soil water (increased nitrate concentrations) is the most common argument against liming practices. In former investigations, a high reduction of N content in the soil was found after liming. However, a mechanical soil manipulation was frequently associated with these liming practices (Seibt and Wittich, 1977; Seibt and Reemtsma 1977).

Ulrich and Keuffel (1970) found that 14 yr after liming without mechanical soil disturbance, the N losses amounted to 450 kg ha^{-1} indicating an average N loss of 32 kg ha^{-1} yr-1. Matzner (1985) found relatively low annual N losses after liming (11 kg ha^{-1} under beech; 50 kg ha^{-1} under spruce). These rates were considerably lower than the annual N deposition. However, decomposition of organic matter is still controversial considering the impact of accelerated soil acidification. Recently, Ulrich (1986) found increasing accumulation of the organic layer in the solling area whereas Zwezschwitz (1985) showed decreasing C/N ratios indicating improved humus layers at different sites in northern Germany. In forest stands of the Black Forest, humus layers also seemed to decrease (Zoettl - pers. communication).

Only a few investigations have been carried out concerning the problem of potentially increased nitrate concentration in seepage water, which might lead to nitrate displacement into the groundwater after soil liming in forest ecosystems. In the solling area, Matzner (1985) found after liming that NO_3-N concentration in the seepage water in 1 m soil depth under beech was 3 Mg L^{-1} and under Norway spruce the maximum value was 10 Mg L^{-1}. These data indicate that the drinking water threshold value was not reached (12 Mg L^{-1}). In the Oden Forest, Evers (1986 a,b) found a nitrate concentration in the seepage water of 20 Mg NO_3-N L^{-1}, but there was no difference between control and limed plots. Evers related these high values to high atmospheric N input.

Kreutzer (1988) showed in the Hoegelwald experiment (Bavaria) that the input of N via the atmosphere is high in dense coniferous forests due to their filtering capacity. This N is partially integrated into the biological cycle. Increased N deposition frequently results in increased growth rates which are more pronounced at sites with a low supply of available N. At these sites which are well-supplied with N, increased N depositions cannot be completely utilized. This will finally lead to higher nitrate losses in the seepage water. On the other hand, at formerly litter-raked sites the nitrate losses are low. Kreutzer also found higher nitrate losses under spruce as compared to deciduous forests. The high NH_4 portion of the N depositions also contribute to the acidity production in the soil leading to enhanced alkaline cation losses. Liming may compensate for these losses, but will probably also result in mobilizing accumulated N contents via humus mineralization. These mechanisms may finally result in nitrate contamination of the groundwater. Four years after liming, Kreutzer measured nitrate values of 250 to 300 Mg L^{-1} in the seepage water in 2 m soil depth of a spruce during winter time. Therefore, Kreutzer postulates that application of lime should be based not only on pH values of the soil and thickness of the humus layer, but also on N input and N uptake capacity. These evaluations have to be carried out for each site and species.

5.2 Mobilization of Heavy Metals in the Humus Layer

Of those heavy metals that are deposited into forest ecosystems, most are accumulated in the humus layer which form organic complexes. However, the humus layer can only be seen as a temporary sink. Therefore, a better understanding of the conditions under which these heavy metals are mobilized and present a danger to decomposer organisms, roots and groundwater, is necessary.

In this context, it is most important to consider that mobilization of heavy metals is, to a large extent, dependent on the pH reaction of the soil. With increasing pH values, the solubility of heavy metals is reduced. In substrates with low pH values, and particularly, in sandy soils, heavy metals can be translocated with the seepage water (Lahmersdorf, 1985). This was also found by Kreutzer (1987). Based on ecosystem balances, Mayer (1981) showed that Ni, Cu and Pb were accumulated in the solling ecosystem whereas Zn and Cd were leached from the soil. Mayer related this to the effects of acid deposition. Also, Shierl and Rodenkirchen (1987) found, after 3 yr of acid irrigation (pH 2.7) in an old Norway spruce stand at Hoegelwald, nearly no decrease of pH in the soil water, but higher concentrations of Al, Cd, Mn and Zn in the mineral soil.

The effects of liming on heavy metal concentrations in seepage water were investigated by Lahmersdorf (1985) in both a beech and a spruce stand in the solling. This research showed different behaviors. For Cu, Cd, Cr and Co, liming reduced losses, whereas losses of the elements Zn, Cr and Ni were higher after liming. Lead concentrations did not change under spruce nor under beech.

Based upon present knowledge, the aim to stop or prevent heavy metal mobilization related to accelerated soil acidification can only be obtained via liming. However, too small amounts of limestone applied to the humus layer will not lead to a significant pH rise and thus the risks of mobilizing organic compounds and complex heavy metals might be increased. In soils with a certain content of loam and clay, the organic compounds will be retained in the mineral soil together with the heavy metals. Only in cases of heavily and deeply acidified mineral soils, and particularly on sandy podzols, heavy metals are transported downward with the seepage water. To decrease the mobilization of heavy metals, Wenzel and Ulrich (1988) suggest the application of lime in combination with phosphate.

5.3 Effects of Compensative Liming on Mycorrhiza

Haung et al., (1986), as well as Ritter et al. (1986), found that limed and healthy looking spruce were marked by vital mycorrhization when compared to spruces of the unlimed control stand. As these results are based on a few qualitative observations, further research is needed to quantify these effects.

5.4 Effects of Compensative Liming on Soil Fauna and Microorganisms

The reaction of the soil fauna and microorganisms to liming practices is different for particular species and population frequencies (Funke, 1986). For instance, most investigations on earthworm activity indicate that liming causes a significant increase in earthworm populations. Funke and Jans (1986) found a dramatic decrease of important decomposer species in limed spruce stands where earthworm populations had increased as a result of the liming. This decrease in decomposer species resulted in a disruption of litter-decomposition. After an adjustment period, litter-decomposition was again increased on these plots. Finally, Funke (1986) showed that most of the investigated species reacted positively to the application of lime.

Positive reactions due to modest lime applications have also been found for microbial soil populations (fungi, bacteria). Under certain conditions, this increased microbial activity resulted in increased mineralization rates, but not necessarily in increased N contents, because N immobilization might be more favored than N mineralization. Furthermore, liming may lead to enhanced nitrification. This had to be considered positive at sites where nutrient cation uptake is low; for NO_3 uptake stimulates uptake of cationic nutrients (Beese and Waraghae, 1985). Kreutzer and Zelles (1986) found a much increased bioactivity in a Norway spruce stand due to liming. Baath et al. (1980) indicated reduced bacteria and fungi frequencies under spruce after liming. Derome et al. (1986) studied 96 stands that had been limed using the same experimental design 20 to 24 yr ago in Finland. They found that liming had brought about a considerable increase in the amount of organic matter in the humus layer and a corresponding decrease of C contentin the uppermost layer of the mineral soil. Many of the changes in the properties of the soil are dependent on the amount and quality of the organic matter present as a result of liming.

To prevent direct damage to the soil fauna by the application of very fine lime materials, Gussone (1987) suggested the application of granulated lime or limestone with a coarser texture.

5.5 Effects of Compensative Liming on Forest Floor Vegetation

Again, the effects of liming on forest floor vegetation (e.g., mosser and harbaceous plants) are not well investigated. According to Hartmann and Jahn (1959), liming leads to the establishment of nitrophylic and more or less demanding plant species.

Two years after liming a Norway spruce stand at Hoegelwald, Rodenkirchen (1986) observed *Oxalis acetosella* more frequently, but this had no impact on the plant species composition of the trial area compared to the initial status. The investigated species were marked by an improved Ca and Mg nutrition. Acid irrigation led to damages to various moss species and to reduced Mn and Zn uptake. These negative effects were clearly reversed by liming.

5.6 *Effects of Compensative Liming on Tree Growth*

Aldinger (1987) conducted growth investigations via stem analysis in limed and unlimed Norway spruce and silver fir stands in the Black Forest which were treated between 1963 and 1975. In general, Aldinger did not find significant growth differences between limed and control trees. However, in cases where "new type" forest damage had developed since about 1980 (associated with tip-yellowing of older needles typical of Mg deficiency), the earlier application of Mg containing lime prevented the development of damage symptoms. This improved vitality status was correlated to better growth. The application of limestone not containing magnesium-carbonate did not influence forest growth. Derome et al. (1986) found a clear reduction in the annual growth increment during the first 2 yr after liming.

6. SUMMARY EVALUATION OF LIMING EFFECTS

The application of lime will always bring about changes both positive and negative in forest ecosystems. It is therefore important to select the right lime materials (form and granulation) as well as suitable application dosages. Based upon present knowledge, a limited application of lime to the surface of forest soils in the amount of 3,000 kg ha^{-1}, does not lead to nitrate contamination of groundwaters. The risk of heavy metal mobilization is probably not more pronounced for limed soils then for soils that are influenced by accelerated soil acidification. Furthermore, no negative impacts of lime on mycorrhiza or forest floor vegetation are known.

The adverse influences on soil fauna and macroorganisms are probably of temporary nature and, in general, positive effects are found. It is very important to choose the appropriate types of lime to prevent direct damage to certain organisms.

7. POTENTIAL RISKS OF FERTILIZATION WITH SULPHATE-BASED, FAST SOLUBLE MATERIALS

Acid deposition and its relationship to the "new type" forest damage has focused attention to the potential adverse effects of high SO_2 emissions. Due to chemical reactions in the atmosphere, H_2SO_4 is formed and deposited to soils, particularly in forest ecosystems. However, SO_4-s input rates vary to a large extent in West Germany (Umweltbundesamt, 1986).

Even though sulphate based Mg and K fertilizers ($MgSO_4$; K_2SO_4) are neutral salts and despite the fact that they do not add H^+ to the soil system - as is the case with H_2SO_4 deposition - SO_4 can be considered the anion of a strong acid. Therefore, when SO_4 is moved through the soil profile and when it is not absorbed, it causes an equivalent displacement of alkaline cations which may result in a relative increase of the content of acidic cations in the percolated soil layers. Hildebrand (1988) described these mechanisms as being due to exchange processes which result in acidification. To test this theoretical assumption, Hildebrand constructed a percolation apparatus to percolate water through soil columns trying to imitate natural water penetration conditions. Before starting the experiment, some columns of highly acid, base poor, sandy-loamy soils with raw humus layers, were fertilized with Kieserite ($MgSO_4.H_2O$). The experiment proved the assumed mechanisms indicating that the exchange process leads to acidification pulses from the organic layer to the mineral soil. However, it must be emphasized that relatively high amounts of water were applied within a very short time (which contributed to ion exchange rates). Therefore, the results are not too surprising. At least in part, these results are in contradiction to findings of Zoettl (1987), Huettl and Zoettl (1986), and Isermann (1987) who studied the effects of Kieserite fertilization under field conditions at sites with comparable soil conditions and in stands with Mg deficiency. In all cases, $MgSO_4$ fertilization led to a pronounced revitalization of the trees and the negative effects of SO_4 anion displacement in the soil were not found at any dramatic level. Similar results were also obtained by Bosch et al. (1986) and various investigators (Huettl, 1987).

Summarizing this discussion, Ulrich (1986) stated that fertilization can be utilized to eliminate acute nutritional deficiencies in order to prevent irreversible damage in forest trees and stands. To achieve such results, Ulrich suggested chloride-free fertilizer salts. As the application of ferrtilizer salts may cause an acidification pulse in acidic soils, this type of treatment should be preceded by lime application. Soils in which nutritional disturbances occur are generally extremely acidified and therefore have only a small capacity to absorb cations like Mg, Ca and to some extent also K. In order to prevent losses of mobile ions, fertilizer application amounts should be based on uptake rates and the remaining absorption capacity of the particular soil profile.

These somewhat controversial findings and discussions indicate that further research is needed in this area to evaluate potential negative ecological effects of the wide spread application of lime and fertilizers.

8. APPLICATION TECHNIQUES AND COSTS OF FOREST FERTILIZATION AND LIMING

The selection of the proper application method and time, the right fertilizer and lime materials, as well as the appropriate application amounts suited for a particular forest stand and site, depend on the following variables: stand structure (stand age and composition), geographical site conditions, physical soil properties, humus form, chemical analysis data of foliage and soil samples, desired management goals (e.g., increased growth, regeneration, revitalization), distribution quality (width and uniformity of distribution), season, costs, size of fertilization area, undesired ecological effects on ground vegetation, microorganisms, pedosphere, hydrosphere and, most of all, experiences gained from earlier treatments.

In general, forest fertilization can be carried out by: hand; blowers and spreaders; and aircraft (helicopter, plane). Besides economical and ecological considerations, distribution width and uniform distribution patterns should be used as parameters to evaluate the quality of the application. Recent research into these factors was carried out by Denniger (1988). The best distribution quality is obtained from hand application. In general, this technique is very costly. It should therefore be applied when very precise distribution of fertilizers is required (e.g., for research purposes) or when labor costs are not of concern (e.g., small private properties). Occasionally, manual fertilizer application is practiced when replanting forest sites.

Fertilization by aircraft is the most widely practiced application technique in West Germany where it is most effective when larger areas are to be fertilized. Here, application quality depends on the experience of the aircraft pilot, weather conditions (wind, rain), marking of fertilization area, stand structure (e.g., dense vs open canopy), geographical characteristics, etc. The application of different forms of fertilizers is best done separately to improve distribution quality.

In even-aged older forest stands on rather flat terrain, fertilizers are commonly applied by blowers or spreaders in West Germany. Using this technique, fertilizer mixtures should only be applied when all materials show similar distribution patterns. As blowers do not generally provide good distribution quality, this method is mainly used to apply materials in larger quantities (e.g., lime > 2,000 kg ha^{-1}).

In conclusion, the application of fertilizers in forests is a well-established management practice. When choosing the right application method (based on economical and ecological considerations), fertilizer application can be very efficient and good distribution quality can be achieved.

Due to the many variables involved in forest fertilization, only very general assumptions can be made when considering costs of fertilization. At present, costs of fertilization/liming applied in West Germany for revitalization/restablization of declining forest ecosystems range between \$180 and 420 ha^{-1} and \$75 and 175 ha^{-1}, respectively.

9. CONCLUDING REMARKS

When reviewing literature in the field of forest fertilization, it becomes clear that the desired gains from the fertilizer treatments in forest ecosystems will be obtained when the right fertilizers are used in the proper amounts at the appropriate times on specific sites.

Evers (1986a,b) evaluating the potential effects of forest liming and fertilization stated the following conclusions:

> "The present knowledge of forest fertilization allows application of fertilizers in all cases where forest damages are associated with nutrient deficiencies or nutritional disturbances. In cases where soil damages are present, restabilization via fertilization is inevitable. For healthy soils, as well as in stands with good nutrient supply, fertilization is unnecessary and may not (based upon up-to-date knowledge) improve the occasionally occurring damages (e.g., premature foliage losses)."

> "In all cases where forest fertilization is needed and where its type can be determined via investigations, fertilization should be carried out and not be refused due to the vague apprehension that damages might result from this treatment. The risks of forest fertilization can be estimated, but the risks of omitted fertilizations cannot be foreseen and are by far much higher."

SECTION II: LAKES

The earliest attempts in aquaculture also involved the use of limestone. Whether through experimentation or accidental discovery, it was observed by our ancestral fish culturists that in some ponds neutralized with liming materials, the fish grew faster and attained a greater size. As fisheries science continued to develop its knowledge base, it became common wisdom that fish had water quality requirements and these requirements had reasonably narrow ranges under which the fish could survive and reproduce. This basic understanding of the water quality requirements of biological organisms, coupled with increasing change in water quality has led to the manipulation of surface water quality as a management tool. The following papers in this section describe many different methods and motivations to neutralize acidic lakes.

CHANGES IN THE FISH COMMUNITY OF A LIMED LAKE NEAR SUDBURY, ONTARIO: EFFECTS OF CHEMICAL NEUTRALIZATION OR REDUCED ATMOSPHERIC DEPOSITION OF ACIDS?

J.M. Gunn
M.J. McMurtry
J.M. Casselman
W. Keller
M.J. Powell

ABSTRACT

Nelson Lake, a moderately acidic (pH 5.7), metal-contaminated (Cu 22 *u*g L^{-1}; Zn 18 ug L^{-1}) lake, 28 km from the smelters at Sudbury, had a degraded fish community in the early 1970's, with lake trout (*Salvelinus namaycush*) scarce, smallmouth bass (*Micropterus dolomieui*) extinct, and the littoral zone dominated by the acid-tolerant yellow perch (*Perca flavescens*). Liming of the lake in 1975-76 increased pH to 6.4, and decreased metal concentrations. Chemical conditions have remained relatively stable in the 10 yr following base addition. Initially, it appeared that neutralization produced dramatic changes in the resident fish community. Yellow perch abundance declined rapidly after neutralization, lake trout abundance increased to the extent that 3.26 kg ha^{-1} were caught in the winter of 1980, and reintroduced smallmouth bass reproduced and established a large population. However, these changes in the fish community can not be directly attributed to liming, as water quality and the sport fisheries of an unlimed nearby lake also improved. Reduced emissions from Sudbury smelters were responsible for improvements in the untreated lake. Recovery of the lake trout population in Nelson Lake appears to have begun prior to liming. Of the lake trout sampled during the 1980 winter fishery, 65.8% were present prior to the chemical treatment. Predation by lake trout was the likely cause of the perch decline. Our results suggest that chemical conditions producing population level responses in fish have abrupt thresholds and that neutralization of lakes above these thresholds may not produce distinguishable effects.

1. INTRODUCTION

Calcium carbonate and other neutralizing agents have been used to rehabilitate a large number of acidified lakes throughout the world (Bengtsson et al., 1980; Booth et al., 1986; Schofield et al., 1986). Lake neutralization has generally improved water quality and allowed

Water, Air, and Soil Pollution **41** (1988) 113–136.

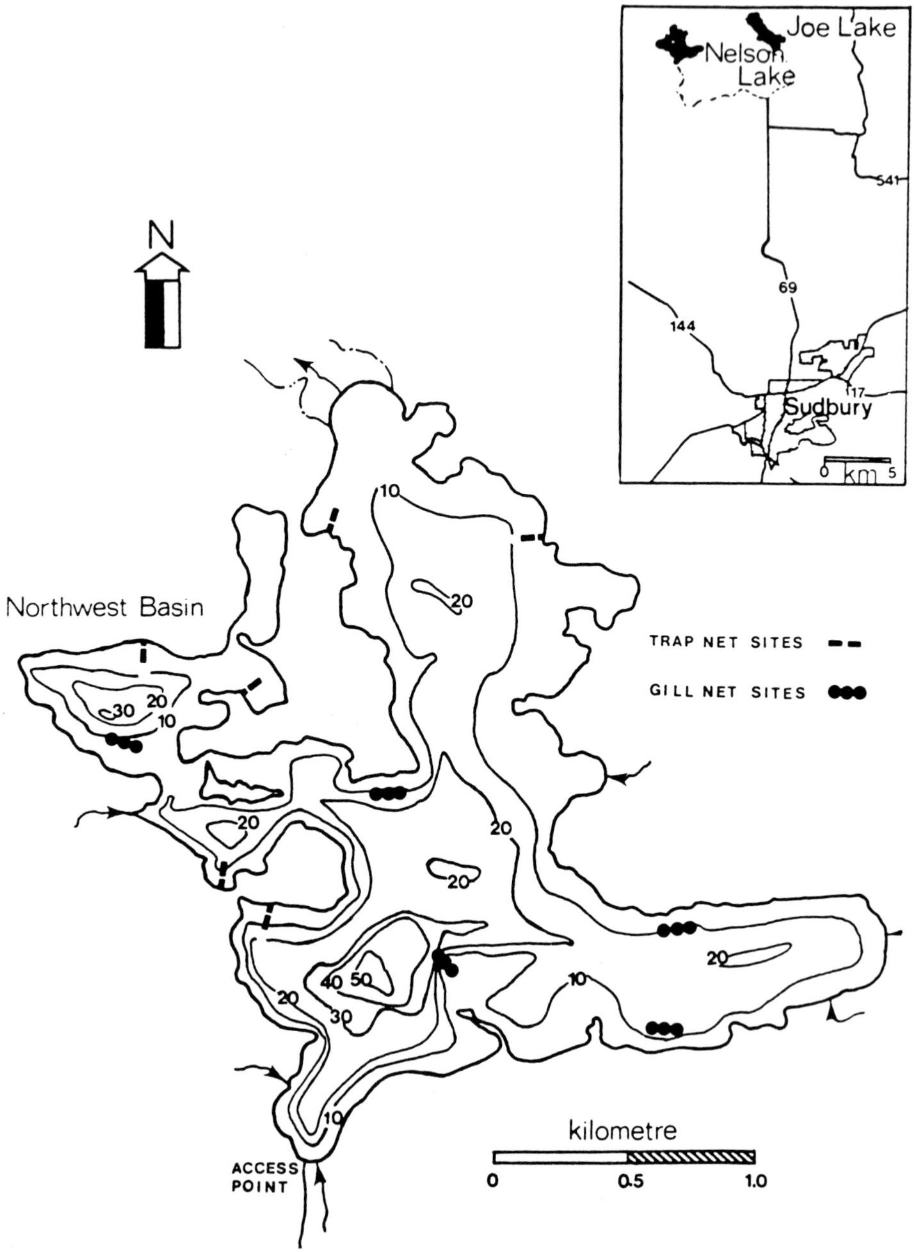

Figure 1. Depth contour map of Nelson Lake (depths in m) showing index netting sites.

for the re-establishment of healthy populations of various aquatic species, including acid sensitive fish (Anderson et al., 1984; Nyberg, 1984; Hamilton et al., 1988). Unequivocal evidence that neutralization has improved conditions for fish has usually come from experiments in which the initial chemical conditions were severe, and *in situ* bioassays could be used to demonstrate that the pre-treated water was acutely lethal to fish (eg. Schofield et al., 1986; Hamilton et al., 1988; Gloss and Schofield, 1988).

Interpretation of the effects of neutralization in less acidic systems with existing fish populations is more difficult. Natural fluctuations in abundance and growth of the resident species and their prey can obscure the effects of the chemical treatment (Eriksson et al., 1983; Eriksson and Tengelin, 1987; Booth et al., 1987). Intensive, long-term studies are needed to determine the biological responses to neutralization in moderately acidic systems.

In this paper we describe changes in the resident fish community of Nelson Lake (initial pH 5.7) during a 10 yr period following neutralization. Nelson Lake is located 28 km from the metal smelters of Sudbury, Canada. The short-term effects of neutralization on the physico-chemical characteristics and plankton assemblages of this lake have already been described in detail (Yan et al, 1977; Yan and Dillon, 1984). To determine the cause of the observed changes in the fish community of Nelson Lake, comparison is made to Joe Lake, a nearby, untreated lake.

2. STUDY LAKES

2.1 Physical and Chemical Characteristics

Nelson Lake and Joe Lake are located 5 km apart, in an area of deciduous forest approximately 30 km from the smelters at Sudbury (**Figure** 1). They are large deep oligotrophic lakes with similar morphometric characteristics (**Table I**). Their watersheds are underlain with granitic bedrock, a part of the nickel eruptive bedrock which forms the perimeter of the Sudbury Basin. Human usage of the lakes is primarily recreational (boating, swimming and fishing). Several cottages are located in the watersheds of both lakes.

Table I

Physical features of study lakes near Sudbury, Ontario. Data are from Yan and Dillon (1984) and Yan and Miller (1984). Drainage basin area and turnover time were not determined for Joe Lake.

	Nelson Lake	Joe Lake
Latitude	48°44'N	46°44'N
Longitude	81°05'W	81°01'W
Altitude (m)	348	367
Distance from Sudbury (km)	28	29
Lake Area (ha)	309	180
Drainage Basin Area (ha)	798	-
Number of Basins	2	1
Maximum Depth (m)	51.0	34.0
Mean Depth (m)	11.6	11.2
Lake Volume ($10^5 m^3$)	359	201
Turnover Time (yr)	8.1	-
Shoreline length (km)	33.3	14.0

Prior to liming in 1975-1976, Nelson Lake was clear (8 m Secchi depth), moderately acidic (pH 5.7) and had relatively high levels of water-borne metals (Cu 22 *u*g L^{-1}, Ni 17 *u*g L^{-1}, Zn 18 *u*g L-1, Al 87 *u*g L^{-1} (**Table II**). The addition of 6.8 x 10^4 kg of $Ca(OH)_2$ and 5.1 x 10^4 kg of $CaCO_3$ in the fall of 1975 (Sept. 9-23) and the spring of 1976 (May 26-31) raised the pH to 6.4, and substantially reduced the concentrations of Cu, Al, Ni, and Zn, but produced no short-term changes in transparency, nutrient chemistry or assemblages of plankton (Yan and Dillon, 1984). Further details on the limnological effects of the chemical treatment are provided in Yan and Dillon (1984). Nelson Lake has not been relimed since 1976, but has not reacidified (**Table II**).

In 1975 the water chemistry of Joe Lake was very similar to that of Nelson Lake. Chemical sampling has not been as frequent on Joe Lake as on Nelson Lake, however, the data available indicate that the water quality of Joe Lake has greatly improved in recent years (**Table II**). Such changes have been observed in a large number of Sudbury area lakes (Keller et al., 1986).

2.2 Historical Information on Fish Communities

There are few quantitative data on the fisheries of Nelson and Joe Lakes as they existed prior to 1975, the first year of intensive index netting, but the personal observations of cottage owners indicated that both lakes had highly degraded fisheries in the early 1970's.

Smallmouth bass (*Micropterus dolomieui*), were frequently caught in Nelson Lake by anglers in the 1960's, but had disappeared by the early 1970's, while yellow perch (*Perca flavescens*), an acid tolerant species (Magnuson et al., 1984), had greatly increased in numbers. Lake trout (*Salvelinus namaycush*) also appeared to have declined in abundance, with the result that during the early years of our study (1975-1979) Nelson Lake received very little fishing pressure.

Joe Lake once supported good brook trout (*Salvelinus fontinalis*) angling but the population had disappeared during the late 1950's or early 1960's. Hatchery introductions of brook trout in 1959 and 1966 failed to re-establish the population (Ontario Ministry of Natural Resources (OMNR), unpubl. data). In 1975, Joe Lake contained a very small population

Table II

Chemical characteristics of Nelson and Joe Lakes, 1973 to 1986. Data up to 1979 are from OMOE (1982). All samples were collected during the ice free period. Samples from 1975 to 1979 and 1985 to 1986 were collected as volume-weighted whole-lake composites. Samples from 1973 to 1974 were surface grabs, and samples from 1981 to 1983 were epilimnion/metalimnion Tygon tube composites. Alkalinity values are based on Total Fixed Endpoint (TFE) measurements before 1978 and on Total Inflection Point (TIA) measurements from 1978 on. TFE measurements were converted to TIA values (mg L^{-1}) using an approximate theoretical conversion of TFE - 1.5 = TIA (OMOE, 1982, Appendix).

Nelson Lake	1973[a]	1974[a]	1975[a]	1976	1977	1978	1979	1981	1982	1983	1985	1986
pH	5.8	5.6	5.7	6.4	6.3	6.5	6.3	6.7	6.4	6.5	6.4	6.4
Alk. (mg L^{-1})	0.3	0.2	0.5	2.2	1.7	2.5	1.7	1.5	1.8	1.6	1.5	1.8
Ca^{++} (mg L^{-1})	5.0	4.4	4.1	5.3	5.0	4.7	4.7	3.8	4.1	3.6	3.3	3.5
$SO_4^=$(mg L^{-1})	16.4	16.1	15.8	15.1	15.3	14.9	14.4	11.5	12.4	12.4	11.8	11.6
Al (ug L^{-1})	-	78	87	30	21	17	19	28	5	27	17	14
Cu (ug L^{-1})	17	29	22	15	4	3	2	4	1	2	-	3[b]
Ni (ug L^{-1})	24	18	17	10	7	7	6	6	5	4	-	4[b]
Zn (ug L^{-1})	18	24	18	14	6	4	4	4	5	4	-	3[b]
n	3-7	4-8	8-17	4-21	9-15	11-28	5-28	1	1	1	4-5	6

[a]before neutralization
[b]from a single mid-summer collection in 1987

Table II (con't)

Joe Lake	1973[a]	1974[a]	1975[a]	1976	1977	1978	1979	1981	1982	1983	1985	1986
pH	-	-	5.7	5.6	-	-	-	6.3	6.3	6.2	6.2	6.3
Alk. (mg L^{-1})	-	-	0.4	0.2	-	-	-	0.8	0.8	0.9	1.0	1.2
Ca^{++} (mg L^{-1})	-	-	4.9	4.6	-	-	-	3.6	-	3.4	3.3	3.5
SO_4=(mg L^{-1})	-	-	16.3	15.5	-	-	-	13.5	13.2	12.4	12.3	12.0
Al (ug L^{-1})	-	-	8.1	58	-	-	-	13	52	9	25	14
Cu (ug L^{-1})	-	-	20	17	-	-	-	3	10	2	-	4[b]
Ni (ug L^{-1})	-	-	23	25	-	-	-	14	14	10	-	10[b]
Zn (ug L^{-1})	-	-	19	21	-	-	-	6	8	5	-	5[b]
n	-	-	8-20	4-5	-	-	-	1	1	1	5-6	6

[a]before neutralization
[b]from a single mid-summer collection in 1987

of white sucker (*Catostomus commersoni*), abundant prey species (Iowa darters (*Etheostoma exile*), lake chub (*Couesius plumbeus*) and brook stickleback (*Culaea inconstans*)), but no sport fish species.

3. METHODS

The following procedures were used to assess the effects of neutralization on the fish community of Nelson Lake: 1) index netting to examine changes in the species abundance and size composition, 2) a winter creel survey to assess angler effort and harvest, and 3) stocking of smallmouth bass to assess suitability of altered water quality for reintroduction of this species.

3.1 Index Netting

Small-mesh trapnets (0.64 cm square mesh, 1.2 m pot, 15 m lead) were the principal sampling gear. Trapnets were set for approximately 24 hr. Trapnetting was done during the month of July, except in 1987, when netting extended into August. From 1975 to 1978, netting sites around the perimeter of the lake were selected randomly. In 1979, 1984 and 1987, trapnets were set at 6 sites that had been sampled in all previous years of the survey. The location of these index sites is shown in Figure 1.

Plexiglas minnow traps (0.5 m pot, 3.6 m lead), design of Casselman and Harvey (1973) were used in the littoral zone. Trap sites along the shore were selected randomly.

In addition to the live capture gear, five gangs of small-mesh gillnets (consisting of one 15 m panel of 1.9 cm mesh and three 15 m panels of 2.5 cm mesh) were set during the July sampling periods of 1975, 1980, 1984, and 1987 to capture small lake trout. Nets were set at or just below the thermocline (depth approx. 10 m) and left overnight. Gillnet sites are indicated in Figure 1.

Length (fork length and total length) and weight were measured and scale samples collected for all sport fish species: lake trout, brook trout, northern pike (*Esoc lucius*) and smallmouth bass. Length and weight of white sucker and brown bullhead (*Ictalurus nebulosus*) were measured in 1975 and 1976 but lengths only were measured in subsequent years. Scale samples were removed from white suckers but no aging structures were taken from brown bullheads. Cyprinids were counted and released. Subsampling was usually required for yellow perch. A random sample of approximately 100 yellow perch was taken from the total catch. Lengths, weights and scale samples were obtained for these fish (weights were measured in 1975 and 1976 only).

All trapnet catches were adjusted to correspond to a standard duration of 24 hr to allow a comparison among years. Biomass of perch was estimated by extrapolating the weight of the random sample to the entire catch. For perch samples with length measurements only, weights were calculated using the weight-length regression developed for the 1975 sample. Missing weights for other species were calculated using the weight-length relationship for 1975 or for the year nearest 1975 with complete data.

3.2 Creel Survey

A creel survey was conducted on Nelson Lake from January to March 1980 in response to a rapid rise in observed fishing effort. The survey was stratified according to day type (weekdays vs weekends or other holidays), month (January, February, March) and time of day 6:00 to 12:00, 12:00 to 18:00). On weekends and holidays the survey clerk stayed at the only point of access to the lake (**Figure 1**). On weekdays, when fishing effort was much reduced, the survey clerk drove his truck around the lake (on the ice surface) counting and interviewing anglers. When possible, lengths, weights and aging structures (see list below) were obtained from the catch. Stomach contents were periodically examined.

3.3. Age Assessment

Initially, the scale method was used to determine the age of sampled lake trout. To reduce variability and bias, the scales were assessed independently by two experienced interpreters. When these results differed, an additional examination was conducted to obtain a final

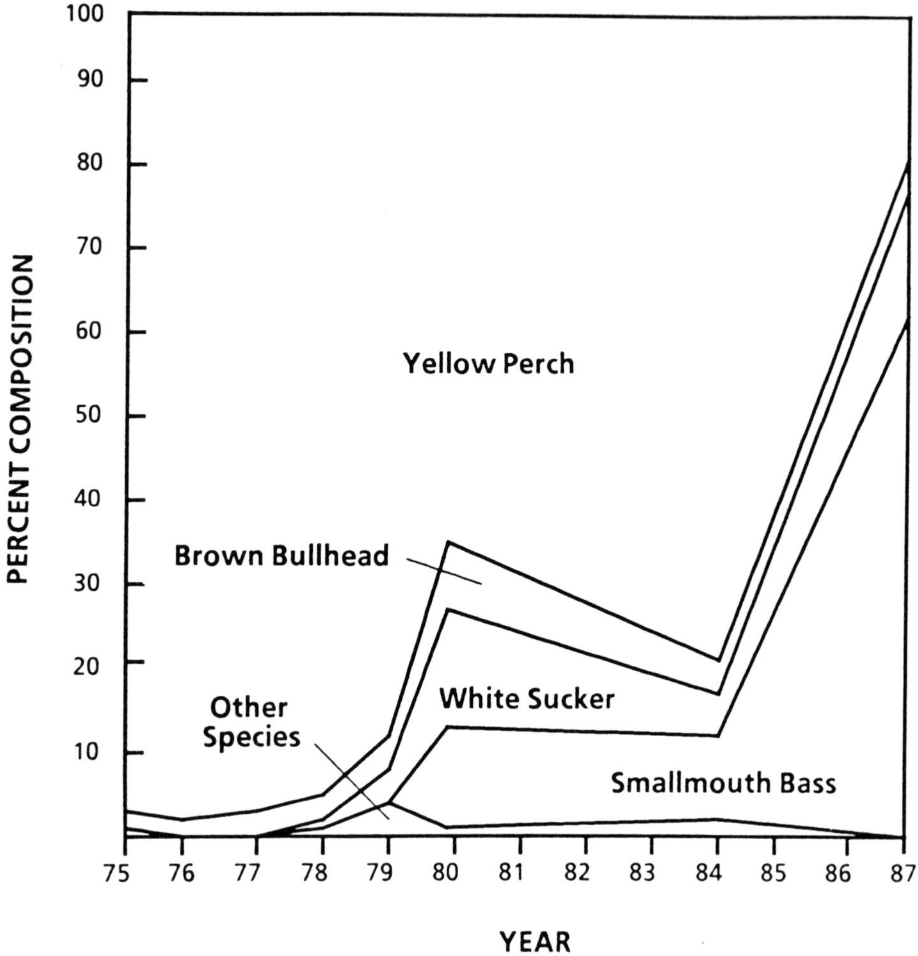

Figure 2. Species composition (% by number) of trapnet catch at the index sites in Nelson Lakes, 1975 to 1987.

interpretation. For 35 lake trout, ranging in assessed age from 4 to 16, scale results were compared with interpretations made on other structures from the same fish. Several flat bones such as opercula, subopercula, and cleithra, were examined, as well as sections of pectoral fin rays and otoliths. Detailed methods are described in Casselman (1987).

3.4 Stocking of Smallmouth Bass in Nelson Lake

Thirty adult bass were transferred to Nelson Lake in the spring of 1978 from McFarland Lake (46°25'N, 80°59'W). Approximately 7,000 bass fry (age approx. 2 mo only) from Sandfield Hatchery (45°42'N, 82°00'W) were also stocked in Nelson Lake during the summer of 1978.

3.5 Stocking of Brook Trout in Joe Lake

On April 21, 1977, 2,000 yearling brook trout (mean FL 17.9 cm, mean weight 71.6 g, left pelvic fin clip) from the North Bay Hatchery (46°19'N, 79°28'W) were stocked in Joe Lake. An additional 1,990 yearlings (mean FL 15.9 cm, weight 52.3 g, right pelvic clip) from Sanfield Hatchery were stocked on May 3, 1977. Trapnetting and gillnetting were conducted in May and September 1977 and June-July 1979 to assess survival and growth of stocked fish.

4. RESULTS

4.1 Nelson Lake Inshore Fish Community

The composition and abundance of species caught in trapnets changed in Nelson Lake after neutralization (**Table III**). Perch constituted 98% by number (**Figure 2**) and 87% by weight of the trapnet catches in 1975. After neutralization, yellow perch declined in importance, representing only 18% by number and 0.3% by weight of the trapnet catch in 1987.

Catch per unit effort (CUE) at the index trapnet sites, indicated that perch numbers and biomass declined continuously, from a mean of 852 perch 24 hn^{-1} (12.53 kg 24 hn^{-1}) in 1975

Table III

Trapnet catch at six index sites on Nelson Lake during July netting period, 1975 to 1987. A small mesh trapnet was set once overnight at each site. The same sites and nets were used throughout the period. The lake was limed between the 1975 and 1976 sampling periods.

	Number Caught							
Species	1975	1976	1977	1978	1979	1980	1984	1987
Yellow Perch	4831	3215	2588	1345	765	135	116	22
Brown Bullhead	80	60	67	38	23	14	4	4
White Sucker	33	1	12	20	39	30	8	16
Lake Chub	-	2	-	2	2	-	-	-
Northern Pike	1	2	1	1	2	1	2	-
Central Mudminnow	-	-	-	4	-	-	-	-
Smallmouth Bass[1]	-	-	-	1	-	26	14	82
Common Shiner	-	-	-	2	32	-	-	-
Netting Duration (hr)	135.83	135.17	140.67	127.67	145.17	140.88	138.50	136.42

[1]Smallmouth Bass were reintroduced in 1978.

Table IV

Plexiglas trap data for July 1975 and 1976 sets in Nelson Lake. CUE is number caught per 24 hr. Seven overnight sets were made in 1975 and 14 in 1976.

	1975		1976	
Species	Number Caught	CUE	Number Caught	CUE
Yellow Perch	685	99.79	294	21.77
Brown Bullhead	24	3.50	18	1.33
Iowa Darter	3	0.44	4	0.30
Central Mudminnow	3	0.44	1	0.07
Brook Trout	1	0.15	-	-
White Sucker	-	-	-	-

to a mean of 4 perch 24 hn^{-1} (0.01 kg 24 hn^{-1}) in 1987 (**Figure 3**). Brown bullhead CUE also decreased significantly after neutralization (R^2=0.76, P=0.0045 for regression of CUE (by number) on year). The biomass of brown bullhead declined from 1.37 kg 24 hn^{-1} in 1975 to 0.13 kg 24 hn^{-1} in 1987.

Plexiglas trap data corroborate the 1975 to 1976 decline in abundance of yellow perch and brown bullhead shown by trapnet data (**Table IV**).

The mean size of yellow perch in the trapnet catches also declined during the study period, while the average size of brown bullhead and white sucker increased significantly (**Figure 4**). The simultaneous decline in both mean size (**Figure 4**) and CUE (**Figure 3**) suggests that size-specific predation was controlling the perch population. The increase in mean size of brown bullhead with declining CUE may be explained by reduced intraspecific competition, however, interpretation of trends in mean size is limited by the lack of a complete record of age composition of the catch.

Adult smallmouth bass stocked in Nelson Lake in 1978, 2 yr after neutralization, reproduced successfully. One of the introduced adult bass was observed guarding a nest on July 10, 1979 and on July 19 some emerging fry were collected. A large reproducing population of smallmouth bass subsequently became well established in Nelson Lake (**Figure 5**), and dominated the littoral zone in the latter years of the study (**Figure 2, 3**). In 1987 smallmouth bass constituted 66% by number and 69% by weight of the trapnet catches.

4.2 Lake Trout

Gillnet catches in the offshore area indicated that lake trout numbers increased during the period when perch and bullhead abundances were declining (**Table V**). In 1975 only one lake trout (1.6 kg) was collected in 5 net sets. In 1980, 1984, 1987 the nets averaged 4 to 7 lake trout per set. Natural reproduction alone produced the increased abundance of lake trout; no hatchery-reared lake trout were planted in Nelson Lake during the study period.

Angling activity on Nelson Lake changed dramatically during the study period. From 1970 to 1978, few anglers used the lake, an indication that angling success was low in this easily accessible lake. In 1979, and particularly 1980, angling activity increased rapidly. In

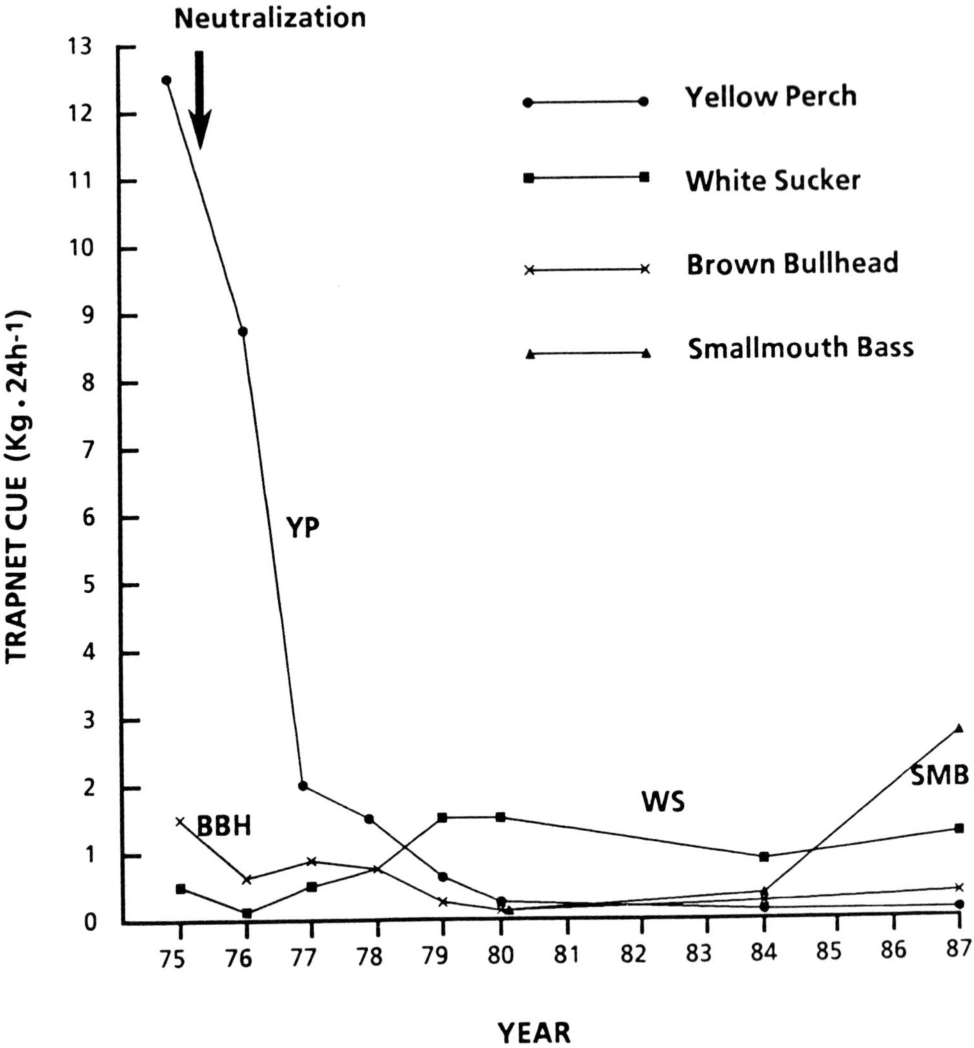

Figure 3. Trapnet catch-per-unit-effort (kg.24 h-1) for index trapnet sites on Nelson Lake during 1975 to 1987.

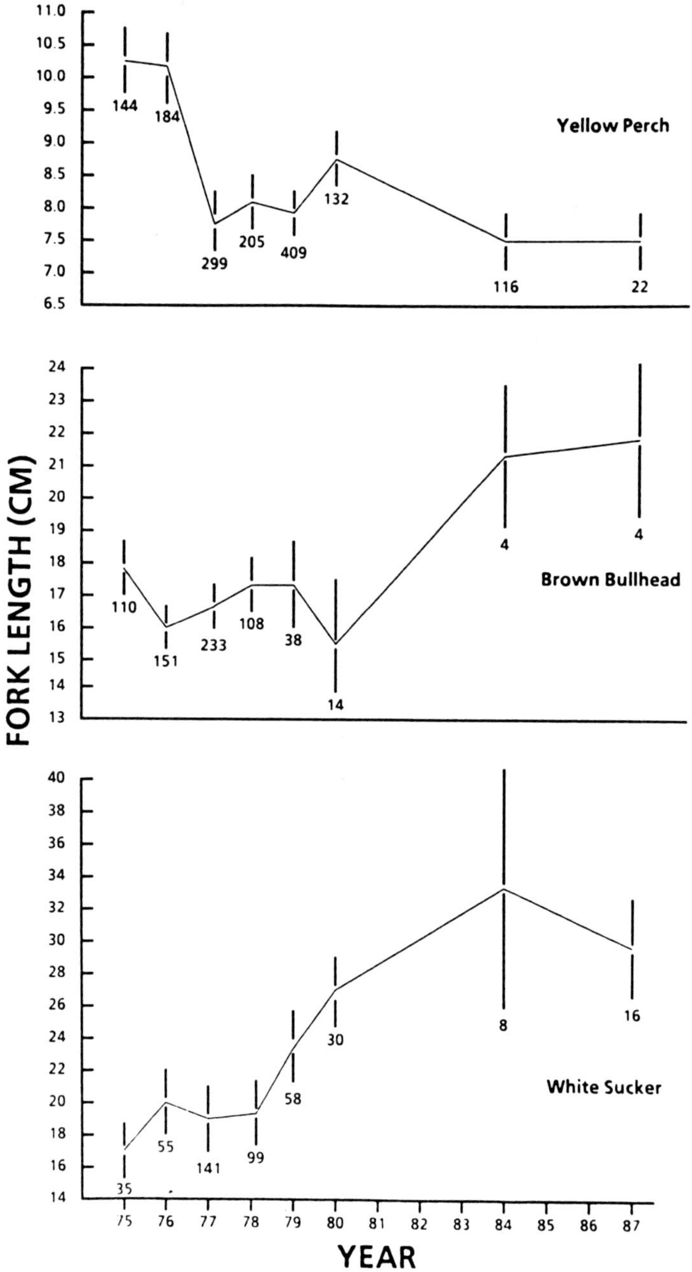

Figure 4. Mean fork length (with 95% CI and N) of yellow perch, brown bullhead, and white sucker in Nelson Lake, 1975 to 1980. Fish were collected with small-mesh trapnets during the July assessment period.

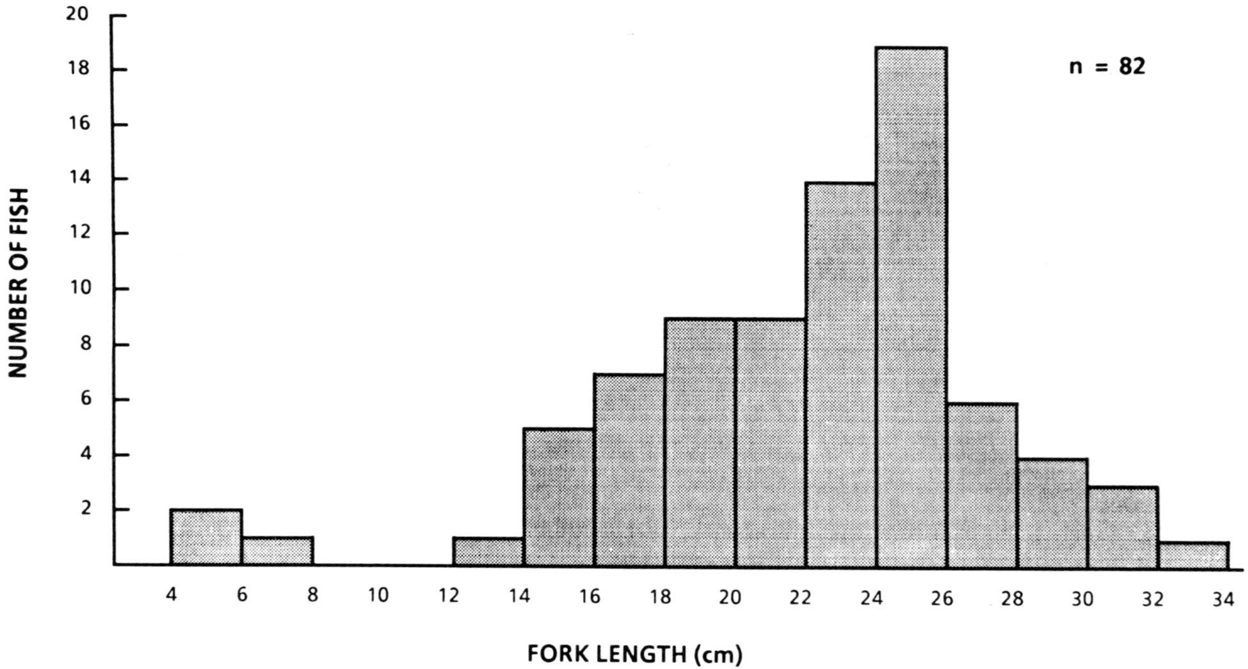

Figure 5. Fork Length frequency of smallmouth bass in trapnet catch in Nelson Lake, 1987.

Table V

Capture records for small mesh gillnets (1.9, 2.5, 2.5, 2.5 cm mesh panels; each 15 m long) set overnight at 5 index netting sites in Nelson Lake.

	Total number from 5 gillnet sets			
Species	1975	1980	1984	1987
Lake Trout	1	37	20	31
Lake Chub	5	43	5	38
Yellow Perch	-	137	-	4
Northern Pike	-	-	-	1
Brown Bullhead	-	1	-	-
Smallmouth Bass	-	1	-	-
Ling	-	-	-	1
Total	**6**	**219**	**25**	**75**

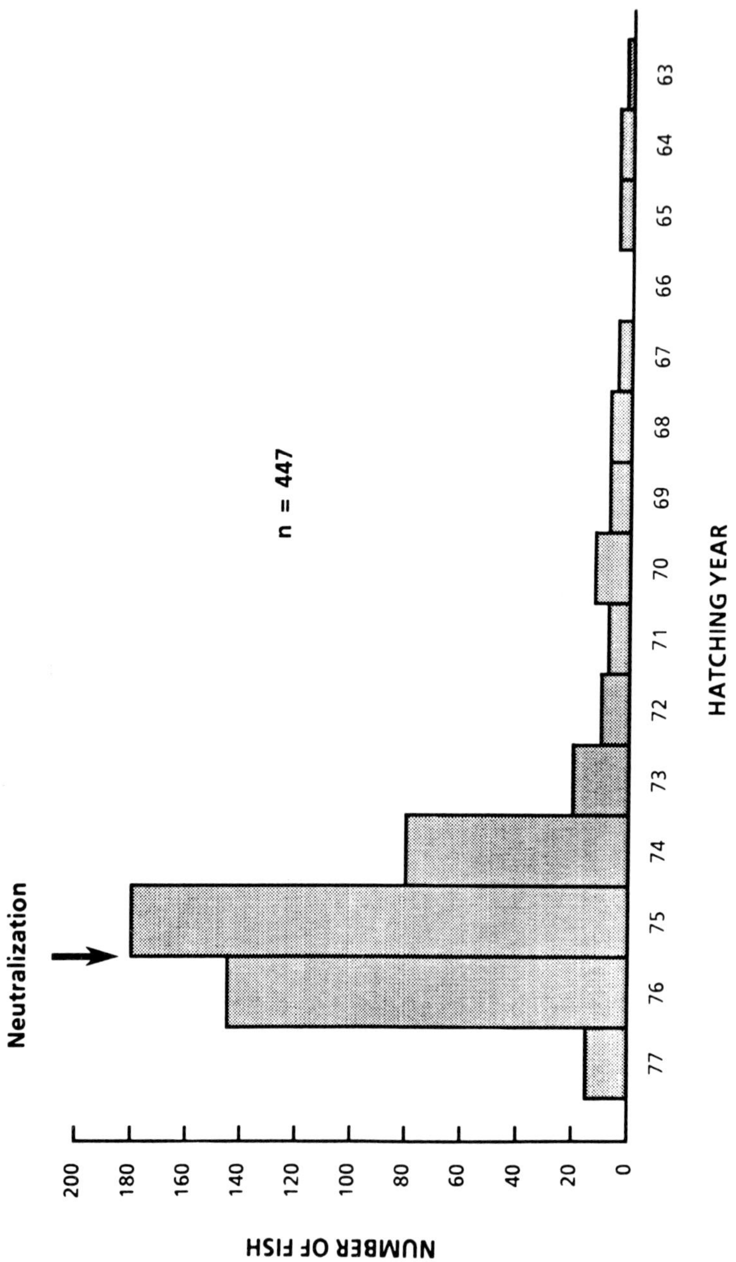

Figure 6. Year of hatching of Nelson Lake lake trout sampled in winter creel census, 1980.

Table VI

Catch statistics for angling fishery for lake trout in Nelson Lake (309 ha) during the winter of 1980.

	Effort (angler-hr)	Harvest		CUE (# per angler-hr)	Mean Fork Length (cm)	Mean Weight (kg)	Yield (kg ha^{-1})
		#	kg				
January	15452	1376	698	0.089	34.6	0.51	2.21
February	9808	559	248	0.058	32.2	0.44	0.79
March	2877	197	82	0.063	30.6	0.42	0.26
TOTAL	28137	2132	1028	0.076	33.1	0.47	3.26

Table VII

Growth of stocked brook trout in Joe Lake. The lake was stocked in the spring of 1977 with yearlings from North Bay (April 21) and Sandfield (May 3) hatcheries.

	NORTH BAY		SANDFIELD	
DATE	Fork Length (cm)	Weight (kg)	Fork Length (cm)	Weight (kg)
Initial	17.9	0.072	15.9	0.052
May 31, 1977	20.4	0.112	18.4	0.085
Sept. 1, 1977	30.7	0.414	30.6	0.408
June-July, 1979	45.0	2.900	49.9	2.675

the winter of 1980, anglers flocked to Nelson Lake (as many as 300 d^{-1}) because of reports of excellent fishing. The low snowfall in 1980, and a winter road to the lake shore also probably contributed to the extremely high pressure. An estimated 2,132 lake trout were taken during this 3 mo period by anglers exerting 28,137 angler-hours of fishing effort (**Table VI**). The survey clerk observed 809 lake trout and obtained biological samples from 447 of these. The estimated biomass harvested during Jan-Mar 1980 was 1,028 kg (3.26 kg ha^{-1}). The mean size (FL $\pm$ SD) of lake trout caught declined during the winter from 34.6 $\pm$ 0.5 cm in January to 30.6 $\pm$ 1.1 cm in March (**Table VI**).

Most of the lake trout (65.8%) sampled during the 1980 creel survey predated neutralization (**Figure 6**). Fish produced in the spring in 1975, approximately 6 mo prior to neutralization, formed the strongest year class (40.2%). Relatively strong year classes were also produced in 1976 (31.6%) and 1974 (18.8%).

4.3 Brook Trout in Joe Lake

Improvements in the sport fishery of Joe Lake were also detected during the study period. Stocked brook trout grew rapidly (**Table VII**) and reproduced without chemical treatment of Joe Lake. Spawning was first observed in the fall of 1978, when the fish were 2^+. Naturally-produced fry were first observed in July and August, 1979 and have been collected at sites near the spawning area throughout the 1980's (Ontario Ministry of Natural Resources, unpubl. data).

5. DISCUSSION

The increased catch of lake trout in small-mesh gillnets, the rapid rise in angling activity, and the collapse of the perch population are the direct and indirect evidence that lake trout abundance rose in Nelson Lake during or shortly before the study period. By 1980 the lake trout population was large enough to permit an estimated winter harvest of 3.3 kg ha^{-1}, one of the highest harvests of lake trout by an angling fishery in Ontario. Goddard et al. (1987)

summarized angling statistics from lake trout fisheries in 87 Ontario lakes, and found that annual harvest ranged from 0.01 to 1.90 kg ha^{-1} in these lakes.

The expansion of the lake trout population appears to be largely unrelated to neutralization as the dominant year classes were produced prior to the chemical treatment. Neutralization may have produced indirect effects, through altered food abundance or changes in growth rate and survival, but based on available data, liming does not appear to be the primary causal factor leading to the enhanced fishery harvest.

Our findings suggest that abrupt thresholds for population level responses exist, and that neutralization of lakes with initial chemical conditions above these thresholds may produce no discernible effects in the fish population of interest. Nelson Lake had an initial pH (5.5 to 5.7) just slightly above the level considered toxic to lake trout (pH 5.5 to 5.6) (Beamish, 1976; Beggs et al., 1985; Mills et al., 1987), but strong year classes of lake trout were produced before neutralization. A threshold effect was also evident from stocking experiments in lakes of varying pH; good growth and survival of lake trout occurred in lakes with pH as low as 5.7 (Gunn et al., 1987).

Lake trout appear to act as a "keystone predator" (Evans et al, 1987) in altering the abundance of other fish species in Nelson Lake, particularly yellow perch. Lake trout prey heavily on yellow perch when these species coexist (Martin, 1952; Gunn, unpubl. data). During the creel survey of 1980, approximately 20% of the stomach samples (those containing food other than bait fish used by anglers) contained yellow perch. The inverse relationship between abundance of yellow perch and lake trout has also been shown in other studies. Hamilton et al. (1988) found that yellow perch abundance declined after lake trout were stocked in their study lake, and Beggs et al. (1985) found that dense yellow perch populations existed in lakes where native populations of lake trout were extinct.

The re-establishment of a reproducing smallmouth bass population in Nelson Lake does appear to be related to improved water quality. However, it is not certain that neutralization was necessary to allow survival and reproduction of bass in Nelson Lake in 1975. There were no bioassays or stocking attempts prior to neutralization, and little information exists to establish a precise pH threshold for this species (Matuszek, 1987), particularly when metal concentrations are also elevated. Beamish (1976) reported that smallmouth bass were more acid-sensitive than lake trout and that smallmouth bass populations disappeared below pH 5.5 to 6.0.

If neutralization did not cause the fish population changes in Nelson Lake, what did? One explanation is that water quality was already improving in Nelson Lake prior to 1975, as was observed in the untreated reference lake, Joe Lake, during the study period.

There are no reliable historical data on the chemistry of Nelson Lake prior to 1973, but emissions from local smelters declined steadily from a peak of approximately 2.5×10^6 t yr^{-1} of SO_2 in the 1960's to $< 1.0 \times 10^6$ t yr^{-1} of SO_2 in the 1980's (Dillon et al., 1986). Data on emissions of metals are less complete, but it has been suggested that the trend follows that of SO_2 (Keller and Pitblado, 1986). The construction in 1972 of the 381 m "super stack" in Sudbury may also have altered the pattern of deposition, further reducing input of acids and metal particulates to lakes near Sudbury. Recent surveys have shown that acid and metal contamination in Sudbury area lakes declined concurrently with reduced emissions of SO_2 and metals (Dillon et al., 1986; Hutchinson and Havas, 1986; Keller and Pitblado, 1986).

An alternate hypothesis to explain the fishery changes in Nelson Lake is that water quality did not change prior to neutralization, but that lake trout had been heavily exploited in the past, and the depleted stock was left to slowly build up to produce the strong 1974-1976 year classes under conditions of abundant food (perch). This "pulse fishing" hypothesis would not account for the extinction of smallmouth bass, an acid-sensitive species (Magnuson et al., 1984), but bass in this area are generally introduced populations, near the northern limits of their range, and prone to year class failure due to fluctuations in climatic conditions (Shuter et al., 1980). We are examining growth rates and year class strength of some of the older lake trout in more detail to test this hypothesis, but at present, water quality changes through emission reductions appears to be the more plausible explanation for the fish community changes.

Dramatic improvements in the fishery of Nelson Lake after neutralization appear now to be largely unrelated to the chemical treatment. Our results and those from other studies (e.g., Dillon et al., 1986; Keller and Pitblado, 1986; Keller et al., 1986), indicate that rehabilitation of some acidic and metal-contaminated lakes in the Sudbury area can occur through emission reductions alone.

ACKNOWLEDGEMENT

This study was begun as part of the Sudbury Environmental Study, a joint project of the Ontario Ministry of the Environment and Ontario Ministry of Natural Resources. R. Bradley, D. Van Exan, B. Morrison and numerous other summer students assisted with the field studies. Age assessment was done by R. McLaughlin, M. Smith and I. Weir. D. Cruikshank conducted the winter creel survey. C. Olver, N. Yan and D. Wales reviewed an earlier draft of the manuscript.

DOSE-RESPONSE RELATIONSHIPS FOR THE ADDITION OF LIMESTONE TO LAKES AND PONDS IN THE NORTHEASTERN UNITED STATES

Timothy B. Adams
Robert W. Brocksen

ABSTRACT

A titration based model (DeAcid) has been used to predict treatment dose and times for reacidification for $CaCO_3$-treated lakes in the Living Lakes, Inc. (LLI) aquatic liming program. Water quality constitutents (pH, ANC and Ca) were used to measure the effectiveness of the dosing model and reacidification rates. Data from 22 lakes or ponds in 5 northeast states have been collected since June 1986. With few exceptions, pH and ANC values ranged from 4.5 to 6.5 and -30 to +65 $u$eq L^{-1}, respectively, in untreated sites and 6.5 to 7.8 and 120 to 300 $u$eq L^{-1}, respectively, in sites approximately 30 days after treatment. Changes in Ca concentration levels have been used to evaluate the utility of the dose model for treatment of both inland and coastal waters. For coastal, seepage lakes application of a single-box mass transfer model to observed post-liming changes in ANC and Ca adequately simulates lake response.

1. INTRODUCTION

Living Lakes, Inc. (LLI) was established in December 1985 to demonstrate the effectiveness of aquatic liming techniques in an applied fisheries management program. As a complement to research programs (Fraser et al., 1982; Saunders et al., 1985), investigating the effects of liming on surface water acidification, LLI focused on technology development and field demonstrations. The primary objective of the LLI program is to develop liming technologies and techniques which may be used to neutralize acidic waters and provide a water quality conducive to the restoration and maintenance of important fisheries. Regular water quality monitoring and fisheries evaluation before and after liming provide the basis to assess the treatment effectiveness. In an earlier article (Brocksen et al., 1987), initial pre- and 30-day post-liming water quality data were presented on 15 lakes in two states. In 1987, the LLI program expanded to include 22 lakes in five states and ten streams in three states. This

Water, Air, and Soil Pollution **41** (1988) 137–163.

Table I
LLI Project Site Background Information

Name	Lake Surface Area Acres	(ha)	Treatment Type	Application Type	Fishery
Baker Pond, MA	29.6	(12.0)	Maintenance	Helicopter	P+T Trout SS Smallmouth Bass
Curlew Pond, MA	43.2	(17.5)	Maintenance	Helicopter	SS Largemouth Bass
Laurel Lake, MA	50.9	(20.6)	Maintenance	Barge/ Slurry Box	SS Largemouth Bass P+T Trout
Lawrence Pond, MA	137.8	(55.8)	Mitigative	Helicopter	SS Largemouth Bass
Lout Pond, MA	12.6	(5.1)	Maintenance	Helicopter	P+T Trout
North Pond, MA	24.7	(10.0)	Maintenance	Helicopter	P+T Trout
Round Pond, MA	21	(8.5)	Mitigative Technology- Transfer	Slurry Box	
Schoolhouse Pond, MA	21.0	(8.5)	Mitigative	Helicopter	P+T Trout SS Smallmouth Bass
Shubael Pond, MA	54.8	(22.2)	Maintenance	Helicopter	P+T Trout SS Smallmouth Bass
Snake Pond, MA	83.0	(33.6)	Maintenance	Helicopter	SS Largemouth Bass
South Pond, MA	25	(10.1)	Mitigative	Slurry Box	SS Largemouth Bass
Barnes Lake, NY	7.2	(2.9)	Mitigative	Slurry box	P, G+T Brook Trout
East Pond, NY	57.3	(23.2)	Mitigative	Helicopter	SS Brook Trout SS Lake Trout P, G+T Rainbow Trout
Heavens Lake, NY	22.2	(9.0)	Research	Helicopter	SS Brook Trout

report presents data on the model dose predictions, effective dissolution of the applied limestone and water quality response in 20 of the 22 treated lakes. In addition, changes in pH, ANC and Ca are discussed for post-liming intervals of up to 20 mo.

2. MATERIALS AND METHODS

Sampling protocols and chemical analyses of lake water were conducted in accordance with the LLI Field Operations Program Implementation Manual (Living Lakes, 1987). The protocols were modifications of those used in the National Surface Water Survey (NSWS) (Linthurst et al., 1986) which are reviewed in an earlier publication (Brocksen et al., 1987). Based upon recommendation of the LLI Scientific Advisory Committee during the first 2 yr of the program, sampling frequency and analyses were adjusted to focus on those parameters which exhibit quantitative changes (10 times required detection limits) as a result of treatments. Although each site is monitored for a full suite of analytes prior to treatment and again 30 days post-liming, a core suite of analytes, pH, ANC, Ca, Al (total dissolved), conductivity, total nitrogen (TN), and total phosphorous (TP) are monitored at least twice a year. This semi-annual sampling is conducted at ice-out and again during late summer, normally prior to turnover. For samples with total dissolved Al content >60 *u*g L^{-1} (vol. wt. mean), analysis for total monomeric and organic monomeric forms is also performed. In lakes with low mean hydraulic retention times (less than 0.5 yr) more frequent sampling is conducted to adequately monitor decreases in ANC, pH, and Ca levels. These lakes will be retreated when any one of three reliming criteria levels are met; pH $\leq$ 6.5, ANC $\leq$ 50 *u*eq L^{-1} or CA/ H^+ ratio (*u*eq L1-^{1}) $\leq$ 100.

2.1 DeAcid Model Predictions and Practical Aspects of Treatment

Various grades of limestone slurry were applied to 22 lakes during 1986 to 1987. Treated lakes or ponds included 11 seepage, 9 drainage, 1 seepage/drainage, and 1 reservoir (**Table I**). Although helicopter applications were used exclusively in 1986, barge and slurry box/jon boat applications were used exclusively in 1987. In all cases limestone was applied as a slurry with

Table I (con't)
LLI Project Site Background Information

Name	Lake Surface Area Acres	(ha)	Treatment Type	Application Type	Fishery
Lake Kanacto, NY	9.9	(4.0)	Mitigative	Helicopter	SS Smallmouth Bass
Lake Madeleine, NY	303.8	(123.0)	Mitigative	Helicopter	SS Brook Trout SS Lake Trout
Lake Tejekna, NY	4.9	(2.0)	Mitigative	Helicopter	P, G+T Brook Trout
Little Simon Pond, NY	155.6	(63.0)	Mitigative	Helicopter	SS Brook Trout SS Lake Trout
Woods Lake, NY	56.8	(23.0)	Mitigative	Helicopter	SS Brook Trout
White Deer Lake, PA	48.9	(19.8)	Maintenance Research	Barge	SS Largemouth Bass
Wallum Lake, RI-MA Trout	322	(130)	Maintenace	Barge	SS Trophy Brown P+T Trout
Flat Top Lake, WV	300.1	(121.5)	Maintenance Research	Slurry Box	SS Largemouth Bass

[1]P+T - put and take;
P, G + T - put, grow and take;
SS - self-sustaining

solids content ranging from 5 to 70%. Limestone was supplied by Pfizer, Inc.'s Minerals, Pigments and Metals Division or in one case by U.S. Gypsum Company, Lime Division. Median particle sizes (*u*m equivalent spherical diameter), included 4, 14, 18, 45, and 200 *u*m.

A titration-based computer model, DeAcid (Warfvinge et al., 1984; Saunders et al., 1985), was used to estimate a dose for each site. The titration-based model uses lake area, mean depth and pre-treatment pH to determine a dose. Simulated titration of lake water is performed for the range of limestone particles in the applied material. At pH $\leq$ 6.5 limestone dissolution is a function of the rate of diffusion of H^+ to the particle surface. At pH $\geq$ 6.5 dissolution is governed by reactions of $CaCO_3$ with water (Sverdrup, 1985). The model also provides for a sediment dose if required for the site. Table II summarizes the DeAcid designed dose, the applied dose, and the short-term dissolution of slurried limestone based upon observed changes in Ca concentration. Increases in Ca, Δ Ca_{exp}, as a result of treatment were determined from re-treatment and 30-day post-treatment samples. Values (**Table II**) are volume weighted means for routine and duplicate samples collected from the epilimnion and mid-hypolimnion. If the relative volume of the hypolimnion was <10%, values were determined solely from epilimnetic samples. These Ca concentration increases were compared to Ca available in the applied dose (Ca applied). Sources of error present in these calculations include errors in lake volume (sometimes > 10%), unknown influences of transevaporation between monitoring periods, and both groundwater and tributary flow changes between monitoring.

2.2 Conductivity Changes

A long-term objective of LLI is to recommend appropriate monitoring for a fisheries management liming program. Constituents such as pH, ANC, dissolved inorganic carbon (DIC), conductivity and dissolved Ca are regularly used to quantify changes in lake water quality as a result of liming. However, measurements such as DIC, which are dependent upon short holding times and rigorous laboratory analysis, normally would not be used in an operational program conducted by a state resource management agency. Therefore, LLI is attempting to identify more cost-effective, reliable methods and instrumentation which could be used by

TABLE II
EVALUATION OF DeACID DOSE MODEL

NAME	LIMESTONE SIZE, um	DeACID DESIGNED DOSES TONNES	APPLIED DOSE, TONNES	APPLIED DOSE, mg/l	APPLIED CALCIUM mg/l	Δ Ca EXP mg/l	% DISSOLUTION DeACID PREDICTION	% DISSOLUTION OBSERVED
1. BARNES LAKE	18	4.5	4.5	38.0	14.7	6.06	35.3	41.1
2. EAST POND	18	17.3	17.4	11.0	4.25	1.79*	42	45.6
3. HEAVENS LAKE	18	4.9	6.4	21.8	8.44	3.18*	35.5	37.6
4. LAKE MADELEINE	(14 & 18)	89.3	90.5	11.5	4.40	2.38*	46	54
5. LAKE KANACTO	18	2.3	4.6	22.6	8.70	4.26	48	40.3
	200	1.0	1.0	4.9	1.87			
6. LAKE TEJEKNA	18	1.8	1.6	10.0	3.87	1.65	54	34.2
	200	0.5	0.4	2.5	0.96			
7. LITTLE SIMON POND	14	22.7	23.5	3.5	1.36	3.82	37.5	30.0
	18	45.6	40.5	6.0	2.32			
	200	22.8	21.5	3.2	1.22			
8. WOODS LAKE	18	10.8	11.1	13.9	5.38	6.64	16-17	33.0
	200	21.8	23.9	29.7	14.71			
9. SOUTH POND	18	8.2	8.2 (38)	15.2	15.2	6.87	26.5	44
10. WHITE DEER LAKE	4	5.0	4.1	8.6	3.44	4.47	85	100+
	18	11.1	12.0	2.5	9.2		28	11

* DATA COLLECTED ≃ 90 DAYS POST-LIMING

+ OBSERVED VALUE REPRESENTS 100% OF DISSOLUTION OF 4u MATERIAL AND 11% OF 18u MATERIAL IN SECOND DOSE

TABLE II
EVALUATION OF DeACID DOSE MODEL

NAME	LIMESTONE SIZE, um	DeACID DESIGNED DOSES TONNES	APPLIED DOSE, TONNES	APPLIED DOSE, mg/l	APPLIED CALCIUM mg/l	Δ Ca EXP mg/l	% DISSOLUTION DeACID PREDICTION	% DISSOLUTION OBSERVED
11. BAKER POND	14	11.0	10.3	11.3	4.38	4.40	48	100
12. CURLEW POND	14	12.5	12.9	13.4	5.20	5.12	49	99.2
13. LAWRENCE POND	14	35	33.7	13.2	5.11	4.56	48	89.0
14. LOUT POND	14	4.8	5.6	15.4	5.96	3.95	48	66.3
15. ROUND POND	14	5.0	5.0	14.9	5.73	3.95	44	69.0
16. SCHOOLHOUSE POND	14	6.1	7.6	16.3	6.31	5.20	49	82.4
17. SHUBAEL POND	18	10.0	13.1	12.0	4.65	3.26	36	70.0
18. SNAKE POND	14	71.1	15.7	11.2	4.34	3.31	42	75.0
19. WALLUM LAKE	18	108.9	99.5	9.5	3.64	2.87	42	70.0
20. NORTH POND	14 200	5.5 3.6	5.5 3.6	13.8 9.0	4.99 3.46	5.17 -	32 -	100.0 -
21. LAUREL LAKE	18	18.2	18.2	25.8	-	-	-	-
22. FLAT TOP RESERVOIR	45	25.4	28.9	27.9	10.70	1.09	41	10.2

fisheries managers in the field to monitor various phases of aquatic liming. One such effort evaluated the use of treatment induced changes in conductivity to estimate the fraction of limestone dissolved 30 days after application. It was assumed that treatment related changes in the conductivity of the three parameters, ANC, Ca and H^+ (pH) could be used to calculate the conductivity change associated with dissolved $CaCO_3$. If sufficient agreement exists between conductivity changes determined from dissolved calcite components (Ca and ANC) and measured conductivity change, measured conductivities may be used to approximate short-term calcite dissolution. For pH values which increase from ~ 5.0 to ~ 7.5 due to liming, it was also assumed that bicarbonate (HCO_3^-) was equivalent to measured ANC. Calculated changes in conductivity were determined from equivalent conductivities of each species and changes in analyte concentrations measured before treatment and 30 days after treatment:

$$\text{Calc. cond } (uS\ cm^{-1}) = \text{Ca } (mg\ L^{-1}) \times 2.6 + \text{ANC } (ueq\ L^{-1}) \times 43.6 - \text{H } (mol\ L^{-1}) \times 3.5 \times 10^{+5}$$

2.3 Post-Treatment Time Dependent Changes in Ca, ANC and pH

A primary objective of the LLI aquatic liming program is to maintain a water quality conducive to the growth and reproduction of a fishery. Although sampling frequency (2 or 3 times yr^{-1}) was not sufficient to quantitatively relate site specific temporal and hydrologic factors to the effects of liming, it has provided adequate data to document that critical liming criteria limits (pH > 6.5, ANC $\geq$ 50 *u*eq L^{-1}, Ca/H^+, *u*eq L^{-1}, > 100) were maintained. The monitoring program has also documented post-treatment decreases in monomeric Al species in five sites where pre-treatment levels of total dissolved Al (> 60 *u*g L^{-1}) were observed. The effects of treatments on Al levels have been characterized in recent research reports (Driscoll et al., 1988a; Driscoll et al., 1988b). Other metals, nutrient parameters and anions have not yet exhibited quantifiable changes (10x required detection limits) 18 mo after treatment. Because of these observations, early data analysis has been limited primarily to evaluation of ANC, Ca and pH.

In order to relate analyte levels to time, data were converted to whole-lake, volume-bathymetry weighted data. This was done by using field stratification profiles, bathymetry data and average values for epilimnetic and hypolimnetic routine and duplicate samples. If the volume of the hypolimnion was <10% of the lake volume or the relative difference between epilimnetic and hypolimnetic analyte values were < 10%, the average epilmnetic concentration of routine and duplicate samples was considered representative of the whole lake.

For lakes where limestone dissolution exceeded 70% (coastal, seepage lakes) whole-lake data were sufficient to apply a simple single-box mass transfer model (Imboden and Schwaraenbach, 1985) to analyze observed changes for Ca and ANC.

$$C_t = (C_o - C_b)e^{-kt} + C_b$$

C_t = conc. at time, t
C_b = background conc.
C_o = initial conc., post-treatment
k = water exchange rate, day^{-1}

For inland lakes where a sediment dose was used, long-term dissolution calcite was expected to contribute to Ca and ANC changes. In these cases the DeAcid reacidification model (Sverdrup, 1985) was applied.

3. RESULTS AND DISCUSSION

3.1 DeAcid Predictions

A majority of treated inland lakes or ponds in the Adirondacks, Berkshires, or Poconos showed good agreement between observed and model predicted short-term dissolution of limestone (**Table II**). The relative precision error for eight of the first ten sites was <20%. For the other two lakes, the utility of the model was limited due to dosing strategy or unusual physical properties of the site. Woods Lake received an initial water column dose of 18 *u*m limestone followed by a large sediment dose (>1.0 t ha^{-1}) of 200 *u*m material. The

TABLE III
CONDUCTIVITY CHANGES IN LIMESTONE TREATED LAKES

NAME	OBSERVED* CONDUCTIVITY uS/cm	CALCULATED** CONDUCTIVITY
BAKER POND	19.3	20.6
CURLEW POND	23.4	24.0
LAWRENCE POND	20.9	22.0
LOUT POND	16.2	17.5
NORTH POND	22.7	24.9
SCHOOLHOUSE POND	14.3	14.9
SHUBEAL POND	18.7	20.1
SNAKE POND	14.9	15.6
WALLUM LAKE	12.9	13.6
ROUND POND	17.7	18.9
SOUTH POND	31.8	31.5
WHITE DEER LAKE	23.0	23.6
BARNES LAKE	20.5	26.5
EAST LAKE	8.3	8.9
HEAVENS LAKE	16.8	14.4
LAKE MADELEINE	13.0	11.8
LAKE KANACTO	10.2	9.8
LAKE TEJEKNA	-4.2	-3.4
LITTLE SIMON POND	16.1	19.1
WOODS LAKE	27.3	26.8

CALCULATED CONDUCTIVITY ($uScm^{-1}$) = ΔCa (mgL^{-1}) x 2.60 + ΔANC ($meqL^{-1}$) x 43.6 - ΔH (MOLES L^{-1}) 3.5 x 1^{+5}

* *OBSERVED CONDUCTIVITY CHANGE DETERMINED FROM PRE- AND 30-DAY-POST SAMPLES. VALUES ARE VOLUME-WEIGHTED MEANS*

** *CALCULATED CONDUCTIVITY DETERMINED FROM VOLUME-WEIGHTED MEAN INCREASE IN Ca^{++} AND HCO_3- AND DECREASE IN H^+*

DeAcid model is designed to simulate the results of single application with a single material. Therefore, the ability to predict the effects of two consecutive applications with two different materials is severely limited. The other site, South Pond, exhibited a mean depth $\leq$ 2 m and CO_2 infusion may have increased the dissolution of applied calcite according to the reaction:

$$CaCO_3 + CO_2 + H_2O = 2HCO_3 + Ca^{++}$$

In coastal regions the observed limestone dissolution in lakes and ponds was approximately 40 to 100% greater than model predictions. Since the effect was observed with different materials applied with different techniques, it seems increased dissolution may be related to lake water composition. Compared to the water chemistry of treated inland sites, coastal lakes exhibited higher conductivities and higher total ionic strengths. In particular, chloride (Cl) levels ranged from 4 to 21 mg L^{-1} in pretreatment coastal ponds (Nos. 11-19, **Table II**) but never exceeded 1.0 mg L^{-1} in inland sites. Small differences between these regional groups existed for other anions. It is hypothesized that in the pH range, 5.9 to 6.5, anions such as Cl act at the calcite particle surface to substantially increase the rate of $CaCO_3$ dissolution. The effects of anion concentrations on dissolution characteristics of limestone are currently being evaluated in the laboratory.

3.2 Conductivity Changes

Observed and calculated conductivity changes are presented in Table III. The general agreement (% cond. diff. < 15%) would indicate that in low conductivity waters (<120 *u*S cm^{-1}) conductivity can be quantitatively related to dissolved $CaCO_3$ resulting from treatment. In fact, for treatments where pH changes in the range from 5.5 to 7.5, the H^+ term becomes insignificant. Under these conditions dissolved $CaCO_3$ in mg L^{-1} can be determined from measured change in conductivity by the simple relationship below. The conversion factor of 1.47 represents the contribution of Ca (2.6 *u*S cm^{-1} $mg^{-1}L$. 0.40 mg Ca mg^{-1} $CaCO_3$) and ANC (0.71 *u*S cm^{-1} mg^{-L} . 0.6.mg HCO_3 mg^{-1} $CaCO_3$) to conductivity.

$$\textbf{Dissolved } CaCO_3 \textbf{ (mg } L^{-1}\textbf{)} = \frac{\textbf{conductivity (uS } cm^{-1}\textbf{)}}{\textbf{1.47}}$$

TABLE IV
FIRST ORDER RATE CONSTANTS FOR DECREASES IN ANC AND Ca

$$C_t = (C_o - C_b)e^{-kt} + C_b$$

SITE	k_{ANC}, $DAY^{-1} \times 10^{-3}$(R)*	k_{Ca}, $DAY^{-1} \times 10^{-3}$(R)	TIME TO ANC = 50 UEQ/L DAYS x 10^2
SNAKE POND	-1.74(0.98)	-1.34(0.98)	7.6
CURLEW POND	-1.44(0.98)	-1.33(0.98)	11.4
LOUT POND	-1.01(0.96)	-0.81(0.94)	14.4
LAWRENCE POND	-1.23(0.98)	-1.09(0.98)	12.9
SHUBAEL POND	-0.96(0.93)	-0.85(0.97)	13.9
SCHOOLHOUSE POND	-0.95(0.95)	-0.69(0.97)	15.7
BAKER POND	-0.97(0.97)	-0.81(0.98)	16.6

* *R VALUES ARE CORRELATION COEFFICIENTS FOR LEAST SQUARE FIT OF IN ANC OR IN Ca VERSUS TIME*

Increased confidence in this relationship is obtained from the fact that a decreased conductivity was calculated and observed at Lake Tejekna, a site where target pH and ANC objectives were not achieved. In this case, H^+ change was the dominant term in the conductivity calculation.

The use of this expression seems restricted to low conductivity waters. In Flat Top Reservoir, where pre-treatment conductivity was >200 uS cm^{-1}, the addition of limestone was associated with a marked decrease in conductivity. In these type waters the addition of calcite particles probably caused extensive coprecipitation of ionic lake constituents. This factor is not accounted for in the above expression.

3.3 Time-Dependent ANC and Ca Changes

Application of the model to data from seepage ponds and lakes in Massachusetts (**Table IV**) demonstrated ($R > 0.90$) that the hydrologic dilution effect caused by lake flushing is the primary influence on observed decreases in both Ca and ANC. The rate constants for loss of ANC and Ca indicate slightly greater rate of decrease for ANC compared to Ca. This effect is influenced both by the increased reactivity of ANC and increased rate of Ca inflow relative to ANC (Godfrey et al., 1985). The excellent linear correlations between Ca and ANC (**Table V**) for the Massachusetts and New York sites indicate Ca and ANC are primarily influenced by dilution through lake flushing.

Four of the Massachusetts ponds, Curlew, Lawrence, Shubael and Snake (**Figures 1 through 4**) were also monitored as part of the Massachusetts Acid Rain Management Program (ARM I, II and III) (Godfrey et al., 1985). For these sites, 9 to 11 timed data points were available. Considering the widely different sampling protocols in the LLI and ARM programs, the analytical data presented in Figures 1 thru 4 indicate both programs have produced mutually comparable data for Ca and ANC. An assessment of interlaboratory bias for these two parameters is currently being performed.

The pH changes over the same time period have been minimal. At Snake Pond, which has shown the most rapid post-treatment loss of ANC in these seepage systems, pH has decreased from a post-liming high of 7.46 to 6.71. Until ANC has dropped below 50 $u$eq L^{-1},

TABLE V
LINEAR CORRELATIONS OF POST-TREATMENT ANC AND Ca
$Ca(UEQ/L) = k\ ANC(UEQ/L) + C_0$

SITE	NO. OF VALUES*	k	CO., ueq/l	R
BAKER POND	11	0.882	66.71	0.917
CURLEW POND	17	0.961	35.53	0.971
LOUT POND	8	0.941	45.66	0.931
SCHOOLHOUSE POND	7	0.860	102.41	0.912
SNAKE POND	17	1.010	26.80	0.89
LAWRENCE POND	9	1.063	41.90	0.978
NORTH POND	6	0.871	80.04	0.920
EAST POND	6	0.836	112.59	0.973
SHUBAEL POND +	7	0.972	38.84	0.877
LAKE KANACTO	7	0.745	87.39	0.989
LAKE TEJEKNA	7	0.495	79.87	0.814
HEAVENS LAKE	5	0.887	100.52	0.991
LAKE MADELEINE	8	0.663	137.19	0.901

** VALUES ARE EITHER MEANS OF ROUTINE AND DUPLICATE SAMPLES OR INDIVIDUAL VALUES*

+ EPILIMETIC VALUES ONLY

ANC/Ca CHANGES AT SNAKE
(UEQ/L VS. DAYS)

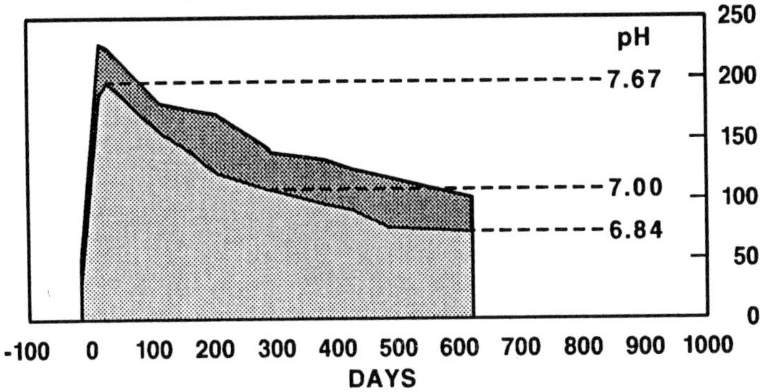

SUMMARY OF MEAN VALUES OF ANC, Ca, AND pH.

TIME, DAYS (POST-LIMING +) (PRE- LIMING -)	ANC UEQ/L	Ca UEQ/L	pH
-16	26.0	50.4	6.46
17	186.0	229.0	6.82
29	197.0	226.0	7.67
116	156.0	180.0	6.37
161	141.0	175.0	7.04
206	122.0	171.0	6.48
290	108.0	142.5	7.02
294	108.0	139.0	7.00
381	96.0	133.0	6.95
427	91.0	125.0	6.71
487	76.0	---	6.90
624	72.6	101.5	6.84

ANC/Ca CHANGES AT CURLEW
(UEQ/L VS. DAYS)

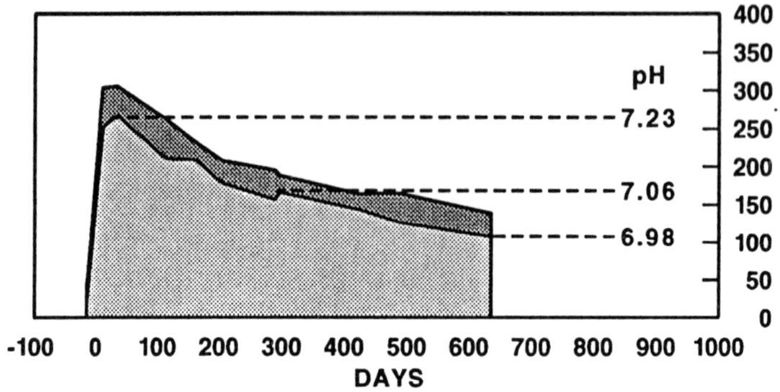

SUMMARY OF MEAN VALUES OF ANC, Ca, AND pH.

TIME, DAYS (POST-LIMING +) (PRE- LIMING -)	ANC UEQ/L	Ca UEQ/L	pH
-16	17.0	45.5	5.99
17	252.0	303.5	6.40
34	267.0	305.5	7.23
115	210.0	262.5	7.20
162	209.0	233.0	7.24
205	178.0	208.0	6.70
290	156.0	195.0	6.70
293	166.0	188.0	7.06
431	143.8	164.0	7.00
487	126.0	---	7.10
635	107.0	138.5	6.98

ANC/Ca CHANGES AT LAWRENCE
(UEQ/L VS. DAYS)

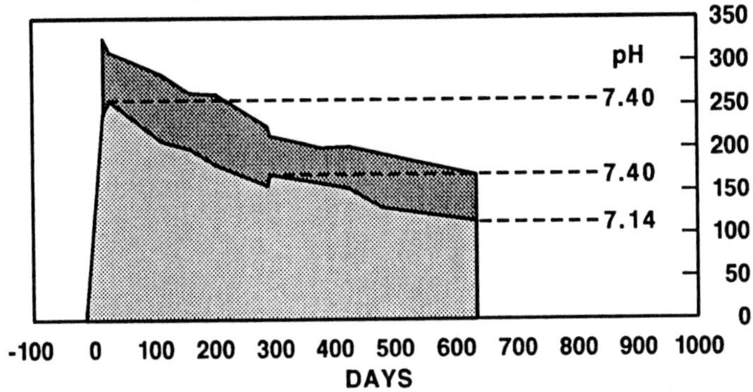

SUMMARY OF MEAN VALUES OF ANC, Ca, AND pH.

TIME, DAYS (POST-LIMING +) (PRE- LIMING -)	ANC UEQ/L	Ca UEQ/L	pH
-13	8.7	---	6.02
17	240.0	327.0	7.29
28	255.0	311.5	7.40
115	208.0	285.0	6.60
161	199.0	264.0	7.08
205	180.0	262.0	6.51
290	156.0	225.0	6.84
294	169.0	214.0	7.14
381	158.8	200.0	7.29
426	153.0	201.5	7.04
481	130.0	---	7.05
634	114.2	168.8	7.14

ANC/Ca CHANGES AT SHUBAEL
(UEQ/L VS. DAYS)

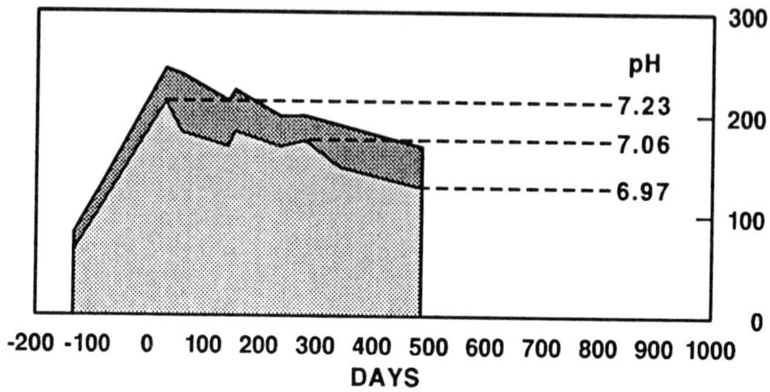

SUMMARY OF MEAN VALUES OF ANC, Ca, AND pH.

TIME, DAYS (POST-LIMING +) (PRE- LIMING -)	ANC UEQ/L	Ca UEQ/L	pH
-134	64.1	81.5	6.72
29	212.0	244.5	7.23
58	182.0	239.0	8.40
142	168.0	213.0	7.14
154	183.0	223.5	8.40
235	168.0	198.0	7.08
277	174.0	198.5	7.06
339	148.0	---	---
485	124.7	168.0	6.97

pH is expected to remain above 6.5. In bicarbonate systems, rapid pH decreases are expected when ANC levels decrease below 20 *u*eq L^{-1} (Linthurst et al., 1986). A goal of the LLI Program is to avoid these concentrations. For our maintenance purposes, ANC and Ca concentrations are used to schedule retreatment.

In Massachusetts sites where sediment limestone dissolution is minimal, determination of rate constants for the loss of Ca can presumably be extended to determination of mean hydraulic retention time (HRT). Assuming that Ca is a conservative ion and that minimal sediment dissolution occurs, the rate constant for volume-weighted loss of Ca can be directly related to HRT by k_{ca}^{-1} = HRT, days. For extremes in Table IV, Snake and Schoolhouse Ponds would exhibit HRT's of 752 and 1,490 days, respectively. The utility of using added Ca to accurately determine HRT's in seepage systems will be presented more completely in a future paper (Adams and Brocksen, unpublished data).

Unlike the Massachusetts seepage pond, the New York sites, Lakes Kanacto and Tejekna, received two different limestone treatments. First, a water column dose (18 *u*m - median particle size) and then a sediment dose (200 *u*m - median particle size). Kanacto was inadvertently overdosed by 70% because of an error in lake volume. However, expected increases in ANC and Ca were not realized probably because extensive *Sphagnum* mats consumed added $CaCO_3$ *via* cation exchange with the organic substrate (Wetzel, 1983). Significant increases in dissolved organic carbon (2.6 mg L^{-1} pretreatment increasing to 6.2 mg L^{-1}, 180 days post-treatment) supports partial neutralization of dissolved calcite by *Sphagnum*. The presence of *Sphagnum* was probably also responsible for the fact that treatment of Lake Tejekna did not achieve water quality objectives, pH > 6.5 and ANC > 100 ueg L^{-1} (**Table VI**).

Because of sediment dosing, concentrations for hypolimnetic samples were as much as four to five times epilimnetic concentrations. The effect of second year turnover at Kanacto gave sufficient redistribution of ANC to raise the spring 1988 epilimnetic ANC values above 100 *u*eq L^{-1} (**Table VI and Figures 5 and 6**). The beneficial effects of sediment dosing have been well documented in recent research programs (DePinto et al., 1988; Sverdrup et al., 1984).

Of the six drainage lakes treated in 1986, Woods Lake and Little Simon Pond are EPRI LAMP research sites and liming results are reported in more depth elsewhere (Porcella, 1988).

TABLE VI
ANC, Ca AND pH CHANGES AT LAKES KANACTO(K) AND TEJEKNA(T)*

TIME, DAYS +POST-LIMING -PRE-LIMING	DEPTH	ANC, UEQ/L K	ANC, UEQ/L T	Ca,UEQ/L K	Ca,UEQ/L T	pH K	pH T
-29	E	-7.0(6)	-26.0(6)	48.7(6)	33.5(6)	4.83(6)	4.51(6)
	H	-	-	-	-	-	-
+33(9/86)	E	152.7(2)	-52.3(2)	206.0(2)	120.0(2)	687(2)	6.70(2)
	H	289.8(2)	35.7(2)+	253.8(2)	79.1(2)	6.34(2)	5.50(2)
+186(2/86)	E	201.0(3)	-	263/5(3)	-	6.80(3)	-
	H	195.1(3)	-	255.0(3)	-	6.61(3)	-
+261(4/87)	E	113.4(2)	18.7(2)	177.5(2)	90.0(2)	6.68(2)	5.93(2)
	H	404.6(2)	54.2(2)	406.5(2)	118.5(2)	6.54(2)	5.75(2)
+409(9/87)	E	74.7(2)	4.5++(2)	149.5(2)	74.5(2)	6.39(2)	4.87(2)
	H	404.6(1)	110.(2)	570.0(1)	127.0(2)	6.58(1)	5.40(2)
+617(4/88)	E	138.8(2)	25.8++(2)	200.0(2)	101.5(2)	6.59(2)	5.80(2)
	H	-	-	-	-	-	-

** VALUES IN PARENTHESIS ARE NO. OF SAMPLES USED TO DETERMINE MEAN ANALYTE VALUE*
+ PRECISION ERROR > 10%, RELATIVE STANDARD DEVIATION
++ E = EPILIMNION, H = HYPOLIMNION

ANC/Ca CHANGES AT KANACTO
(UEQ/L VS. DAYS)

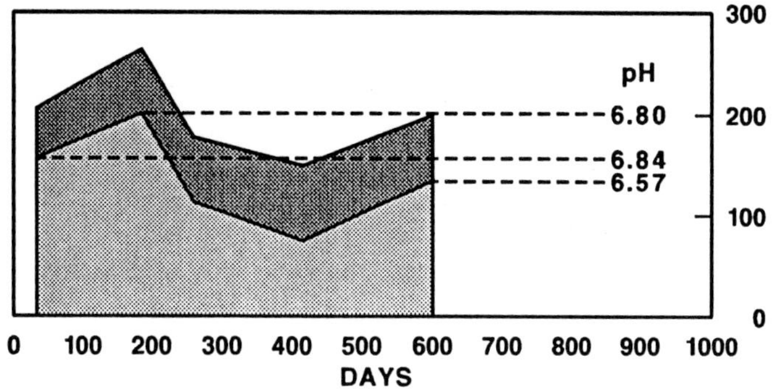

SUMMARY OF MEAN EPILIMNIC VALUES OF ANC, Ca, AND pH.

TIME, DAYS (POST-LIMING +) (PRE- LIMING -)	ANC UEQ/L	Ca UEQ/L	pH
33	156.6	206.0	6.84
186	201.0	264.0	6.80
261	113.4	177.5	6.68
416	74.7	149.5	6.42
600	134.7	199.5	6.57

ANC/Ca CHANGES AT KANACTO
(UEQ/L VS. DAYS)

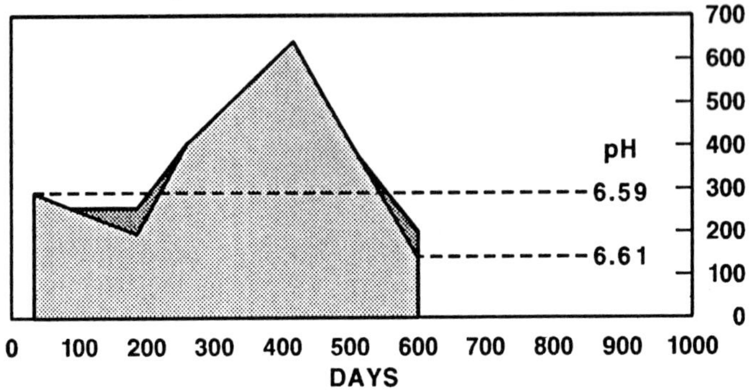

SUMMARY OF MEAN LOWER DEPTH VALUES OF ANC, Ca, and pH.

TIME, DAYS (POST-LIMING +) (PRE- LIMING -)	ANC UEQ/L	Ca UEQ/L	pH
33	289.8	253.8	6.59
186	195.0	255.0	6.61
261	404.6	406.5	6.54
416	638.9	570.0	6.61
600	137.9	200.0	6.61

Three of the remaining four drainage sites are in New York. They provide a range of both surface area and estimated HRT. As expected, Lake Madeleine, the largest site (103 ha) with highest estimated HRT (> 2.0 yr) exhibits the slowest rate of loss for ANC and Ca (**Figure 7**). The loss of ANC and Ca for Madeleine are influenced by both hydrologic dilution and inflow of increased concentrations of ANC and Ca from Heavens Lake, a treated site which feeds Madeleine. Presumably more frequent treatments of the rapidly flushing (0.4 yr, HRT) Heavens Lake will prolong circumneutrality of Madeleine.

Heavens Lake (9 ha; 0.4 yr HRT) and North Pond (10 ha; 1.4 yr HRT) are small and rapid flushing systems which received substantial sediment doses. More than 60% of the 21.8 mg L^{-1} of material applied to Heavens Lake was deposited on the sediment while North Pond received a separate sediment dose (**Table II**). Semi-annual collection of data (**Figures 8 and 9**) indicates that after 2 yr ANC levels remain above the retreatment criteria limits (50 ueg L^{-1}). In light of the reported rapid flushing of Heavens Lake, it seems sediment dosing has increased the duration of neutralization. The application of a sediment dose and its subsequent effects on reacidification rate are the result of research efforts by DePinto (1988) and Sverdrup (1984).

4. CONCLUSIONS

Results indicate the DeAcid model has been an adequate tool to predict the quantitative aspects of limestone dissolution in inland drainage and seepage sites. However, the model has consistently underestimated the short-term dissolution of limestone applied to coastal seepage systems under maintenance conditions (pH = 6.0 to 6.5). Although the effects exerted on dissolution of calcite in these coastal waters need to be better characterized, the DeAcid model has always underestimated dissolution in these systems. Therefore, minimum criteria objectives for ANC (>100 ueg L^{-1}) and pH (> 6.5) were always achieved.

Water quality data indicate that, for the most part, only those parameters directly associated with liming (ANC, Ca, pH, DIC) changed significantly in the early post-treatment phases.

ANC/Ca CHANGES AT MADELEINE
(UEQ/L VS. DAYS)

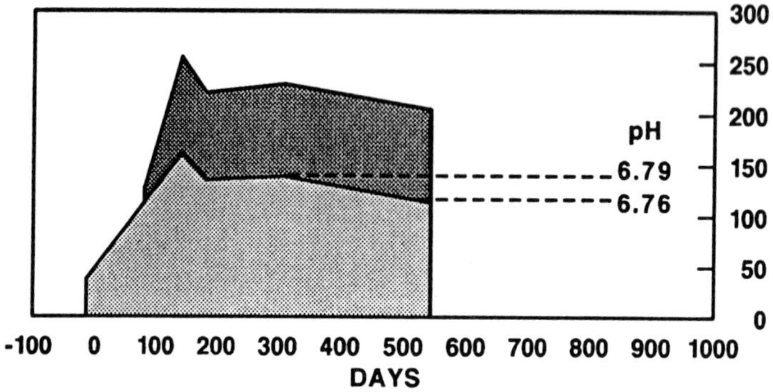

SUMMARY OF MEAN VALUES OF ANC, Ca, AND pH.

TIME, DAYS (POST-LIMING +) (PRE- LIMING -)	ANC UEQ/L	Ca UEQ/L	pH
-39	37.2	128.0	6.25
110	160.7	255.0	6.82
182	135.2	220.5	6.90
327	138.6	229.0	6.79
545	113.4	205.0	6.77

ANC/Ca CHANGES AT HEAVENS
(UEQ/L VS. DAYS)

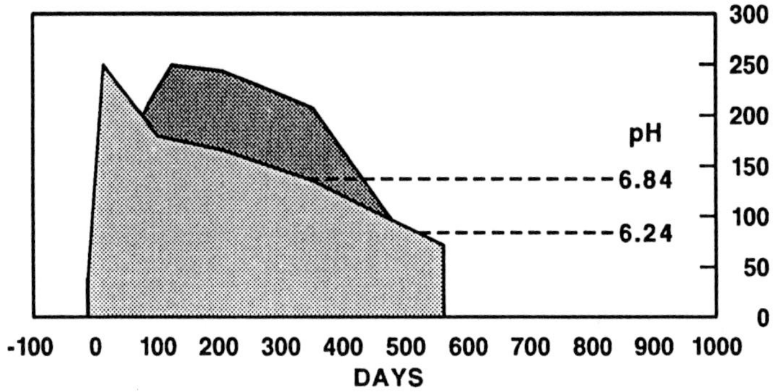

SUMMARY OF MEAN VALUES OF ANC, Ca, AND pH.

TIME, DAYS (POST-LIMING +) (PRE- LIMING -)	ANC UEQ/L	Ca UEQ/L	pH
-36	38.6	115.0	6.12
1 *	249.9	412.0	----
112	179.5	249.9	6.77
185	165.6	244.0	6.78
329	135.4	207.5	6.84
545	71.2	178.0	6.24

* CALCULATED VALUES DETERMINED FROM FIRST ORDER RATE CONSTANTS FOR DECREASE IN ANC AND Ca.

ANC/Ca CHANGES AT NORTH
(UEQ/L VS. DAYS)

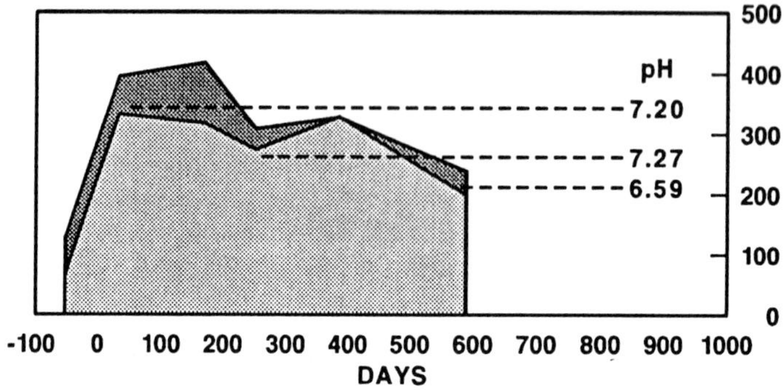

SUMMARY OF MEAN VALUES OF ANC, Ca, AND pH.

TIME, DAYS (POST-LIMING +) (PRE- LIMING -)	ANC UEQ/L	Ca UEQ/L	pH
-70	65.6	129.0	6.65
32	332.4	394.3	7.20
176	317.5	416.5	7.11
254	273.5	308.4	7.27
384	328.1	327.5	7.39
597	199.0	259.0	6.59

Post-treatment monitoring of ANC and Ca has provided sufficient basis for predicting the duration of treatment and the scheduling of retreatment to sustain water quality conditions consistent with LLI's fisheries management goals. An accurate determination of mean hydraulic retention time from rate data for loss of Ca is an additional benefit of limestone treatment, especially for seepage ponds where no sediment dose was administered.

ACKNOWLEDGMENTS

The authors gratefully acknowledge the advice and consultation of the members of the Living Lakes, Inc., Scientific Advisory Committee (R.A. Beck, R.C. Bratcher, D.J.A. Brown, J.G. Holsapple, P. Nyberg, D.B. Porcella, G.W. Redfield, J.L. Regens, R.K. Schreiber and H. Sverdrup) and technical assistance from the professional staff of the primary field contractor, International Science and Technology, Inc. We are especially indebted to Dr. Gerald Filbin, who through consistent dedication, provided a verified data set. We are also indebted to Dr. Paul Godfrey, his laboratory staff, and Acid Rain Management volunteers, for providing additional data for treated sites in Massachusetts.

LIMING OF LAKES, RIVERS AND CATCHMENTS IN NORWAY

Bjorn Olav Rosseland
Atle Hindar

ABSTRACT

Despite a slight reduction in the level of acidic deposition in Norway, acidification of lakes and rivers continues. The Norwegian Liming Project (1979-84) demonstrated that lime treatment can be an effective measure against acidification of watercourses given appropriate adaptation to local conditions. Liming in Norway is difficult because of (1) large amounts of precipitation, (2) short retention time of lakes, and (3) episodic changes in water chemistry. In 1988 NOK 14 mill. has been allocated to operational liming and research. We report here on chemical and biological responses from lime treatment of a lake, a river and a catchment. Lake Store Hovvatn was limed in 1981 and successfully stocked with brown trout. Before reliming in 1987, fish growth had ceased, but increased post liming. The River Audna has been continuously limed since 1985. Sea trout fisheries have improved, and the stocking of Atlantic salmon smolts at the mouth of the river in 1986 has already resulted in the return of spawners. Liming of the entire terrestrial catchment to the pond Tjønnstrond in 1983 by helicopter was also successful; stocked brown trout have survived to the present.

1. INTRODUCTION

In Norway, liming of acidic waters started in 1920's (Sunde, 1926; Dahl, 1927) and has since continued in several areas of Norway (Rosseland and Skogheim, 1984a). As the loss of fish populations from acidic lakes increased in the 1960's and 1970's (Sevaldrud et al., 1980; Sevaldrud and Skogheim, 1986), the Norwegian Liming Project (1979-1984) was initiated. The project evaluated different neutralizing agents, their chemical and biological effects, and different strategies for neutralizing lakes, rivers and catchments. In 1983, operational liming funded by the Norwegian environmental authorities was started, based on the recommendations given by the Norwegian Liming Project (Liming Project, 1987). We report here on three main liming projects; the liming of River Audna (from 1985), the whole-lake liming of Lake Store Hovvatn (from 1981), and the catchment liming of Tjønnstrond (from 1983) (**Figure 1**). These

Water, Air, and Soil Pollution **41** (1988) 165–188.

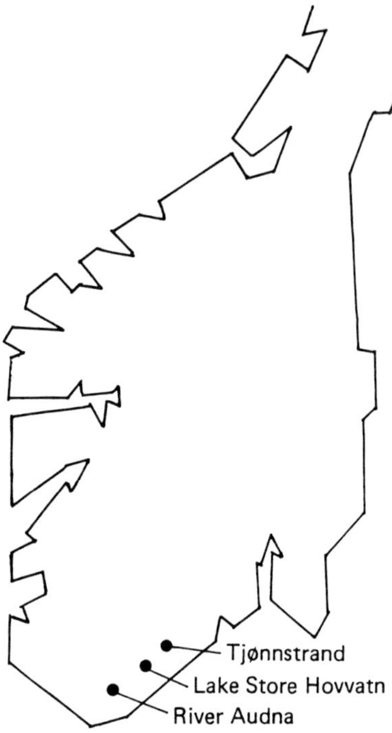

Figure 1. Map of southern Norway with the sites of River Audna, Lake Store Hovvatn and Tjønnstrond marked out.

treatments are all a part of the ongoing operational liming activities in Norway (Hindar et al., 1987; Hindar and Rosseland, 1988).

2. METHODS

Different lime application methods have been used in the reported liming projects and will be described separately for each project. The chemical data include total Al and species of Al. The separation method for Al-speciation follows the analytical scheme of Driscoll (1984). The fraction of monomeric inorganic Al, termed "labile Al" (LAl), is connected to the toxicity of aquatic biota (Driscoll et al., 1980; Rosseland and Skogheim, 1984b). The total (TAl) and the reactive Al (RAl) fractions contain both the organic and inorganic species of Al (Driscoll, 1984). The other analytical methods were identical to those described by Skogheim and Rosseland (1984).

3. LIMING OF RIVER AUDNA

River Audna, situated in the country of Vest-Agder in southern Norway (Sorlandet), has a catchment of 450 km^2 with a mean run-off of 1500 mm and a mean flow of 22 $m^3 s^{-1}$ (**Figure** 2). This river lost its Atlantic salmon (*Salmo salar*) population by the early 1970's (Leivestad et al., 1976), and by the 1980's had only a minor sea trout (*Salmo trutta*) fishery left. Brown trout (*Salmo trutta*), Arctic charr (*Salvelinus alphinus*) and perch (*Perca fluviatilis*) were

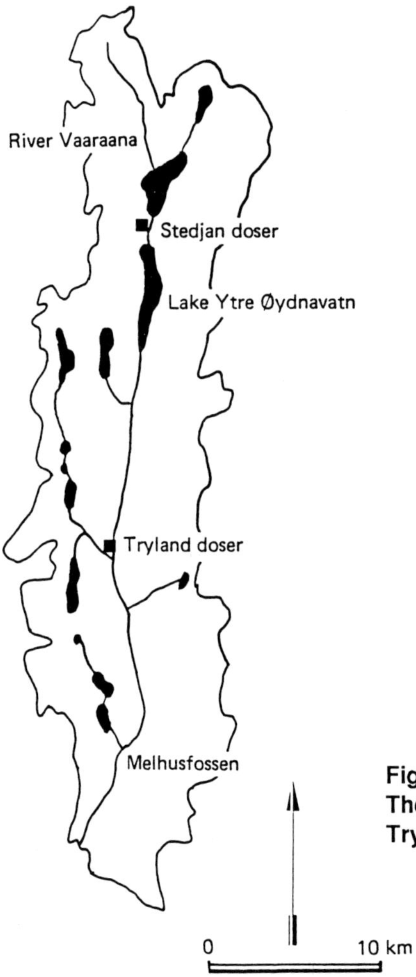

Figure 2. The River Audna catchment area. The position of the two dosers at Stedjan and Tryland is indicated.

common in the watershed. Lake Ytre Oydnavatn, the largest lake in the catchment area, has been a part of the liming strategy for River Audna. It has a mean depth of 32 m, surface area of 3.3 km^2 and a volume of $105 \cdot 10^6$ m^3. The retention time is 0.6 a yr. The 50 km of river downstream from this lake represents the former salmon stretch.

During the Norwegian Liming Project (1979-84) a tributary, River Vaaraana, was successfully limed by means of limestone slurry (Abrahamsen and Matzow, 1984; Rosseland and Skogheim, 1984a; Rosseland et al., 1986a; Liming Project, 1987). Since 1985, another strategy has been used (Hindar et al., 1987).

At Stedjan, by the inlet of Lake Ytre Oydnavatn (**Figure** 2), a limestone doser (**type NORCEM**) using finely powdered limestone (NORCEM type SR, 80% as $CaCO_3$, 2% as $MgCO_3$, 14% as SiO_2, with particle size 90% <0.064 mm, 50% <0.010, and 20% <0.002 mm) was installed. To ensure maximum dissolution, the limestone powder was mixed with water to form a slurry prior to dosing the stream. A similar doser has been installed at Tryland, ca. 40 km from the river mouth. At the start of the liming (October 1985), Lake Ytre Oydnavatn was limed using 890 t of NORCEM SR limestone spread by boat.

The monitoring of the effects of liming included water chemistry, invertebrates and fish. In addition to test-fishing, tagged Atlantic salmon smolts have been stocked near the river mouth. Both hatching and survival experiments with salmon smolts have been performed before and after liming. Reported here are the major results from these studies.

3.1 Chemical Response

Prior to liming, typical water chemistry in the main river was pH 5.0 to 5.1, Ca 1 to 1.5 mg L^{-1}, RAl = 200 to 250 *u*g L^{-1} and LAl = 100 to 150 *u*g L^{-1}. Two weeks post-liming (November 1985), the pH of Lake Ytre Oydnavatn had increased to 6.7 to 6.8, Ca increased to 3.5 mg L^{-1} in the surface waters with alkalinity increasing from 0 to 90 *u*ekv L^{-1}. RAl showed a minor change from 200 to 260 *u*g L^{-1} prior to liming to 230 *u*g L^{-1} post liming. A LAl concentration of 80 *u*g L^{-1} at the surface and 160 *u*g L^{-1} below 15 m depth prior to liming, decreased to 20 to 35 *u*g L^{-1} and 55 to 70 *u*g L^{-1} in the surface and >15 m depth, respectively. The minor change in RAl indicated little precipitation of Al post liming. Using the dissolution model of Sverdrup (1985), a theoretical dissolution rate of 72% was calculated. Two weeks after liming, 72.5% had dissolved in agreement with the theoretical value.

The automatic dosing systems at Stedjan and Tryland are controlled by water flow only. In 1985, 330 t of limestone powder were dosed, whereas 3320 t were used in 1986. Both dosers have stopped for periods of time due to technical problems. However, due to dissolution of previously sedimented limestone on the riverbed, the water chemistry in these periods (hours or days) changed only moderately.

Before the liming, even large changes in water flow had little effect on the stable low pH of the River Audna. Due to acidic tributaries (pH 4.7), the pH decreased from the outlet of Lake Ytre Oydnavatn (pH 5.20) and downstream (pH 5.15 at Melhusfossen 5 km from the river mouth). After liming, pH downstream of the doser at Stedjan has varied between 6.0 and 7.5 except for a few episodes (pH 5.6 in May 1986). Calcium has varied between 2.0 and 4.5 mg L^{-1}. At Tryland, the pH has been above 6.0 and Ca between 2 to 3 mg L^{-1}. The effects of liming have been satisfactory down to Melhusfossen (pH > 6.0) and the river mouth (**Figure 3**).

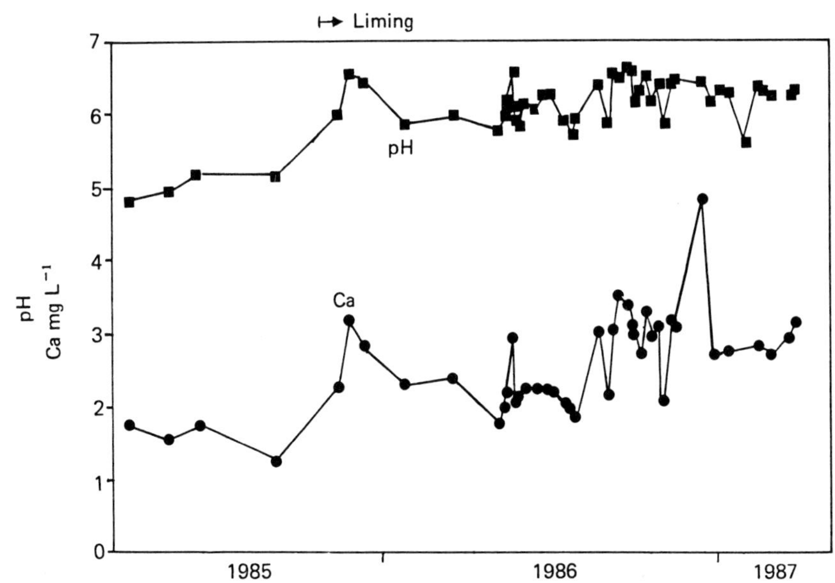

Figure 3. The pH and Ca in River Audna at Melhusfossen (see Figure 2) in the period March 1983 to march 1987. The start of the liming is indicated.

3.2 Biological Response

3.2.1 Invertebrates.

Invertebrates have been sampled at 9 to 13 stations in the main river and tributaries since 1985, by means of the *kicking method* (Frost et al., 1971). The invertebrate community contained acid tolerant species of Ephemeroptera, Plecoptera and Trichoptera. Until 1987, few acid sensitive populations of stoneflies and caddisflies were observed in the river. Using the acidification index of Raddum and Fjellheim (1984), the River Audna as a whole had a value of 0.3 in 1985-86. In fall 1987, however, the first specimens of the acid sensitive mayfly *Baetis rhodani* was observed, increasing the index to 0.5 (G.G. Raddum-pers. com.), indicating a stable neutralized ecosystem.

3.2.2 Fish Populations.

In June 1985, prior to the liming, 3400 fin-clipped Atlantic salmon smolts were stocked at Rødberg in brackish water at the river mouth. Survival experiments were performed using 50 salmon smolts in fish cages at Tryland (pH 5.3 to 5.4, LAl 63 to 87 $ug\ L^{-1}$). All smolts died within 6 days at Tryland compared to 11 dead smolts at Rødberg. In spite of the high survival rate at Rødberg, no returns of the fin-clipped salmon were observed in 1986 (Hindar et al., 1987). As a part of a national sea ranching program for Atlantic salmon, River Audna and River Lygna (acid control) were stocked with 3000 smolts at the river mouth and 3000 smolts 5 km upstream, all tagged by Carlin tags. In 1987, Atlantic salmon were caught for the first time in many years on their return to River Audna (L.P. Hansen-pers. com.). Also, the sea trout fishery had increased tremendously after liming (O.E. Larsen-pers. com.).

Hatching experiments using freshly fertilized eggs were conducted upstream of the liming at Stedjan (pH 5.11, Ca 1.1 mg L^{-1}, LAl 115 $ug\ L^{-1}$) and on the limed river stretch downstream Lake Ytre Oydnavatn (pH 6.3, Ca 2.9 mg L^{-1}, LAl 25 $ug\ L^{-1}$). These experiments gave a 83% and 11% mortality until hatched in the acidic and limed areas, respectively. At Stedjan, 84% of the alevins died whereas all survived on the limed river stretch (Hindar et al., 1987).

3.2.3 Conclusion.

So far, the liming of River Audna has been a success. A yearly funding of NOK 1.5 x 10^6 covers the expenses for lime and the leasing and operational cost of the dosers. The chemical and biological program receives NOK 0.2 x 10^6 annually. Increased salmon fisheries, being the only salmon river at Sorlandet, might in the future more than balance these budgets.

4. LIMING LAKE STORE HOVVATN

Lake Store Hovvatn, situated in the county of Vest-Agder (**Figure 1 and 4**), has a surface area of 1.1 km^2 and a total catchment area of 7 km^2. With a specific discharge rate of 32 L s^{-1} $km^{2\ -1}$, the theoretical retention time is 0.9 yr. Pollen, a smaller water body (surface

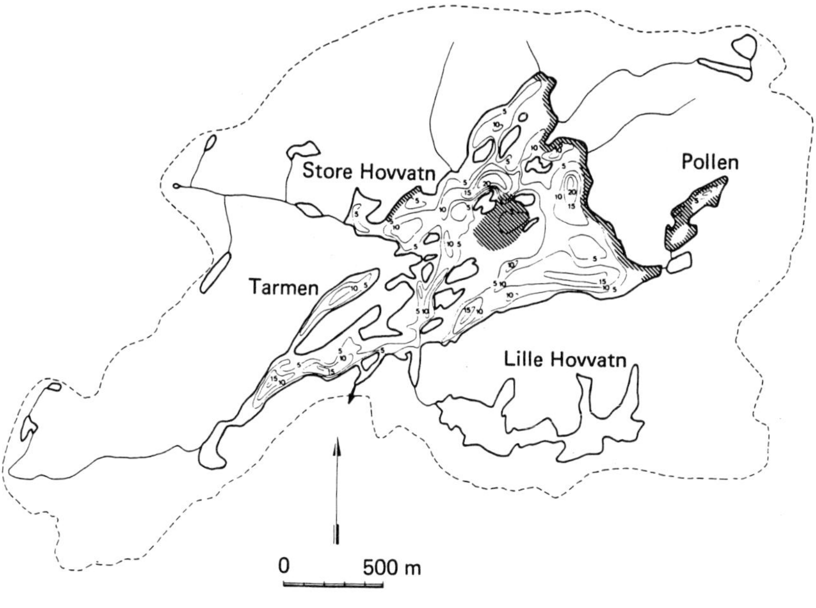

Figure 4. Lake Store Hovvatn and its catchment. The pond Pollen, Tarmen and the reference Lake Lille Hovvatn are shown. In 1981, limestone powder was spread on the ice in the hatched areas. In 1987, limestone powder was spread on the lake surface.

area of 4.6 ha) is connected with Lake Store Hovvatn in the eastern part (**Figure 4**). Lake Lille Hovvatn is not limed and serves as the acidic control in the experiment (Raddum et al., 1986).

The last brown trout was caught in Lake Store Hovvatn in the 1940's (Matzow et al., 1985). Lake Store Hovvatn has no suitable spawning streams, and the brown trout population was lake-spawning. Early stocking attempts yielded no recaptures after 1966 (Matzow et al., 1985).

Lake Store Hovvatn and Pollen were limed in 1981 as part of the Norwegian Liming Project (Wright 1982, 1984, 1985; Raddum et al., 1986). A total of 200 t of powdered limestone was spread on the ice in a 10 to 20 m wide and 3 km long strip on the shore in the eastern part of the lake and around the islands (**Figure 4**). Forty tons were spread on the ice of the entire Pollen Pond. The limestone powder used was the same as described for River Audna.

In the years 1981-1984, a total of 11,437 fry of brown trout was stocked in Lake Store Hovvatn. The fish were from a nearby hatchery at Byglandsfjorden and were of the Lake Byglandsfjorden strain. This trout population is known to be lake spawning. Test-fishing was performed in the years 1982-1987.

A goal for the research program has been to follow the water chemistry, bottom fauna and fish biology until effects of reacidification were observed. By 1986, reacidification had proceeded to the point where a marked decrease in the trout population occurred.

The lake was relimed in the summer of 1987 as a whole-lake liming of both Lake Store Hovvatn and Pollen, but with a substantially lesser dose compared with the liming in 1981. Sixty-nine t were spread at the surface of Lake Store Hovvatn. An additional 17 t were spread on the surface of a narrow bay (called Tarmen) in the western part of the lake. A total of 4.8 t was spread on the entire Pollen.

The 17 t spread on Tarmen was an attempt to see if the bottom fauna would respond similarly to that observed after liming Pollen Pond with a heavy dose in 1981.

We report here on the major results in water chemistry, bottom fauna and fish biology following liming in 1981 and 1987.

4.1 Chemical Response

Water quality data indicate a background pH of Lake Store Hovvatn in the 5.2 to 5.5 range as being high enough to sustain a brown trout population until the 1940's. Before liming in 1981 pH was about 4.4 and Ca 0.4 mg L^{-1} and repeated stockings with brown trout had failed. Typical RA1-concentrations in the 150 to 300 *u*g L^{-1} range were found in this clearwater lake (TOC of 2 to 3 mg L^{-1}), a large fraction of the available Al was in the labile Al form.

Liming in the winter of 1981 resulted in an immediate increase in pH and Ca at ice-out (**Figure 5 and 6**). During the first summer, pH was in the 6.3 - 6.7 range, and Ca increased to 2.5 mg L^{-1} at the surface (Wright, 1985).

Dilution during periods with heavy rain in the autumn 1981 resulted in dramatic decreases in pH. Following a typical dilution curve for a conservative element (Wright, 1985), Ca should have decreased almost to background levels within 2 yr. However, as can be seen in Figure 6, that was not the case. Periods of decreased Ca concentrations coinciding with increasing flow were observed, while during the summer months Ca concentrations increased.

Limestone powder, sedimented onto the littoral shore bottom or resuspended with subsequent sedimentation at greater depth, supplied the lake with neutralizing capacity. Within the first 2 weeks after ice-out in 1981, 14% of the lime dose dissolved (**Figure 7**). Compared to ordinary lake liming, this figure is low. A momentary dissolution of 65 to 70%, according to dissolution models (Sverdrup, 1985) would have been expected in whole-lake liming of Lake Shore Hovvatn.

Shoreliming of Lake Store Hovvatn did not result in Ca increase at the lake bottom at greater depth. Calcium concentrations at 20 m did not vary from concentrations at 1 m in 1981. Dramatic increases are often seen when limestone powder is allowed to sediment after whole-lake liming. We believe this shows that resuspension of limestone particles from the shoreline with a subsequent transport to greater depth is of minor importance.

In the autumn of 1981 another 12% of the dose at Lake Store Hovvatn had dissolved (**Figure 7**). The dissolution rate was 8% yr^{-1} corresponding to 15 t yr^{-1} during the next 3 yr. In the autumn of 1984, a total of 50% of the dose had dissolved. From autumn 1984 until

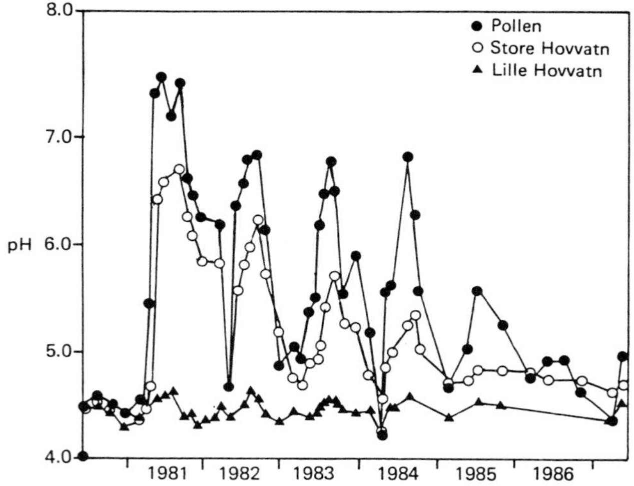

Figure 5. The pH of Lake Store Hovvatn, Pollen and the reference Lake Hovvatn in the period 1981-1987.

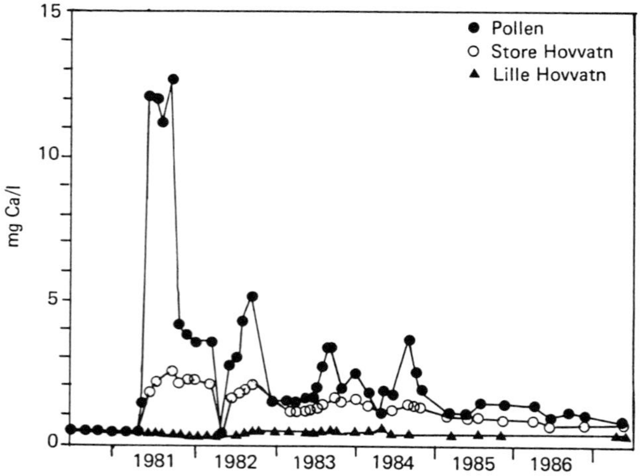

Figure 6. Calcium (mg L^{-1}) of Lake Store Hovvatn, Pollen and the reference Lake Lille Hovvatn in the period 1981-1987).

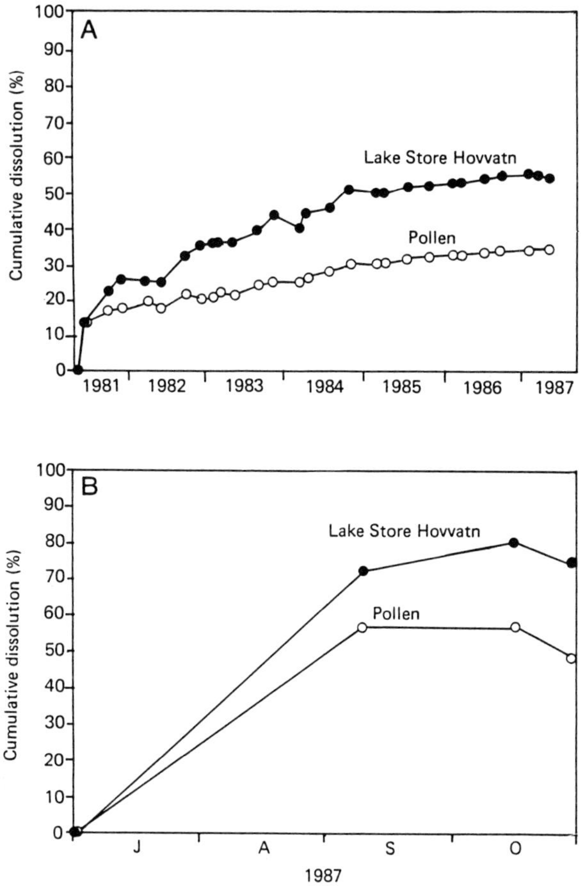

Figure 7. Cumulative amount of limestone in a) 1981 and b) 1987 (as percentage of lime dose) dissolved at Lake Store Hovvatn and Pollen.

reliming in 1987, the dissolution rate was 2% a year for Lake Store Hovvatn, resulting in a total dissolution of 55% for the period 1981-1987.

The total dissolution is 10 to 15% lower than would have been expected as a momentary effect following a whole-lake liming. Dissolution of sedimented limestone powder delayed the reacidification of the lake, however, and was adequate to sustain a brown trout population for the next 6 yr.

In periods with snowmelt, pH-depressions of more than one unit were seen in Lake Store Hovvatn due to acid meltwater flowing beneath the ice-cover. In lakes with an even shorter retention time than Lake Store Hovvatn, the upper 2 to 3 m of limed water may be pushed out during such episodes. No fish kills have been associated with these episodes, however, indicating that this cold water (0-2^{0}C) is of minor importance for fish survival. Marked disturbances attributed to the surges of cold water within shallow areas of these lakes and outlet streams may, however, occur for both fish populations and bottom fauna. No studies of such disturbances have been performed in Norway. Liming of Pollen resulted in dramatic increases of both pH and Ca (**Figure 5 and 6**). In the summer of 1981, pH had increased to 7.5 and Ca to 11 to 13 mg L^{-1} due to the very high dose of 280 g lime m^{-3-1} compared to typical doses of 20- 30 g lime m^{-3-1}. This concentration of dissolved lime was rapidly diluted with acid run-off from the catchment. A decrease in Ca from 12 to 4 to 5 mg L^{-1} and pH from 7.5 to 6.6 in one month (from September to October 1981) makes it clear that most of the momentary dissolved lime (75%) was flushed out of Pollen during this period.

A typical pattern is seen in Pollen in April 1982. Acid runoff reacidified the upper part of the pond (**Figure 4 and 5**). At the same time, dissolution of the high proportion of sedimented limestone powder increased the near-bottom (8 m) - and even the intermediate depth (5 m) - concentrations of Ca from below 4 to 5 to 8 mg L^{-1} (not shown on the figures).

The high sediment dose of the Pollen (7.5 t ha^{-1}) allows the fluctuation in both pH and Ca to be much greater than in Lake Store Hovvatn. The shorter retention time and a more important role played by the sediments in delaying the reacidification should also be emphasized.

As for the Lake Store Hovvatn, 14% of the limestone powder was dissolved in Pollen as a momentary effect. With a typical dose of 20 to 30 g lime m^{3-1}, the dissolution would have been 50 to 55% following the dissolution model of Sverdrup (1985). A pattern similar to that of Lake Store Hovvatn is seen, however, with a total dissolution of 34% (**Figure 7**).

Before reliming in 1987, Ca was still higher (0.7 mg L^{-1}) than preliming values in 1981 (0.4 mg L^{-1}). Liming with 69 t of limestone powder increased pH to 6.8 in Lake Store Hovvatn. The pH of Tarmen was about 7.0 due to a much higher dose. In Pollen, the dose of 4.8 t increased pH to about 7.0.

The whole-lake liming in 1987 resulted in an immediate dissolution of 65 to 68% (45 to 47 t) of the lime dose in Lake Store Hovvatn and 49 to 53% (2.3 to 2.5 t) in Pollen. Liming of Tarmen is not included in these figures. Compared with the shore liming of Lake Store Hovvatn in 1981, a much higher fraction of the lime dose dissolved during the first summer of 1987, but the absolute amount was of the same magnitude.

Johannessen and Hindar (1987) found acceptable agreement between dissolution models for limestone powder in whole-lake liming (Sverdrup, 1985) and results from Norwegian liming projects. Liming of Lake Store Hovvatn and Pollen in 1987 also fits well into these models as far as the momentary dissolution is concerned. As is shown for Lake Store Hovvatn, Pollen and other whole-lake liming projects, the long-term dissolution is important and deserves attention.

We have estimated the long-term dissolution of lime in Pollen to be 0.35 t ha^{-1} yr^{-1} as a mean figure for the first 3 yr. Hindar (1984) reported long-term dissolution of 0.42 t ha^{-1} yr^{-1} the first 15 mo after liming of the pond Lille Finnetjern. This pond received a sediment dose of 1 t ha^{-1} compared to 7.5 t ha^{-1} in Pollen. We have estimated the dissolution from the limed shore area of Lake Store Hovvatn to be as high as 2.5 t ha^{-1} yr^{-1} as a mean for the first 3 yr after liming. This figure corresponds to 0.15 t ha^{-1} yr^{-1} for the whole lake area. The limed shore area of Store Hovvatn (20 meters · 3 km) received 20 to 30 t ha^{-1}.

The pH and Ca in the reference Lake Lille Hovvatn were both unchanged throughout the period (**Figure 5 and 6**). The pH was almost constant at 4.5 as was the Ca concentration at 0.4-0.5 mg L^{-1} both at the surface and at the greatest depth.

The Al concentration of Lake Store Hovvatn was reduced from 200 to 250 *u*g L^{-1} to almost half that level within the first summer following liming. In the summer of 1983, the total concentration was below 50 *u*g Al·L^{-1}. As reacidification continued, Al concentration increased to 150 *u*g Al L^{-1} in 1985 and 1986, and to 175 *u*g L^{-1} immediately before reliming in 1987.

The operationally-defined labile fraction of Al is strongly dependent of pH (Driscoll, 1984). This is also documented for Lake Store Hovvatn (Wright, 1985). Labile Al was the dominant fraction from late 1983, when this parameter was first included in the program. The same trend was also seen in Pollen. In the reference Lake Lille Hovvatn, the total Al has remained relatively high throughout the period.

Rapid alterations in both total levels (total or reactive fraction) and species of Al can be ascribed to factors in the catchment as well as in the lake itself. The catchment of Lake Store Hovvatn is not limed which allows for Al to be transported to the lake according to interactions between soil/bedrock and runoff. Limed lakes serve as sedimentation basins for Al when the pH is over 5.5 to 6.5 (Wright and Skogheim, 1983). The success of removing Al from the water body depends, thus, on the relationship between input rate and sedimentation rate.

4.2 Biological Response

4.2.1 Invertebrates.

Before the lime treatment, the composition of invertebrates was typical for acidic lakes, characterized by a sparse density pattern and few species (Raddum, 1980; Raddum et al., 1986). Highest abundance was found in the littoral zone, while the numbers at 5 and 10 m depths were low (Raddum et al., 1986). Bivalves and snails were lacking.

Liming resulted in changes in the composition of the benthic fauna both in Lake Store Hovvatn and Pollen. The response in Lake Store Hovvatn was delayed both in the littoral zone and at greater depth. In Pollen, the response was more immediate.

Oligochaets were relatively abundant at 0.5 m depth, but were not found at 5- and 10-m depths in Lake Store Hovvatn before liming. The first year after treatment, these animals were not found at any of these depths, but in 1982 they were found at all depths. During 1983, the abundance of oligochaets increased further to relatively high numbers (Raddum et al., 1986).

In Pollen, the oligochaets become present or increased in number at all depths the first summer after liming. The number increased further during 1982 and 1983.

A delay in response was also seen among the chironmidae larvae in both Lake Store Hovvatn and Pollen Pond (**Figure 8**). The greatest relative increase was found in Pollen,

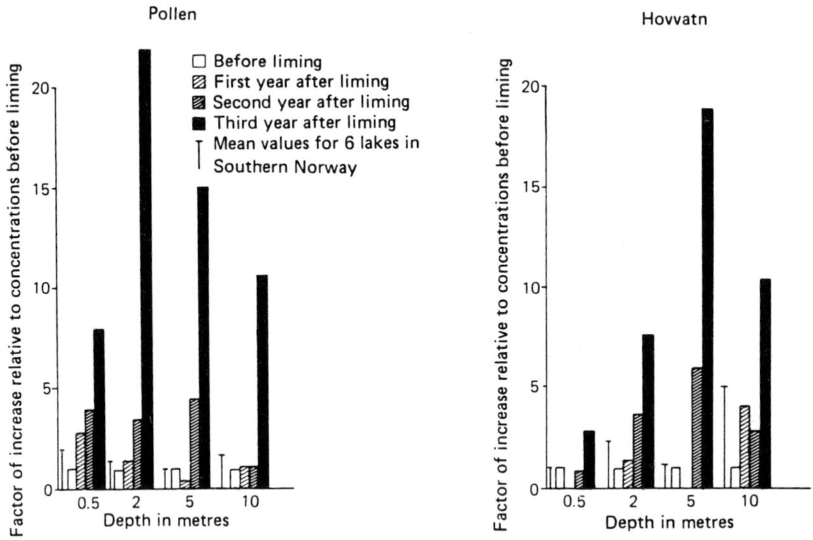

Figure 8. Density of chironomids in the benthos of Pollen and Lake Store Hovvatn before liming and 3 yr after liming, expressed as relative to preliming values. Also shown are the densities found in 6 acid lakes in southern Norway (adapted from Raddum et al., 1986).

where the abundance increased by 2.8 times at 0.5 m depth the first autumn after liming (Raddum et al., 1986). The third autumn, the increase was dramatic at all depths both in Lake Store Hovvatn and in Pollen (**Figure 8**). In Pollen, 80,000 larvae m^{-2} were found at 2-m depth, a 22-fold increase.

Raddum et al. (1986) ascribe the dramatic effects of the benthic fauna after liming to the amount of unexploited organic matter and the lime dose. The increase indicates that accumulated food had become available. Elevated Ca concentrations might also have stimulated the oligochaets. The difference in response between Lake Store Hovvatn and Pollen is ascribed to differences in bottom substrate. The sediments of Pollen consist of soft organic material even in the uppermost part of the littoral zone. Limestone particles at the inorganic shore of Store Havvatn might have delayed the response in the littoral zone relative to Pollen.

In 1985, a dramatic decrease among the chironomids was found. The density at 2, 5 and 10 m depth decreased to 1/14 compared to the figures from 1984. This density is similar to the situation before liming. Depletion of food resources may have been the cause of this decrease. The availability of the organic manner might also have been reduced as a result of lower pH during reacidification (Hindar et al., 1987). The data from 1987 post-liming are not yet available.

4.2.2 Fish Populations.

A summary of the stocking program for Lake Store Hovvatn is shown in Table I. Following liming in winter 1981, a total of 4,797 brown trout was stocked (2,885 of 0+, 980 of 1+ and 932 of 2+). The stocking material was first year progeny. The stocking density was 32 fish ha^{-1} in Lake Store Hovvatn and 244 fish ha^{-1} in Pollen. To compensate for an assumed lack of reproduction and reduction of the population through test fishing, fish were stocked every year until 1985 (Matzow et al., 1985).

The brown trout in Lake Byglandsfjord grow asymptotically toward 33.8 cm (Rosseland and Skogheim, 1984a; Matzow et al., 1985). The hatchery reared fish were considerably smaller than the corresponding age classes living in Lake Byglandsford at the time of stocking (**Figure 9**). After the 1981 stocking in Lake Store Hovvatn, however, the brown trout showed a very rapid growth in the first year (**Figure 9**). The 1-yr-old fish doubled the annual growth rate compared to the parent population. All three age groups were significantly larger in August 1984 than fish of the same age in Lake Byglandsfjord (Matzow et al., 1985). Later stockings have not shown the same growth rate, but appear to approach the same growth pattern as in the parent population (**Figure 9**).

Table I

Number of stocked and recaptured fish at Lake Store Havvatn following liming in 1981.

Year	Stocked	Recapt.	Testfish	Recapt./test-fishing
1981	4797			
1982	2800	1094	3	364
1983	1975	789	3	263
1984	1865	732	3	244
1985	0	415	2	207
1986	0	75	1	75
1987				

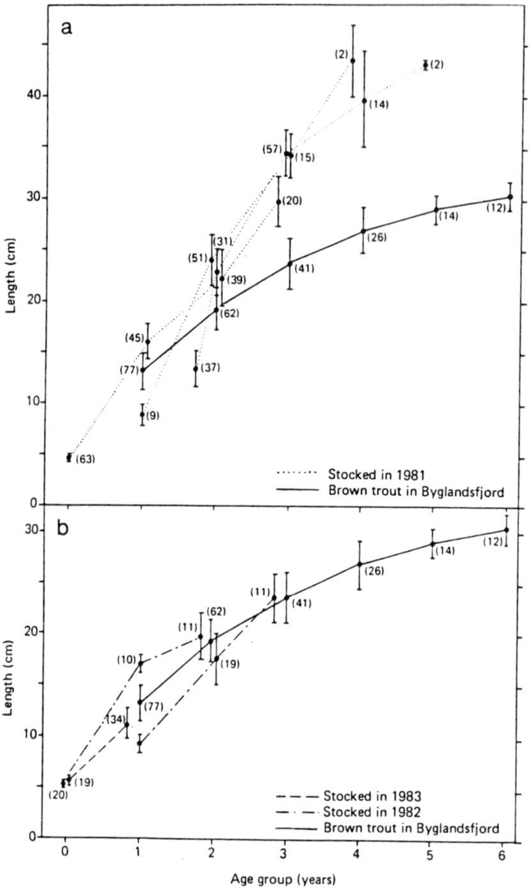

Figure 9. Empirical growth of the brown trout population in Lake Store Hovvatn and in Lake Byglandsfjord (after: Matzow et al., 1985).

The early high growth rate was probably due to the availability of accumulated food in the lake. This effect is well known for stocked brown trout (Nilsson, 1972) and brook trout (Grande et al., 1980), and is a phenomenon in some acidic and sparsely populated lakes (Rosseland et al., 1980; 1986).

Benthic fauna dominated the stomach contents. Feeding was characterized by large, mobile and littoral species (Matzow et al., 1985), typical of acidic lakes without fish, or with sparse fish populations (Raddum, 1979; Rosseland et al., 1980; Hultberg and Andersson, 1982;

Eriksson et. al., 1983). The most important prey were mayfly nymphs (Ephemeroptera, with species *Leptophlebia vespertina* and *L. marginata*) and beetles (Coleoptera, mainly diving beetles). Also, alderflies (Megaloptera) and lesser water boatmen (Corixidae) were important. There were no signs of changing food patterns among the brown trout during the first years after liming.

During the period 1986-1987, the growth of the brown trout population had ceased. The catches in 1986, using standard gillnet series (Rosseland et al., 1980), were low compared to earlier years (**Table I**). During the test-fishing in May 1987, before the reliming, one dead female trout with unshed eggs was found in Pollen, however, no other observations of dead fish were recorded. The catch was low, with 1.8 brown trout per gillnet series and night in Lake Store Hovvatn. After the reliming in 1987, the growth increased for certain age-classes, and the catches exceeded those of 1986, Table I. The mean length growth for the 3+ age-group of 1986 had been 4.3 to 4.5 cm yr^{-1} in 1985-86. During the summer of 1987, after reliming, the length increment for these 4+ fish was 7 cm in mean, and the catches in October 1987 had increased to 10.5 fish pr. gillnet series and night in Lake Store Hovvaatn (G.G. Raddum,-pers. com., **Table I**).

The low growth prior to the reliming in 1987 indicates that the water chemistry of 1986 and spring 1987 had been suboptimal. The reduced catches in 1986 and spring 1987 did obviously not reflect a high mortality, as the catches increased after reliming. In 1986 and spring 1987, the surviving fish seemed, thus, to have sought refuge in other areas which eventually reduced the catches in areas which had earlier provided suitable habitats.

4.2.3 Conclusion.

Liming Lake Store Hovvatn in 1981 and 1987 was successful. The invertebrates and the stocked brown trout population have responded according to food availability and water chemistry, showing a biological response both to the neutralization, the reacidification and the repeated neutralization of the lake. The different liming strategies used and their chemical consequences have given valuable information for future liming operations in Norway.

5. LIMING THE CATCHMENT TO THE POND TJONNSTROND

The Tjønnstrond catchment area is situated in the county of Telemark (**Figure 1 and 10**). The catchment area is 0.25 km^2 and contains two lakes connected by a stream. Both lakes are shallow with a maximum depth of 3.5 m. Combined retention time for the two lakes is 2 mo, typical for many lakes in the area. The terrestrial catchment at Tjønnstrond was limed in summer of 1983. A nearby reference catchment, Storgama, was not limed. Both catchments are well documented from the SNSF-project ("Acid Precipitation - Effects on forest and fish," 1972-1980). They are characterized by barren rock and thin soils covered with peat and poorly-developed podzols. Catchment liming in this area was believed to give rapid response in the runoff.

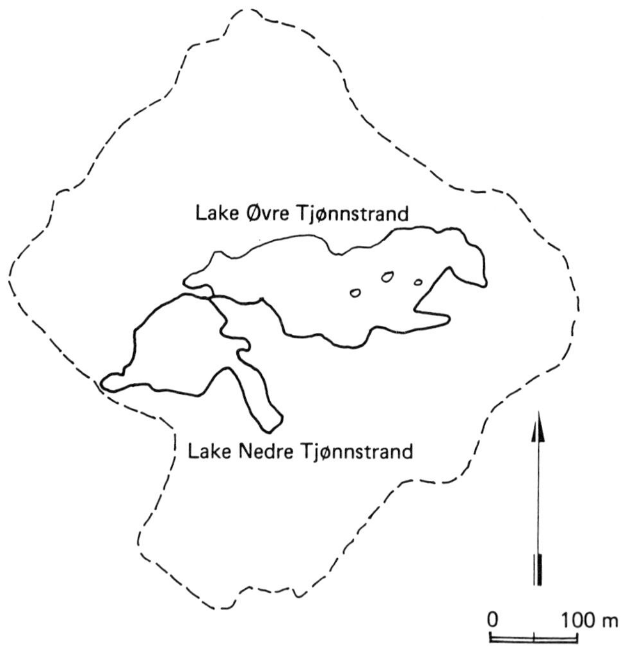

Figure 10. The Tjønnstrond catchment with the two lakes.

Before liming, both small ponds were barren of fish. At Tjønnstrond, the small brook connecting Lake Ovre Tjønnstrond and Lake Nedre Tjønnstrond allows downstream fish migration only. Due to lack of spawning areas, the former good brown trout populations were dependent of annual stocking. Due to an increasing water acidity, the stocking has failed the last 20 yr. Even a 1982 stocking of brook trout (*Salvelinus fontinalis*), a species more acid tolerant than brown trout (Grande et al., 1978; Rosseland and Skogheim, 1984b; Rosseland et. al., 1986b), failed (L'Abee-Lund and Kleiven, 1987).

Prior to liming, the water chemistry was typical for high elevation lakes (600 m above sea level) in the most acidified area of Southern Norway. The pH was 4.4 to 4.8, Ca concentrations about 1 mg L^{-1} and RAl 100 to 400 *ug* L^{-1}.

The liming strategy at Tjønnstrond was whole catchment liming using the same limestone powder as in the River Audna and in the Lake Store Hovvatn. This strategy was chosen because the retention time is too short to allow whole-lake liming once a year or once every second year. Also, Lake Ovre Tjønnstrond has no inlets. Liming the lakes with dosers is therefore impractical.

The lime dose of 75 t was chosen based on earlier experiments in Sweden. The dose corresponds to 3 t ha^{-1} catchment area. This is equivalent to 1600 g m^{3-1}, about 50 times the dose at Lake Store Hovvatn. Lime was brought in and spread over the entire catchment by helicopter. Liming on the lakes surface was successfully avoided.

In October 1983, 4 mo after liming, 125 fry of brook trout were stocked at Lake Ovre Tjønnstrond and 125 fry of brown trout were stocked at Lake Nedre Tjønnstrond. The trout were of the Lake Byglandsfjord strain, the same as used in Lake Store Hovvatn.

5.1 Water Chemistry

The first rain following liming resulted in dramatic changes in water chemistry at the outlet of the lakes. The pH increased from 4.5 to 7.1 (**Figure 11**). No changes were seen in the reference catchment. During the spring flood of 1984 when pH in the runoff from the reference area decreased to 4.36, the pH of Tjønnstrond was 5.55 (**Figure 11**).

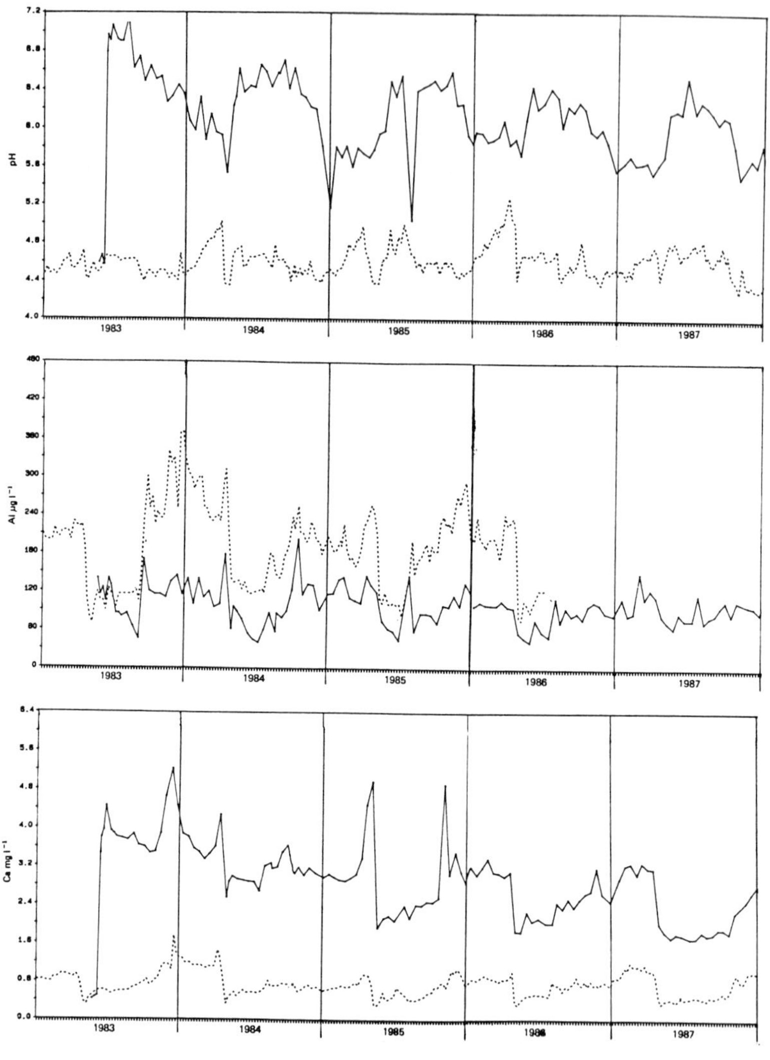

Figure 11. The pH, Ca and Al at the outlet of the catchment Tjønnstrond (solid line), and the reference catchment Storgama (hatched line).

Although drops in pH occurred during spring floods, water quality has so far not reached levels critical for the fish population. As can be seen from Figure 11, Ca increased from a pre-liming value of about 1 mg L^{-1} to above 4.5 and later 5.2 mg L^{-1}. During spring floods, Ca of the limed catchment is markedly higher than in the reference catchment. Al is 50 to 150 μg L^{-1} lower in the limed catchment.

In spite of the short retention time of the lakes and the spring drops in pH, Figure 11 shows a remarkably stable and acceptable water quality. Critical water quality is not yet documented 4 yr after the lime treatment. From our experience, such effects are not possible for lakes of corresponding retention times following a traditional whole-lake liming even at extreme doses.

The marked decrease in Al is also very uncommon for these kinds of lakes. The decrease in Al is a result of both transport into the lake and sedimentation within the lake. Sedimentation is a slow process relative to input rates. Many lake liming projects in Norway show that reactive Al is not markedly changed after liming (Hindar et al., 1987). This is ascribed to short retention times and unchanged input rates. The result seen here for Tjønnstrond is therefore probably an effect of treating the whole catchment with lime.

The lime, a 3.5% dose, was transported out through the lakes the first year. At the end of 1986, a total of 10% had benefited the water mass, and at the end of 1987 this was probably 12 to 13%. This is much higher than typical values of 1 to 3% from the early Swedish liming program (SNV, 1981).

The duration of the acceptable effects depends on target levels of water quality parameters. The lime dose for Tjønnstrond corresponded to a 30-yr-dose of a typical whole-lake liming. Although 30 yr seems to be very optimistic, the results presented here for the first 4 yr indicate a long-lasting effect. Labile Al is now part of the chemical program in order to evaluate the status of the future water quality.

5.2 Fish Response

The stocking of 125 summerold brook trout and 125 summerold brown trout in Lake Ovre Tjønnstrond and Lake Nedre Tjønnstrond, respectively, took place in October 1983. Test-fishing during the autumn of 1984 and 1985 resulted in 17 (14%) and 23 (18%) recaptures of brook and brown trout, respectively.

Although stocking experiments with brook trout in Norway have demonstrated a great migratory activity (Grande et al., 1978; Qvenild, 1986), there were few brook trout captured in Lake Nedre Tjønnstrond. The good and relatively stable water quality in the lakes is suggested to be the explanation for this phenomenon (L'Abee-Lund and Kleiven, 1987). The growth and condition factor was high for both species, although due to species differences in growth potential, the brook trout exceeded the brown trout in specific growth. The mean length, weight and condition factor in 1984, 1 yr after stocking, was 20 cm, 100 g and k = 1.05 for brown trout, and 25 cm, 200 g and k = 1.14 for brook trout, respectively. In 1985, the mean weight of brown trout was 168 g compared to 653 g for brook trout (L'Abee-Lund and Kleiven, 1987).

5.3 Conclusion

Liming of the catchment Tjønnstrond has been a success. We believe the catchment characteristics of Tjønnstrond and probably the high discharge rate accounts for a relatively high liming efficiency on a yearly basis. The thin soils are more saturated with base from the lime treatment than deep soils would have been. The retention time of the ground water is probably very short in this area compared to acidified areas in Sweden. Old groundwater probably does not exist in the catchment. The rapid response seen here also reflects the fact that Tjønnstrond consists of areas of barren rocks.

The stocking of brook trout and brown trout has demonstrated successful survival and good conditions for growth with very low migratory activity of the brook trout. The chemical response has thus provided suitable fishery conditions.

ACKNOWLEDGMENT

We greatly acknowledge the comments given on the manuscript by Dr. Richard F. Wright.

LAKE LIMING IN DIFFERENT TYPES OF ACID LAKES USING VARIOUS TYPES OF CALCITE POWDERS AND METHODS

H. Sverdrup
P. Warfvinge

ABSTRACT

Results from experiments show that instant dissolution efficiency of calcite powder increases with finer particles, and that initial dissolution efficiency decreases with and increased dose. The Sverdrup model predicts and the results show that the dissolution efficiency is dependent on pH, dose, depth and grinding. Dry application can reduce the dissolution efficiency by 40% in relation to a well slurried application. Experiments in two lakes indicate that residual calcite on the lake bottom cannot accomplish bulk neutralization of the lake even if substantial effects on the microenvironment around the calcite and the reacidification time can be seen. Bioassays with Atlantic salmon indicate that fish mortality due to Al toxicity is greatly reduced when the Ca concentration exceeds 3.5 mg L^{-1}. The results show that the pH and alkalinity in almost any acidified lake can be established and maintained by liming at levels where fish populations can be sustained by natural reproduction.

1. INTRODUCTION

In Sweden, the first serious effects of acidification were detected on a large scale in the early 1960's. By then several decades of acidic deposition had changed the base saturation in the soils located on granitic or gneissic bedrock, and the water could no longer be fully neutralized during its passage through the soils of the catchments.

Today approximately 20,000 lakes of a total of more than 200,000 lakes in Norway and Sweden have a pH value so low that naturally reproducing fish populations cannot be sustained in these waters. In several regions of southern and western Sweden and southern Norway, more than 50% of all streams and lakes have been affected in this way. Every year more lakes and streams become acidified due to the continuing depletion of the base cation storage in the soils (Johnston et al., 1986; Renberg et al., 1985).

Water, Air, and Soil Pollution **41** (1988) 189–222.

Liming of lakes, streams and soils are in no way an acceptable ecologically sound solution to the acidification problem. Still, it is the only large-scale remedy for postponing some of the worst effects of acidification on the aquatic environment, until emission of acidifying S and N compounds have been reduced substantially.

Today, lake and stream liming is a well documented and highly developed method for neutralizing acid surface water, and its success in restoring fish populations is well proven. In Sweden, the United States and Norway, liming is carried out in operational programs, as a way to manage the aquatic environment in areas affected by acidification. Britain, Finland and Canada have experimental liming programs, and may possibly expand these to become operational liming programs (Bengtsson et al., 1980; Fiskeristyrelsen, 1982; Sverdrup, 1983, 1985; Hindar et al, 1985; Howells and Brown, 1985; Fraser et al., 1985; Booth et al., 1985; Lessmark and Thornelof, 1986)

Acidification mitigation is carried out by a combination of different mitigative methods to obtain the targets defined:

Lake liming is carried out by boat, barge, helicopter or on the winter ice, and serves to neutralize the bulk water volume of the lake, as well as to deposit a required amount on the lake bottom.

Stream liming is carried out with different types of continuously operating calcite dosers, e.g., water-powered dosers, tipping bucket dosers, diversion wells, rotary drums and fully automated dosers, in order to neutralize the flow at some target zone downstream. These technologies may be used on lakes as well.

Streamside soil liming is a specialized treatment method designed to neutralize the runoff from a discharge area adjacent to the stream. Only modest volumes of runoff can be treated. This method may also be used to prevent leaching of dissolved Al into the stream.

Shoal liming is often used as a complement to lake liming in order to achieve a local neutralization and create special habitats in a lake or stream.

Soil liming is the bulk neutralization of soil in a catchment. Soil liming is used to prevent soil acidification in forest, for long-term mitigation of groundwater acidification, as well as a way to mitigate acidic runoff.

In most surface water liming programs the objectives and target values are originally defined as ecological values, the goal being that a certain ecological balance be restored or that certain types of aquatic populations be protected and sustained. These ecological target values must be reinterpreted in terms of chemical and physical variables before any mitigative measure can be planned. For some systems, like fish populations, this can be fairly straightforward (Brown, 1982).

During mitigation planning and design, the result of a certain treatment is predicted using physicochemical dose-response relationships and mathematical models. The back-estimation of the ecological system response, however, is seldom straightforward. In Figure 1 a flowsheet for this evaluation is sketched. The actual engineering design of the liming will follow a flowsheet as shown in Figure 2 (Sverdrup, 1985; Saunders et al., 1985; Hindar et al., 1985).

The large acidification mitigation programs result in an increased demand for competent selection of objectives to be considered, as well as efficient administration and planning. Competent technical planning procedures and calculation routines are needed in order to produce cost efficient liming operations. With respect to the planning of lake liming operations, crucial information concerning liming is now available (Sverdrup et al., 1983, 1985a,b,c; Sverdrup, 1983, 1985; Sverdrup and Warfvinge, 1984, 1985; Warfvinge et al., 1987; Saunders et al., 1985; Hindar et al., 1985):

- Simple computer models predict the result of a certain lake treatment or the dissolution efficiency of neutralization agents under lake liming conditions, in terms of pH, particle size distribution and other critical parameters.
- Reacidification models predict the duration of a certain lake treatment.
- Overall catchment acidification mitigation strategies optimize mitigation results.
- Cost optimization routines critically evaluate ecologically equivalent acidification mitigation alternatives for the most cost-effective option.

This report will relate how some recent Swedish lake liming results verify the models used for planning and calculation, in particular, the models used to estimate the dissolution efficiency and the chemical response in the lake. An example is given as to how this is implemented in the liming of a complex lake and stream system.

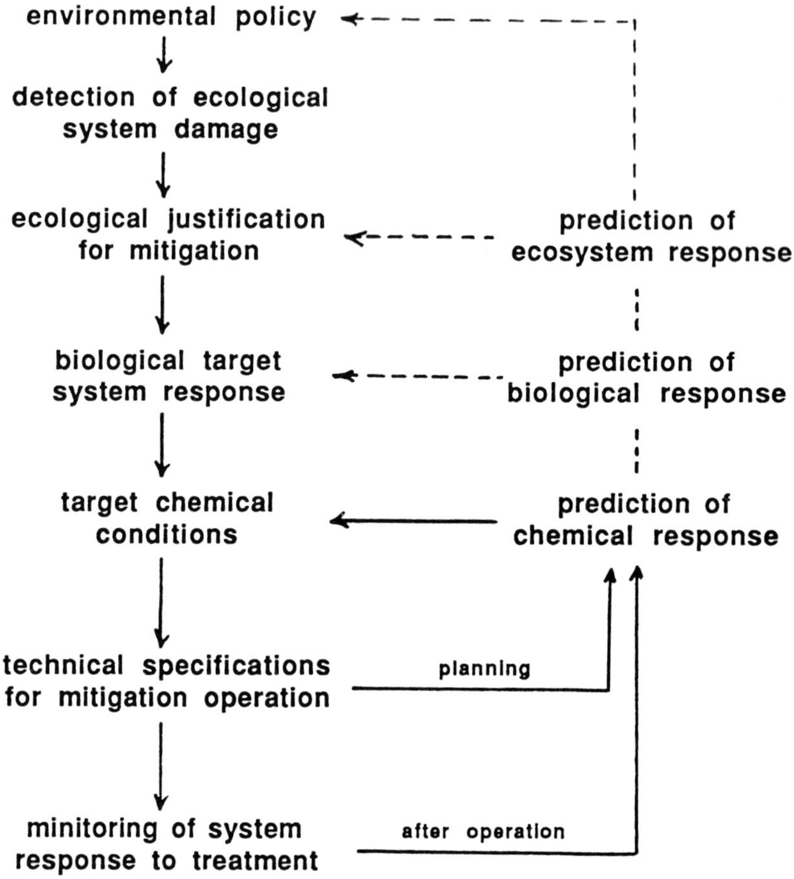

Figure 1: The structure of a liming project planning, and the chain of response prediction. Models for prediction changes in biological systems in response to an aquatic chemical change are not available at present, and exist only as subjective and qualitative judgements. Such dose-response relationships are very much needed since most liming programs are related to the ecological status of an aquatic environment.

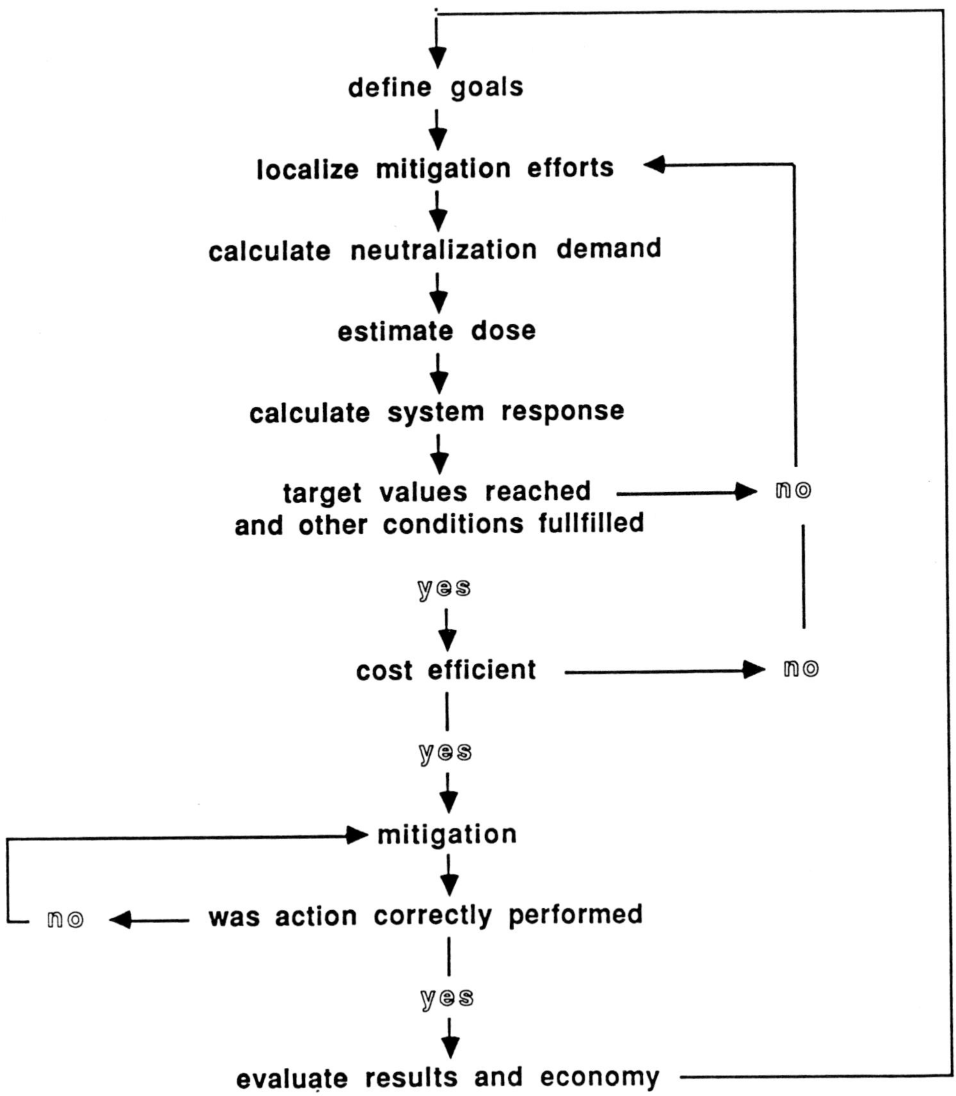

Figure 2: The calculation procedure used in all liming projects planned by the authors.

2. CALCITE DISSOLUTION KINETICS AND LAKE LIMING

Calcite is the most frequently chosen neutralizing agent for lake liming since the dissolution efficiency is sufficiently high to obtain the desired level of neutralization in the lake. The price is reasonable in relation to its dissolution properties and it is available in sufficient quantities on the market. The material has a low content of environmentally harmful impurities such as heavy metals and is safe for manual handling, and thus, does not contain aggressive alkaline components.

Most limestone powders used in lake liming do not dissolve completely during the application. In order to be able to predict the result of the addition of a certain dose to a lake, the dissolution efficiency must be known. The dissolution of calcite particles in water of low ionic strength occurs simultaneously along two reactions:

$$H^+ + CaCO_3(s) = Ca^{2+} + HCO_3^- \tag{I}$$

$$H_2O + CaCO_3(s) = Ca^2 + HCO_3^- + OH^- \tag{II}$$

For solution pH values below pH 4.4, the first reaction involving the reversible reaction with the H^+ will dominate. In this reaction the diffusion of H^+ ions into each calcite particle is the rate determining factor. The partial pressure of CO_2 must be in excess of 0.1 atmosphere for the rate of the reaction with CO_2 to be equal to the rate of the reaction with water at any time. At the partial pressure found in surface waters in contact with the atmosphere, the partial pressures of CO_2 are far too low for the rate of this reaction to be significant (Amrhein et al., 1985; Plummer et al., 1978; Sverdrup, 1985).

The dissolution of a calcite powder is proportional to the total surface area of all particles, and the total dissolution rate is the sum of the rates of all reactions taking place. The total dissolution rate of calcite is thus given by the expression:

Dissolution rate = Exposed surface · Chemical driving force. (1)

Since the exposed surface area per unit mass of liming material increases as particle size is reduced, the dissolution rate is significantly higher for finely ground powders than for coarse limestone products. The chemical driving force given by the kinetic expression which describes the dependence of the reaction rate per unit surface and time as a function of solution concentration of reactants for calcite dissolution in natural lake water is:

$$R = k_{H+} \cdot [H^+] + k_{H2O} - k_B \cdot Ca^{2+}][HCO_3^-] \quad (2)$$

Experiments indicate that the coefficients have values in the range pk_{H+}=3-3.5, pk_{H20}=7.8-8.5 and pk_B=2.4 (Plummer et al., 1978; Sverdrup 1985). Even small amounts of dolomite in the limestone will change these coefficients significantly.

A number of factors do not influence the dissolution efficiency significantly. Dissolved CO_2 and components changing the titration curve of the lake water may change the response to the dissolved substance in the water, but not influence the dissolution efficiency to any large extent. The temperature is not important as the increase in dissolution rate at higher temperature is almost exactly counteracted by the change in viscosity that will cause the particle to sink faster, reach the lake bottom sooner, and therefore be exposed for a shorter period of time. A thorough discussion of limestone dissolution kinetics in lake environments can be found in Sverdrup (1985).

When calcite powder is distributed over a lake surface, the particles are rapidly distributed according to size due to their size related settling velocity, and consequently sink and dissolve simultaneously. The large particles sink more rapidly than the smaller, and partially neutralize the water in front of the smaller particles sinking in their wake.

A differential equation may be formed for the dissolution of a calcite powder sinking in a lake by combining the kinetic expression for the chemical driving force, the rate dependence on the particle surface expressed as the geometric surface area of a powder and the physical law for particles sinking in a water column (Sverdrup, 1983). The equation can be solved numerically, but to interpret the expression in terms of governing parameters, we may rephrase the equation into words:

$$\frac{\partial X}{\partial z} = \frac{(\text{exposed surface}) \cdot (\text{driving force}) \cdot (\text{time of exposure})}{1 + (\text{dose}) \cdot (\text{exposed surface}) \cdot (\text{calcite saturation}) \cdot (\text{time of exposure})} \quad (3)$$

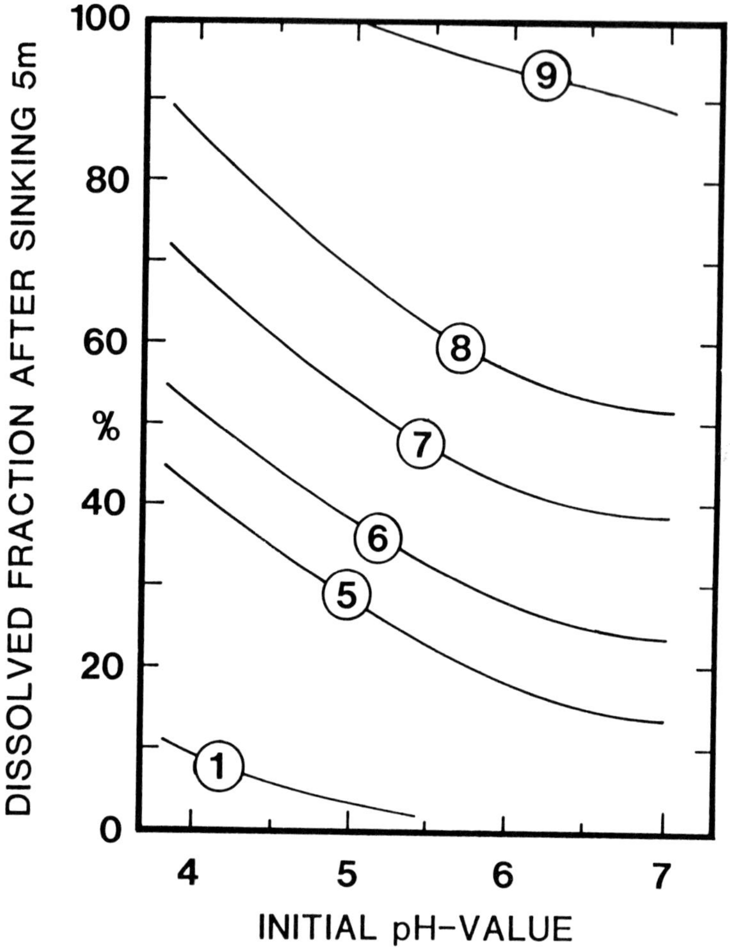

Figure 3: **The dissolution of different limestone powders in 5 m of water, the effect of the dose on the dissoluton of calcitic limestone and the effect of different lake depths on the dissolution efficiency for a powder like no. 7.**

Where $(\partial X/\partial z)$ is the change in dissolved fraction of the dose added with lake sinking depth z. At low dosage, the precipitation reactions may be neglected, as well as the effect of the dose, and the dissolution efficiency is determined mainly by the exposed surface, the chemical driving force for dissolution and the time of exposure to the acid water, the time of exposure being a measure of the particle size distribution and the lake depth at the site of the application. The exposed surface is a function of the particle size distribution and hence the grinding grade of the calcite product used. For high dosages, the dissolution efficiency depends mainly on the chemical driving force, the dose and the degree of calcite saturation. With the model, dissolution efficiency diagrams can be made for a certain lake depth as an aid in hand calculation of the dissolution efficiency. These diagrams apply to the case when the dose does not exceed 20 g m^{-3}. The validity of the model has been thoroughly verified in a large number of laboratory experiments and field studies of lake liming (Sverdrup, 1985). The results from dissolution model calculations are shown in Figure 3, showing the calculated dissolution efficiency of different calcite powders sinking through 5 m of water. The particle size distribution of the different products is shown in Figure 4.

However, with depths different from that of the diagram, the dissolution efficiency can be assessed by changing the pH used to enter the diagram (Sverdrup, 1985);

$$pH(diagram) = pH(lake) - \log_{10}\left(\frac{Lake\ zone\ depth}{5\ meter}\right) \tag{4}$$

Since the material is distributed at the surface of the lake, there may be saturation effects close to the surface which will decrease the dissolution efficiency. The larger the application rate, the more significant such effects will be on the total dissolution efficiency. Greater depths will increase the dissolution efficiency.

With the increasing number of parameters that must be considered, more diagrams are required, and the use of a comprehensive computerized version of the model is the most convenient way to calculate the dissolution efficiency. For the planning of the liming of the lakes presented in this study, the full version of the model developed at the Lund Institute of Technology by Sverdrup (1985) was used. The predicted dissolution efficiencies for lake segments of the large lakes in the study were used to determine the actual dose to be applied.

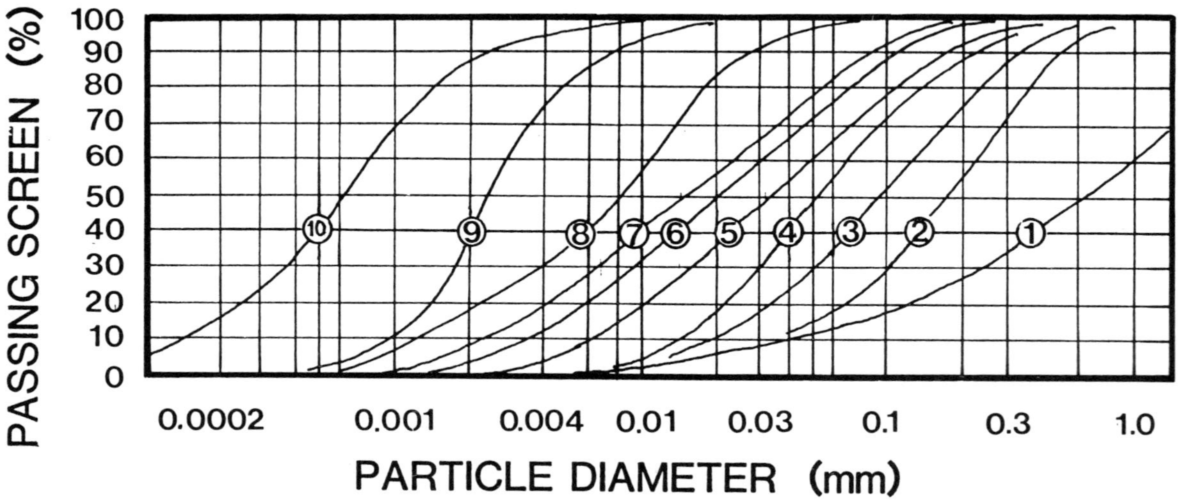

Figure 4: The particle size distributions of calcite powders used in scandinavian liming.

The purity of the product is important to the extent that even low amounts of dolomite may reduce the dissolution efficiency significantly. The content of dolomite in the limestone should preferably be less than 10%. Contents of silicates in the calcite material do not generally affect the dissolution efficiency, as they do not form any solid solution with calcite, and thus do not incorporate in the calcite crystal structure (Sverdrup, 1985).

In general, dolomite dissolves more slowly than calcite. This implies that a dolomite powder must be more finely ground to obtain a comparable dissolution efficiency with respect to calcite, which also implies a higher cost. This always gives dolomite, which is a fully acceptable neutralization agent once it is dissolved, an economic disadvantage in terms of cost effectiveness of neutralization as compared to calcite products.

3. LIMING OF DIFFERENT TYPES OF LAKES: INITIAL NEUTRALIZATION

For the study of liming in different types of lakes, we selected 14 lakes located in the Province of Bohuslan and in the Providence of Hallan (**Table I**). The lakes fell into the following four hydrological categories:

- Seepage lakes;
- Seepage lakes with outlet;
- Drainage lakes with simple form; and
- Multiple basin drainage lakes.

The complexity of the form of the lakes could be distinguished into three groups as shown in Figure 5. In terms of depth and lake size, the lakes fell into four well distinguished categories:

- Large and deep lakes;
- Large and shallow lakes;
- Small and deep lakes; and
- Small and shallow lakes.

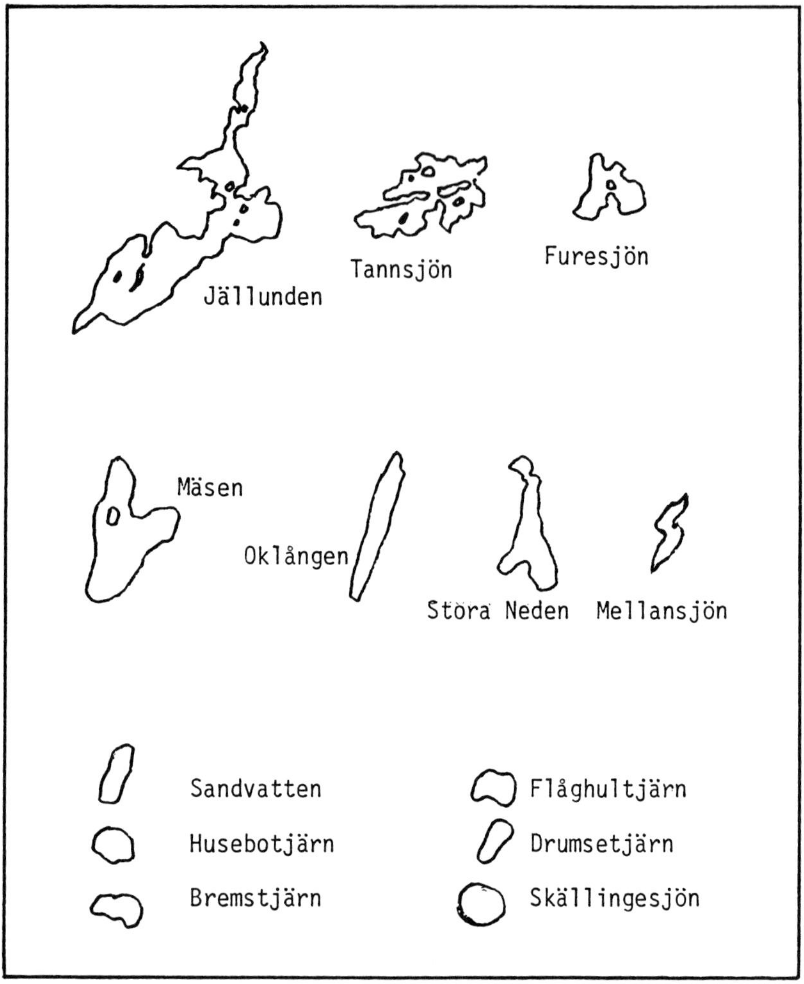

Figure 5: The morphology of some of the lakes in this study, arranged according to complexity into three groups. Each group comprises deep and shallow lakes, clear and colored.

Lake	Lake surface area km^2	Average depth m	Maximum depth m	Retention time yr	Volume 10^6 m^3	Hydrol. type	Water color mg PtL -1
Jällunden	8.60	4.0	13	1.4	37.5	d	30-50
Stora Neden	2.90	21	57	7.1	60.0	d	0-10
Mäsen	4.20	23	60	6.0	96.0	d	10-30
Oklången	2.58	10	26	2.8	27	d	30-50
Skällingesjön	0.30	10	26	11.2	3.1	s	0-10
Furesjön	0.40	8	30	2.8	2.5	d	0-10
Sandvatten	0.26	17	34	14	4.4	s/d	0-10
Gjällarpesjön	0.08	6.4	14	1.4	0.5	s/d	50-100
Husebotjärn	0.24	6	10	1.3	1.4	d	30-50
Brämstjärn	0.21	5	9	0.7	1.1	d	30-50
Flåghult	0.49	6	16	1.4	2.8	d	50-100
Drumsetjärn	0.20	5	10	1.8	1.0	d	50-100
Tannsjön, inner basin	0.60	4.5		2.5	1.8	d	30-50
Tannsjön, outer basin	0.40	2.5		1.2	1.6	d	30-50
Mellansjön	0.50	4.0		2.5	2.0	d	50-100

Table I: Characteristics of the lakes in the first part of this study. Gjallarpesjon has only seepage tributaries, but a defined drainage outlet. Tannsjon is divided into two distinct basins with limited water exchange. d=drainage, s/d=seepage tributaries, drainage outlet lake, s=seepage lake.

Chemically, the lakes selected could be classified in terms of acidity and water color, since the acidity is known to have a profound effect on the dissolution efficiency, whereas water color has been claimed to be an important factor, despite lack of evidence:

- Very acid and very clear lakes;
- Moderately acid and very clear lakes;
- Neutral and very clear lakes;
- Moderately acid and moderately clear lakes;
- Moderately acid and very brown lakes; and
- Neutral and very brown lakes.

The lakes were limed using basic technologies for lake liming, of which the first is now obsolete in Scandinavia (barge dry application, barge dry-wet application, boat wet application, aerial dry application and land stationed dry discharge of powder). Dry application of calcite powder was carried out in the first liming in 1980 of Lake Jällunden. The resulting dustbowl was a problem both for personnel and for the environment, and the practice of dry application from boats has been abandoned. Several lakes were treated with a wet-dry application from a barge, where the calcite powder is pneumatically transported to a flash mixer immediately before discharge into the lake. This produces a semidispersed air-water-calcite mixture. The best technology used today (IMEK-boat) implies very vigorous mixing of calcite and water to produce a well dispersed calcite slurry prior to discharge into the lake. Both barges and boats have a loading capacity of 5 to 10 t of calcite powder, and the daily application capacity is 72 to 200 t day^{-1}.

A number of lakes were treated with the standard Scandinavian technology for remote lakes: dry application by helicopter. The calcite is carried in a container holding 400 to 600 kg of calcite powder under the aircraft. The load can be gradually released from the container during flight over the lake. Skilled pilots can do very accurate applications, but sometimes calcite can get dumped in single piles which, of course, is of no use for neutralization. In North America, wet applications from helicopters are more common, but this technology has not yet been applied in Scandanavia.

Lake liming is sometimes carried out by blowing the dry powder mixed with air out from the calcite transport truck in the direction of the lake where a road runs close to a substantial part of the shoreline. This method is used in a substantial number of projects, even though it is fairly obvious that it is not always capable of neutralizing a lake. It was also used by local liming initiators in preparing sampling stations prior to sampling visits by the provincial authorities in now defunct lake liming projects. The author regards this method as highly unacceptable in terms of dissolution economy and meeting chemistry targets. It was applied on the smallest lake in the study to document its poor performance.

The lakes were treated with calcite powders available on the market, ranging from the regular Swedish run of agricultural limestone with a surface weighted average particle size of 22 *um*, to superfine fillers used in the paper and pulp industry with a surface weighted average particle size of less than 2 *um*. The dose applied ranged from 2.1 to 37 g m^{-3} of calcite, the smallest project involving 10 t of material, the largest 1700 t.

4. RESULTS

The lakes were limed and relimed over a number of years, and a total of 23 lake treatments were closely monitored. Some of the treatments were not carried out according to the original plans, and, hence, did not meet the water quality targets (Oklängen I, II, Slkällingesjön I, Furesjön I, Jällunden II and Tannsjön I) until they had been retreated. However, the results could still be evaluated and used to broaden the experience. In all the lakes treated, conditions were created so that naturally reproducing fish populations could be sustained or reintroduced.

4.1 Oklängen and Mäsen

In Lake Oklängen, a different product, mesa, a $CaCO_3$ precipitate produced by paper and pulp factories as a waste product, was tried. Mesa is a grey sludge, containing approximately 30% water and consists of pure $CaCO_3$ (98% dwt). The first test phase of the application worked well, but in the second phase, the contractor lost control over the application equipment, as well as the supply of material, and allegedly applied 230 t. If this number is correct, mysteriously little of it ever dissolved in the lake. The lake was later relimed by the contractor using the standard lake liming powder and standard technology.

In the large clearwater Lake Masen, 1700 t of calcite were applied by barge, and the dissolution efficiency, observed closely, matched the value predicted by the model.

4.2 Skällingesjön

The liming of Lake Skällingesjön, a seepage lake, was carried out with 70% of the intended dose due to an error by the contractor carrying out the application. The applied powder dissolved almost completely and instantly due to its fineness, and as a result only the epilimnion was neutralized, while the hypolimnion was strongly acidic. The small undissolved residual of the applied dose did not penetrate the thermocline, but stayed on the top of the thermocline for several weeks after the application. This showed that very fine products may be used if it is decided to lime the epilimnion selectively for some reason, and they will not affect the hypolimnion. The acidity of the hypolimnion stayed at pH 4.7 until the second liming, a pH difference of 2 units compared to the epilimnion. The lake was strongly thermally stratified during the application.

4.3 Stora Neden

Lake Stora Neden was treated with a powder somewhat finer than the regular grade, and the dissolution efficiency was correspondingly increased. The powder was applied as a water-air mixture, and the observed dissolution efficiency was slightly lower than the theoretical prediction for a pure water suspension application. In Lake Stora Neden, small amounts of calcite were selectively placed in the littoral zone in order to study the dissolution of calcite into the lake and to possibly detect any positive action of waves on the dissolution rate and transport of material into the lake. The special investigation revealed that the calcite bank created a neutral microenvironment several millimeters above the bottom, but the bulk transport of dissolved calcite into the lake was insignificant. This illustrated that liming of the littoral zone may be an interesting tool to create neutral microenvironments, but it is definitely worthless for neutralizing the lake volume.

4.4 Gjällarpesjön

This small headwater lake to Stora Neden was treated by blowing dry powder from a transportation truck standing on a lakeside road. This method is frequently used in certain parts of the Swedish liming program, and, for this reason, was tested on the smallest lake in the study. The method was extremely inefficient in terms of observed dissolution efficiency. The application precision was bad, and a large dust bowl was created that was a nuisance to the environment. Due to the very poor dissolution efficiency, as well as the unacceptably low application precision, very large amounts of calcite are required to reach the target values (in some cases it is probably physically impossible to reach the target values). This is probably one of the most costly liming methods. We suggest that this method be abandoned in the Swedish operational liming program.

4.5 Furesjön

In Lake Furesjön, another headwater lake to Stora Neden, bioassays were carried out with Atlantic salmon taken from a fish hatchery on a neighboring river system still supporting naturally reproducing salmon. Prior to liming, repeated tests with 110 fish in each test at 4 stations in the lake, showed 50% mortality after 4.5 days and 100% mortality after 7 days. The pH was in the range of 4.4 to 4.7, Ca in the range of 2.5 to 2.8 mg L^{-1} and Al in the range 0.5 to 0.25 mg L^{-1} during the initial bioassay. In Lake Furesjön, the pipeline equipment used by the contractor malfunctioned and most of the calcite was applied directly to the sediment with very little entering the water column. During the 7 mo period prior to reliming, the alkalinity and pH of the lake stayed constant around pH 5.1 to 5.5, thus verifying that calcite on a lake bottom is not capable of neutralizing the bulk water volume of an acid lake. Fish survival was 100% after 400 hr at pH 5.35, 0.14 mg L^{-1} Al and 3.5 mg L^{-1} Ca in the bioassay carried out during the application and after. The lake was later fully neutralized through an aerial (helicopter) application of standard dry lake liming calcite powder. Fish survival was 100% at pH 6.8 and 7 mg L^{-1} Ca after more than 500 hr in the bioassay experiment carried out following the final treatment.

4.6 Jällunden, Tannsjön and Mellansjön

The relatively low effectiveness of littoral zone applications for bulk lake neutralization was illustrated by the consecutive liming of Lake Jällunden. In the first liming, a full neutralization was achieved by distributing 1400 t in the littoral zones of the lake. Three years later, the same degree of neutralization was achieved with approximately 450 t of calcite, distributed proportionally by volume over the lake. This corresponds to 35% of the amount applied in the first lake liming, leaving 90 t available for selective liming to create microenvironments or to reduce costs.

Tannsjön and Mellanjön are headwater lakes above the large Lake Jällunden. Tannsjön, which is divided into two distinct basins, was initially limed in the outer basin only. The water exchange between the two lake parts was too small to permit the water exchange between them to neutralize the untreated basin. The inner basin was limed later for this reason.

4.7 The Sandvatten-Husebotjärn-Bremstjärn Lake System

Sandvatten and Husebotjärn were both limed with a dry application from a helicopter. Even if the dissolution efficiency was lower due to the dry application, good results were obtained. The helicopter crew applied the prescribed amounts evenly over the lake surfaces according to the plan. Both lakes have no access for heavy vehicles, and liming from the air was virtually the only option. Husebotjärn may possibly be treated by boat. In Bremstjärn, the alkalinity and pH values rose due to the liming of the upstream lakes, and Bremstjärn was limed to circumneutral condition with a coarse calcite powder from a boat.

Fläghultsjärn was limed from a boat with a finer calcite powder. There were no problems and the lake was neutralized. Drumsetjärn was initially limed, due to aerial navigation programs, from a helicopter with the dose and product intended for Bremstjärn. The particle size distribution of the powder delivered was not the one originally intended for Bremstjärn, since the contractor did not check the particle size distribution curve specified, but rather used a commercial brand name to define the grinding grade. This is common practice in the Swedish operational liming program, despite the fact that the commercial brand names are severely misleading. Later Drumsetjärn was relimed with the correct type of powder.

4.8 Concluding Remarks

Of the 14 lakes treated, only seven were carried out correctly according to the written instructions given to the different contractors (Mäsen, Gjällarpesjön, Sandvatten, Husebotjärn, Jällunden I and IV and Mellansjön). In seven of the treatments, totally unacceptable errors occurred (Oklängen, Furesjön, Skällingesjön, Drumsetjärn, Bremstjärn, Skällingesjön and Tannsjön). All retreatments were carried out correctly (Jällunden III, Skällingesjön II, Oklängen III, Drumsetjärn II, and Tannsjön II). Thus, it became very clear to us that the cost of not controlling the application is generally several times higher than the actual cost of doing on-site inspection of the application work.

One significant result was that we were not able to detect any problems in arriving at the target water quality in the lakes with medium colored water or in those with strongly brown water (O to 100 mg Pt L^{-1}). As long as the proper titration curve is used, there is no problem.

The results in terms of dissolution efficiencies are listed in Table II, together with some of the crucial parameters considered in this study.

It is possible to show that the particle size distribution, initial pH and dose do influence the dissolution efficiency the way it is predicted by theory. This is visualized in Figures 6 and 7. From the results, it is also evident that a better mixing with water before application of the calcite gives a higher dissolution efficiency. These data indicate that for fully dry application, the dissolution efficiency will be approximately 60% of what full mixing with

Lake	Commercial name	Purity and origin	Method	Amount applied calcite (t)	Amount dissolved calcite (t)	Average particle size (μm)	Application method	pH prior to liming	Depth	Dose (g calcite m^{-3})	Dissolution efficiency %
Måsen	0-0.2 mm	78% Skövde	Barge	1700	885	12	W/D	5.6	22.8	16.0	52
Oklången I	Mesa	95% Delary	Kalkade	57	34	10	W	5.4	10.4	2.1	60
Oklången II	Mesa	95% Delary	Kalkade	227	36	10	W	5.9	10.4	8.5	16
Oklången III	0-0.2 mm	84% Skövde	Barge	328	240	10	W	5.6	10.4	12.2	67
Stora Neden	0-0.05 mm	92% Köping	Barge	852	464	7	W/D	4.9	21.5	15.0	55
Skällingesjön I	0-0.005mm	98% Malmökrita	Boat	18.5	16.5	2	W	4.9	10.4	5.9	90
Skällingesjön II	0-0.2 mm	84% Skövde	Boat	30.4	7.1	10	W	6.9	10.4	9.8	24
Furesjön I	0-1 mm	91% Ignaberg	Dry/Kalkade	20	4.4	18	D	4.6	8.1	6.1	22
Furesjön II	0-0.2 mm	91% Ignaberg	Helicopter	60	20	12	D	5.7	8.1	18.5	34
Gjällarpesjön	0-0.05 mm	92% Köping	Dry/car	10	2	7	D	5.6	6.1	20.0	20
Sandvatten	0-0.2 mm	92% Köping	Helicopter	65	26	12	D	4.6	17.1	14.8	41
Husebotjärn	0-0.2 mm	92% Köping	Helicopter	20	7.3	12	D	5.2	5.7	14.3	36
Brämstjärn	0-2 mm	92% Köping	Helicopter	15.8	3.4	22	D	6.0	5.1	15.8	21
Flåghultstjärn	0-0.05 mm	92% Köping	Boat	47	33	7	W	5.5	5.7	16.5	69
Drumsetjärn I	0-0.2 mm	78% Skövde	Barge	17	7	12	W	5.8	4.9	16.8	42
Drumsetjärn II	0-0.5 mm	92% Köping	Helicopter	15	4.9	14	D	6.6	4.9	14.6	33
Jällunden I	0-0.2 mm	91% Ignaberg	Dry/Barge	1400	300	12	D	5.0	2.0	37.0	21
Jällunden II	0-0.2 mm	91% Ignaberg	Barge	356	160	12	W/D	5.8	4.4	9.5	45
Jällunden III	0-0.2 mm	88% Skövde	Boat	104	53	10	W	6.3	4.4	2.8	51
Jällunden IV	0-0.2 mm	91% Ignaberg	Barge	448	143	12	W/D	6.2	4.4	11.9	32
Mellansjön	0-0.2 mm	91% Ignaberg	Barge	30	15	12	W	5.5	4.0	11.9	44
Tannsjön, inner basin	0-0.2 mm	91% Ignaberg	Barge	50	14	12	W/D	5.5	3.5	28.0	28
Tannsjön, outer basin	0-0.2 mm	91% Ignaberg	Boat	25	11	12	W	5.5	3.5	14.0	42

Table II: The dissolution efficiency observed in the lakes studied; together with some of the governing parameters. W=wet slurried calcite powder prior to application (Boat, Kalkade). W/D=mixing of calcite, air and water prior to application (Barge, kalkade). D=application of dry powder directly on the lake (Truck blower, barge, helicopter).

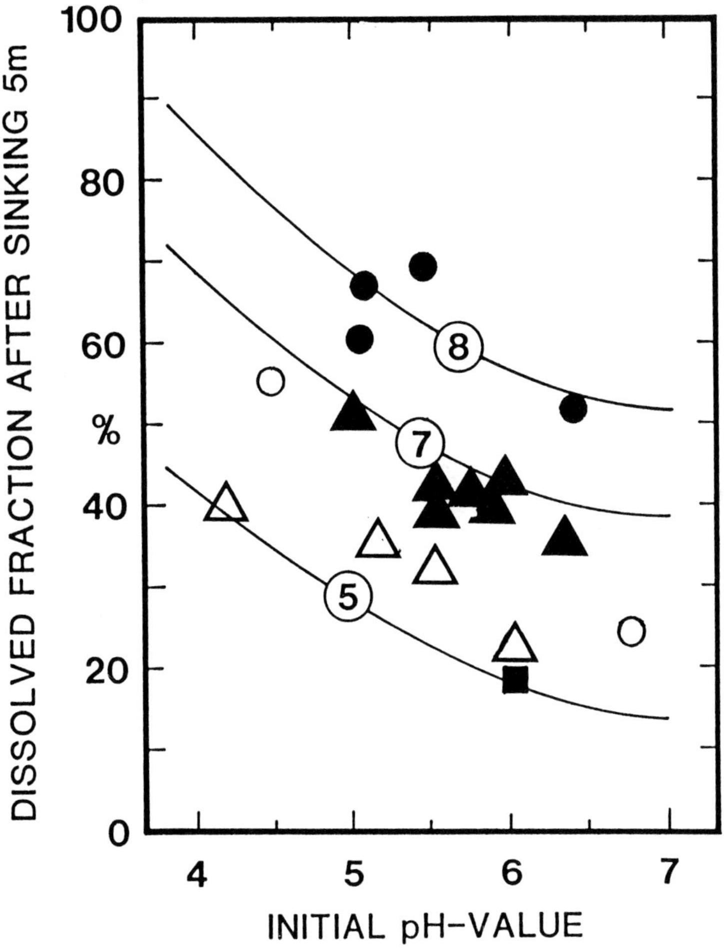

Figure 6: The observed dissolution efficiency in the experimental lake liming carried out by the authorities. The drawn line reflects the calculations with the model. The symbols represents; (O); No. 8, 0-0.5 mm applied wet, (●); No. 8, applied dry, (△); No. 17, 0.02 mm, applied wet, (▲); No. 7, 0-02 mm applied dry, (■); No. 5.0-2mm applied dry.

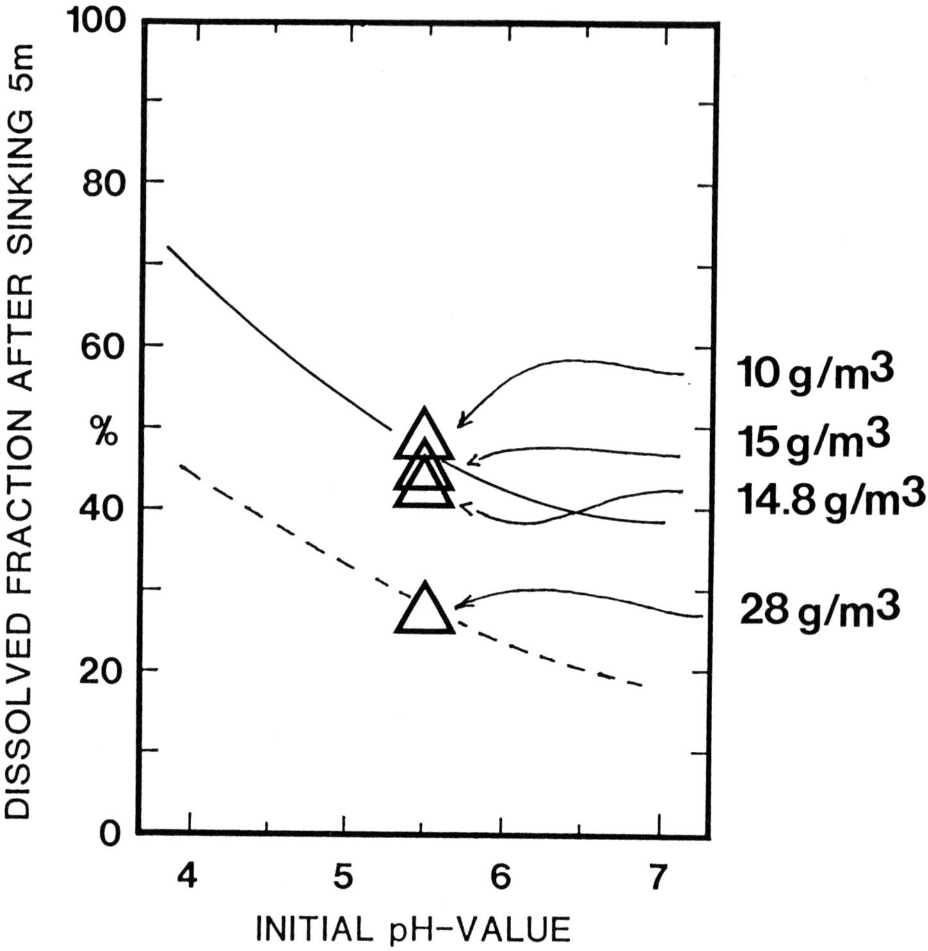

Figure 7: The observed effect of increasing the dose on the dissolution efficiency. The observed field results agree well with laboratory simulations of lake liming and with theoretical predictions.

water would imply, consistent with the approximation that all particles less than 6 mm are tied up on the surfaces of larger particles.

Dry application from a calcite transport truck is a method which should be abandoned due to its poor performance.

The results have some implications worth considering for the present Swedish operational liming program. The results show clearly the importance of the grinding grade for determining the dissolution efficiency, and, accordingly, the actual dose to be applied. However, the present official requirements are that the calcite powder be "90% less than 0.25 mm" or "90 % less than 0.1mm" which would include nearly every powder on the market, from a coarse crush which would be for liming cost efficiency, to superpowders used for paper production.

It is evident that the fixed doses given in today's liming instructions apply only to one particular particle size distribution (No. 7; often called 0 to 0.2 mm), and the given doses may be significantly too large or too small if other powders with other particle size distributions are used. As long as this simplified approach is insisted upon, this will continue to impose large and unnecessary costs on the Swedish liming program. In comparison with the conventionally planned Swedish lake liming, costs can be cut by 10 to 40% by cost optimization, and this is often combined with a better result due to the greater care taken during planning.

4.9 Reacidification and Maintenance Lake Liming

In all the lakes limed in this study, the desired level of neutralization was reached (pH 6.8 to 7.2, alkalinity of 0.12 to 0.2 meq L^{-1}). The dissolution rate of residual calcite at the lake bottom will be several orders of magnitude less than during the phase when it sinks through the water column. Calcite on the lake bottom will not be able to significantly influence the bulk neutralization of deeper lakes, but it may, under specific circumstances significantly prolong reacidification. Depending on the hydrological retention time of the lake and the amount of calcite on the bottom, this level will be sustained in the lake for a time before reacidification (Sverdrup and Warfvinge, 1985). The reacidification time can easily be

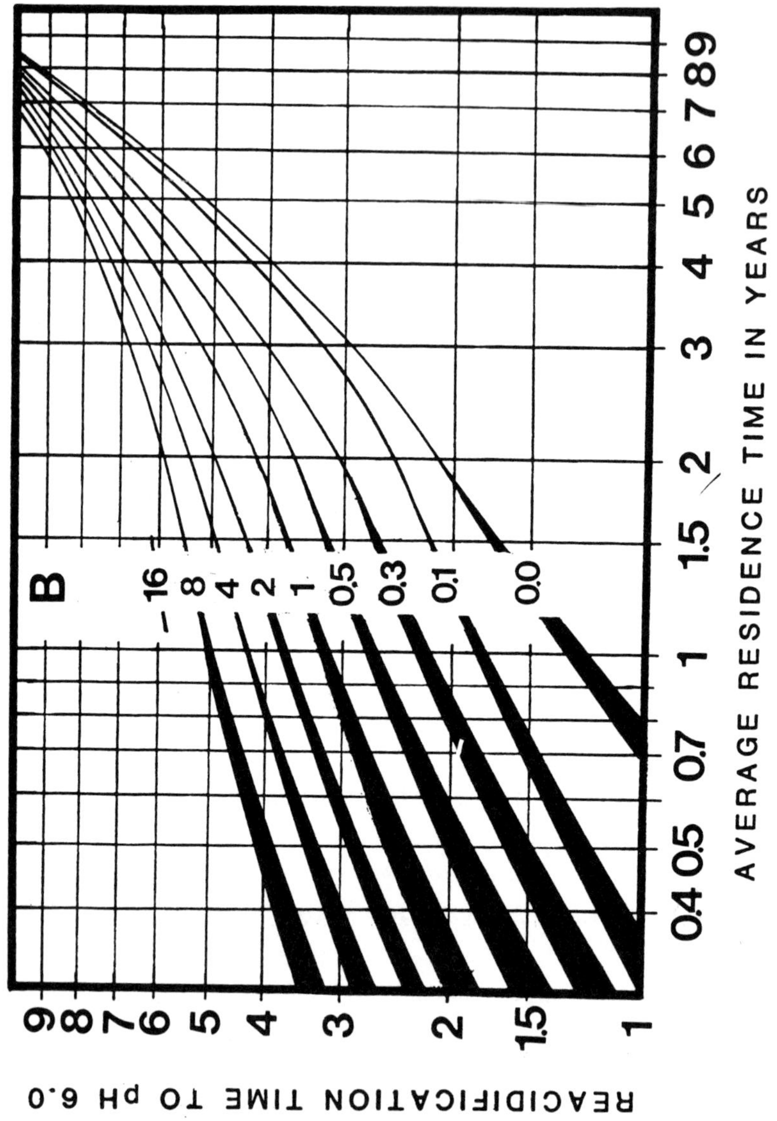

Figure 8: A simple diagram to predict the time required to reacidify a limed lake to pH 6, as a function of the hydraulic retention time and the amount of calcite on the bottom (B) in t ha^{-1}.

estimated with the diagram shown in Figure 8, which is based on field results and the computerized model. The validity of the diagram has been verified on a large number of field observations in Sweden, Norway, Finland, Canada and the United States (Booth et al., 1985; Saunders et al., 1985; Sverdrup et al., 1985a,b,c; Sverdrup 1985a).

Jällunden has been relimed four times in 1980, 1983, 1985 and 1987, and the pH of that lake has been kept constantly above pH 5.8 since 1980. The lake harbors large populations of reproducing fish. The first reliming was carried out in response to reacidification. Due to the lag-time built into the bureaucratic process of initiating a new liming project, the lake reacidified below pH 6.0. Today when the reacidification can be reliably predicted, reacidification can be prevented by initiating the bureaucracy before it has reached the critical point. A critical review of the limings indicates that the optimal amount of liming for Lake Jällunden is 500 to 550 t of 0 to 0.2 mm Ignaberga calcite powder.

The lakes in the Sandvatten-Husebtjärn-Bremstjärn Lake System will reacidify at different rates depending on their hydraulic retention times which are very different. A calculation with the reacidification model indicates that if the two lakes, Sandvatten and Husebotjärn, are kept at an alkalinity of at least 0.06 meq L^{-1}, then Bremstjärn will stay neutral without liming the lake itself.

4.10 Whole Watershed Approach to Lake Liming

We have only discussed liming of individual lakes, so far, and shown how almost any lake can be limed and maintained at circumneutral conditions at any time. However, lakes are only one component in a watershed, and often the purpose is better served if the whole catchment is studied in an effort to develop the optimal liming strategy in terms of meeting multiple target values in a cost effective way.

The ecological target values are reinterpreted in terms of water chemistry, and this results in two parallel approaches to achieving the goals. One is the bulk neutralization of the water passing the target points in the catchments. This will consume the major part of the neutralizing agent in the project. The second part is the creation of neutralized microenvironments through selective liming at specific locations.

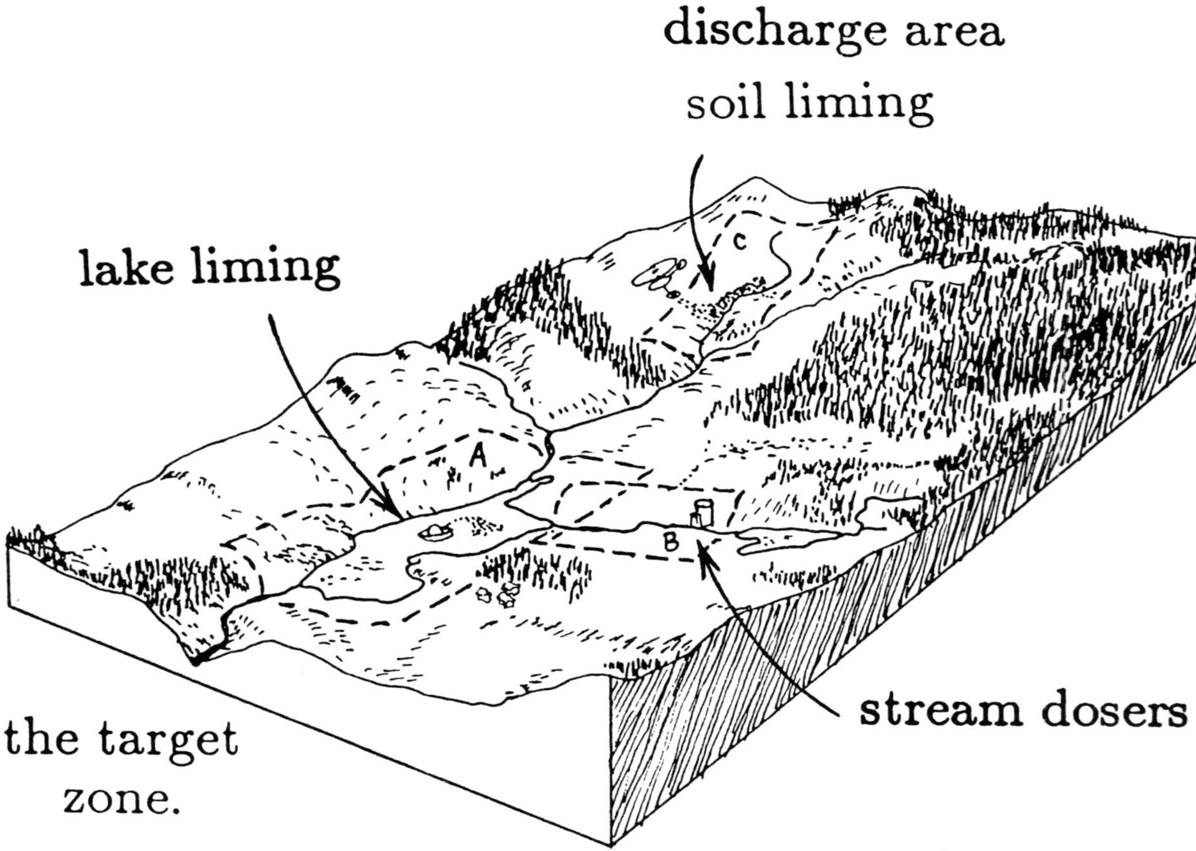

Figure 9: A whole catchment approach means that many different liming methods are combined in the most optimal way in relation to meeting the target values, reliability and minimum long-term operational cost.

Accordingly, the optimal liming project first meets all set target values and, second, is the most cost effective way of doing so. In order to achieve this goal, the optimal method will probably contain several elements such as lake liming, dosers, streamside soil liming, discharge area soil liming and specific liming of aquatic habitats.

This may be illustrated with the drawing shown in Figure 9. Bulk neutralization of the water would imply liming of lakes in the watershed (A), or would be combined with or substituted by stream liming using different types of dosing devices (B). Streamside soil liming may be an effective way of bulk neutralizing smaller streams (C), as well as be a specialized effort designed to stop leakage of potentially toxic Al to the stream. Likewise, lake segments may be limed selectively in spots to create spawning shoals or special microenvironments even if this measure is useless for bulk neutralization of the lake. In a whole catchment approach to acidification mitigation, all these components are used to obtain the best result.

4.11 The Högvadsän River and Lake System as an Example

The Salmon River Högvadsän in the Province of Hallard in southwestern Sweden illustrates the principle of whole watershed liming. The river is a natural breeding locality for Atlantic salmon; the good catches made there were both appreciated by a large number of enthusiastic fishermen, and were documented in the files of the local administration. During the late 1960's and 1970's, it became evident that the river had become acidified, and the stock of Atlantic salmon was dwindling to the point of virtual extinction. A large liming program was started to save the salmon river.

The program was to have one main objective, to sustain a naturally reproducing population of Atlantic salmon in the river. Later, the objectives were broadened to include self-sustaining fish populations in some of the lakes, such as Barken, Hjartaredsjon and Svarten.

A schematic representation of the river system is shown in Figure 10, where some of the characteristic data have been included. A short description of the lakes in the system is given in Table III.

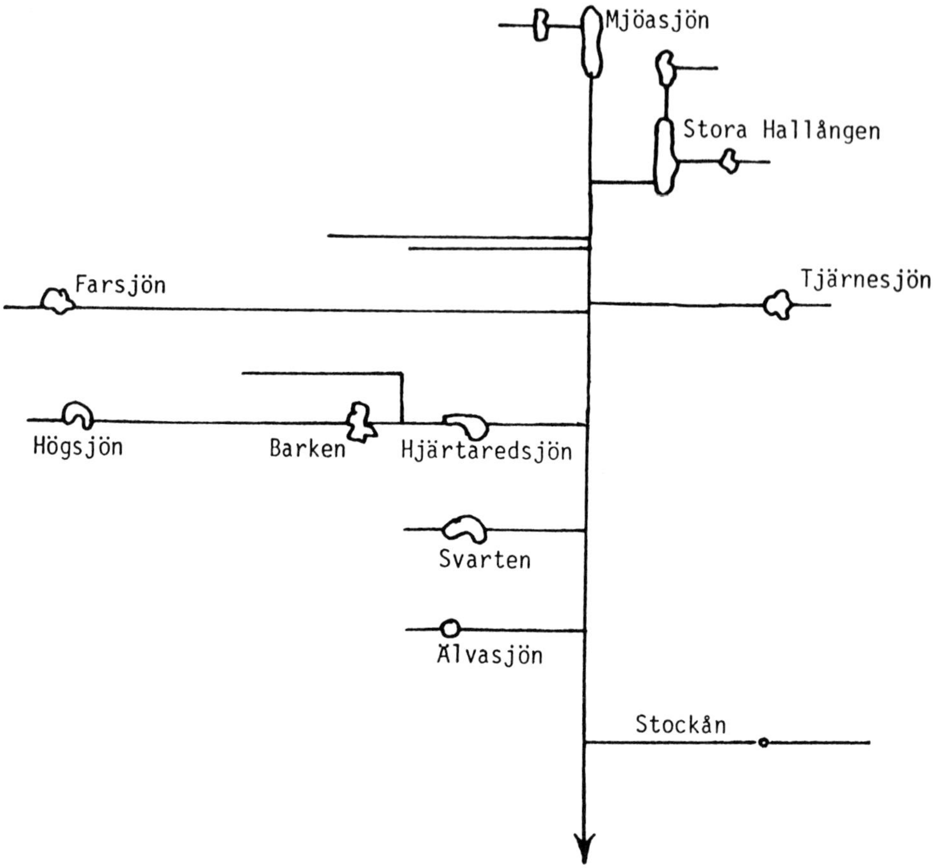

Figure 10: A schematic representation of the Hogvadsan river system. The system is severely acidified, but fish are sustained by liming. There are water quality requirements on the lower parts of the river due to salmon and separate water quality requirements in several of the lakes for preserving the lake trout populations.

Lake	Volume 10^6 m^3	Retention time yr	Average depth m	Runoff contribution %
Bredasjön	4.0	2.4	5.0	(0.3)
Mjöasjön	5.6	1.2	7.0	1.4
Lilla Hallången	6.0	1.6	10.0	(2.4)
Stora Hallången	13.1	1.0	10.0	6.0
Farsjön	2.7	0.9	2.5	1.1
Högsjön	10.0	2.4	4.3	(2.2)
Barken	12.5	0.42	10.0	(11.0)
Hjärtaredsjön	5.7	0.11	5.7	25
Björsjön	1.9	1.9	7.0	(1.2)
Svarten	16.2	2.1	11.3	2.9
Ålvasjön	3.3	1.65	5.0	1.4
Tjärnesjön	24.4	2.2	6.1	5.0

Table III: Characteristics of the lakes in the Hogvadsan river system. All lakes are drainage lakes, most have access for heavy vehicles. A parenthesis indicates that the lake is upstream from the next lake on the list.

The ecological target was interpreted to imply that the pH of the main course of the river from Älvsered down to the confluence with the semineutral river Ätran should, at all points, never go below pH 6.0, corresponding to an alkalinity of approximately 0.05 meq L^{-1}. The subobjectives later adopted for some of the lakes imply similar values. The given values will automatically restrict the concentration and toxicity of dissolved Al in the river, and hence, no limit for Al concentrations was defined.

4.12 The Optimal Strategy for Högvadsån

For the Högvadsån River system, we shall look at four different approaches to meeting the required water quality in the target zone, and we will examine which of these alternatives meet our criteria:

Alternative 1: Lake liming of all lakes with access in the system; these lakes provide 43% of the runoff to the target area.

Alternative 2: A stream doser installed at Älvsered and at the outlet of Tjärnesjön. The lakes Barken, Hjärtaredsjön, Svarten, Bredsjön, Mjöasjön, Högsjön, Stora Hallängen and several very small lakes are limed. These lakes contribute 34% of the runoff to the river. This represents the actual liming program carried out in the catchment today.

Alternative 3: A stream doser installed at Älvsered and at the outlet of Hjärtaredsjön. The lakes Barken, Hjärtaredsjön, Tjärnesjön, Svarten and Stora Hallängen are limed. These lakes contribute 37% of the runoff to the river.

Alternative 4: A stream doser installed at Älvsered and at the outlet of Hjärtaredsjön. No lake liming.

Lake liming of all the lakes in the system is the conventional approach in rivers with a substantial number of lakes in the catchment. There are, however, a number of complications to the approach which may present difficulties. During the high flow due to spring snowmelt, a situation may occur where the cold, very acid, meltwater will float on top of the slightly warmer water of a lake and take a short cut to the outlet entering the stream. If the distance across the lake is large, this is less likely to occur accordingly the phenomenon is most often observed in small lakes. In our case, the lakes contribute 43% of the runoff to the river, and water quality targets can be held if there are no acid short-cuts and if the lake alkalinity is kept constantly above 0.12 meq L^{-1} in the spring.

A few of the lakes are large and they have their major inlets far from the lake outlet; this will decrease the likelihood of an acid short-cut. Such lakes are Stora Hallängen, the combination of Barken and Hjärtaredsjön and Svarten. This is also probably the case in Tjärnesjön, but, a substantial amount of runoff is contributed between the lake and the main course of the river. Runoff from the lakes just mentioned will not be able to neutralize the river to target values during spring flow conditions.

At present, two dosers are installed in the system. One at Älvsered, which is a few kilometers above the highest point where salmon occur in the river, and one at the outlet of Lake Tjärnesjön. Tjärnesjön is not limed since it was chosen to be a reference lake when the liming program was still experimental. For an operational liming program in River Högvadsän, a reference lake is not required. The doser downstream from Tjärnesjön works on a small volume of water, severly limiting the calcite output to the main river from this source. Accordingly, the doser at Ävsered may be sped up to compensate for a standstill at Tjärnesjön, but not the opposite. During a power failure at Älvsered, acid pulses may flush through the system when acid short-cuts occur at the same time. This scenario is not unlikely, and has occurred several times in the past. As a backup to the doser system, nearly every lake in the catchment is limed, except Tjärnesjön. Acid pulses may go through the river if there is an acid short-cut and one doser is not operating.

Alternative 3 implies that the dosers are upgraded with automatic control units and wet slurrying units for the calcite powder. The doser at Tjärnesjön was moved downstream of Hjärtaredsjön, which substantially increased the capacity to deliver calcite to the main stream. A failure at the lower doser can be compensated for by speeding up the upper doser, and vice versa. In a failure of the upper doser, the upper part of the lake will be safeguarded partially by the lake, whereas the lower doser will be able to neutralize the lower half of the river. To meet the target values in the main course of the river, only Stora Hallängen, Mjöastjön, Hjärtaredsjön, Tjärnesjon and Svarten are limed. Other lakes are limed if separate target values exist for these lakes only. Lake liming in the other lakes is not cost efficient in relation to their contribution to the neutralization of the main course of the river. The lakes picked for liming are the ones which have the most significant contribution to the river in terms of runoff, and also which are most like to show acid short-cuts. Only when all adverse events coincide, when there is an acid short-cut in all lakes and both dosers do not operate, will the river be subject to acid pulses through the full length.

Alternative 4 implies two dosers as described in alternative 3, but with no lake limings as backup in case of a failure of the equipment. The configuration of the dosers will permit the slower doser to fail without any effect on the neutralization of the main course, but a failure at the top doser will affect half the target zone.

In Table IV we have listed the approximate water alkalinity at the target zone under some different situations in order to evaluate the different ways of neutralizing the river system against each other.

5. DISCUSSION

Table V indicates, that under acid short-cut conditions which are known to occur frequently in the system, only alternative 2 and 3 will satisfy the water quality criteria for the target zone, provided that the dosers have 100% availability. The presently adapted alternative 2 will probably prevent fishkills under most single doser failure conditions, but reproduction may be occasionally interrupted. At lower doser availability, only alternative 3 will be able to meet the requirements.

There are, of course, widely different costs involved in carrying out the different alternatives. Alternative 1, which really does not work, would cost approximately 55% of what the present implemented alternative 2 costs today to run on a long-term basis. Alternative 3 would cost approximately 80% of the present cost, if a slightly different choice of products, a fine tuned lake-liming and upgraded dosers were used. Alternative 4 which would work well if there were no doser failures (involving a higher level of risk) and would cost approximately 50% of the presently implemented program.

Scenario number	Event occurring	Lake liming alt.1	Dosers alt 2.	Dosers alt. 3	Dosers alt. 4
				alkalinity at target site meq L^{-1}	
1	No dosers operating Acid throughbreak in all lakes	-0.05	-0.05	-0.05	-0.05
2	No dosers operating Acid throughbreak in all lakes except Hjärtaredsjön and St. Hallången	-0.01	-0.01	-0.01	-0.05
3	Doser at Älvsered not operating Acid breakthrough in all lakes except Hjärtaredsjön and St. Hallången	-0.01	0.01	0.10	0.10
4	Doser at Älvsered in operation other doser not operating Acid breakthrough in all lakes except in Svarten, Tjärnesjön, Hjärtaredsjön and St. Hallången	-0.01	0.03	0.05	0.05
5	All dosers operating Acid breakthrough in all lakes	-0.05	0.02	0.10	0.10

Table IV: The water quality at the target in the River Hogvadsan under different operational liming scenarios. We have assumed the conditions in the river to be similar to the spring high flow, due to snowmelt, and being acid.

ACKNOWLEDGMENT

This study was financed by the Swedish Government on the recommendation of the National Board of Fisheries at Goteborg. Project officer at the National Board of Fisheries is Fil. Lic. Bo Bengtsson. Many persons contributed significantly to this study. The sites were selected and local contacts established with the help of Lisbeth Nilsson at the Province of Hallands Bureau of Environment, Per Olsen at the Province of Goteborg and Bohus Bureau of Environment, Inge Lund and Key Hoglind at the Uddevalla Countys Board of Fisheries, Kurt Claesson at the Vargerg Countys Bureau of Environment and Municipal Health. We were assisted in sample collection by Robin Wellington, Stromstad, Province of Bohuslan and Henry Johansson, Jallunden. Jim Hamilton and Dr. Lewis Molot, Booth Aquatic Inc, on behalf of the Ontario Ministry of the Environment, proposed the bioassy design and assisted us in installing the experiment on Lake Furesjön.

EFFECTS OF LAKE LIMING ON PHYTOPLANKTON PRODUCTION IN ACIDIC ADIRONDACK LAKES

Paul A. Bukaveckas

ABSTRACT

Phytoplankton community characteristics were monitored prior to and following $CaCO_3$ addition to two small, highly acidic lakes (Cranberry Pond and Woods Lake) and one larger, less acidic lake (L. Simon Pond). Data were also collected from a control site (Dart's Lake) exhibiting chemical characteristics similar to the pre-treatment conditions observed at the experimental sites. In the two small, most acidic lakes, base addition was associated with higher chlorophyll levels during the first summer following treatment. Woods Lake was maintained at a circumneutral pH for 3 yr and exhibited increased phytoplankton abundance throughout the post-treatment period. In contrast, Cranberry Pond reacidified within 1 yr following based addition. Reacidification was accomplished by a decrease in lake chlorophyll levels to pre-treatment levels. At the larger, less acidic lake (L. Simon Pond), liming was associated with lower chlorophyll levels during the first summer after treatment. Reductions in chlorophyll levels at L. Simon Pond reflect the absence of the spring phytoplankton peak and a decrease in phytoplankton growth below the depth of the thermocline. At Cranberry Pond, annual differences in phytoplankton production did not correspond to changes in lake acidity and phytoplankton abundance. Productivity in Woods Lake exhibited an increasing trend during the 3 yr following treatment. Interpretation of treatment effects on productivity was confounded by high between-year variability at the control site.

1. INTRODUCTION

Atmospheric deposition of strong acids has caused acidification of surface waters in sensitive regions of North America and Europe. Acidification is accompanied by changes in the biological characteristics of impacted lakes and streams. Reductions in species richness and changes in the species composition of phytoplankton, macrophyte and zooplankton communities (Baker, 1984; Stokes, 1986; Jackson and Charles, 1988) have been reported. In chronically acidified systems, fish populations are eliminated and succeeded by macroinvertebrates as top predators (Schofield, 1976; Ericksson et al., 1980; Bendall and McNicol, 1987).

Water, Air, and Soil Pollution **41** (1988) 223–240.

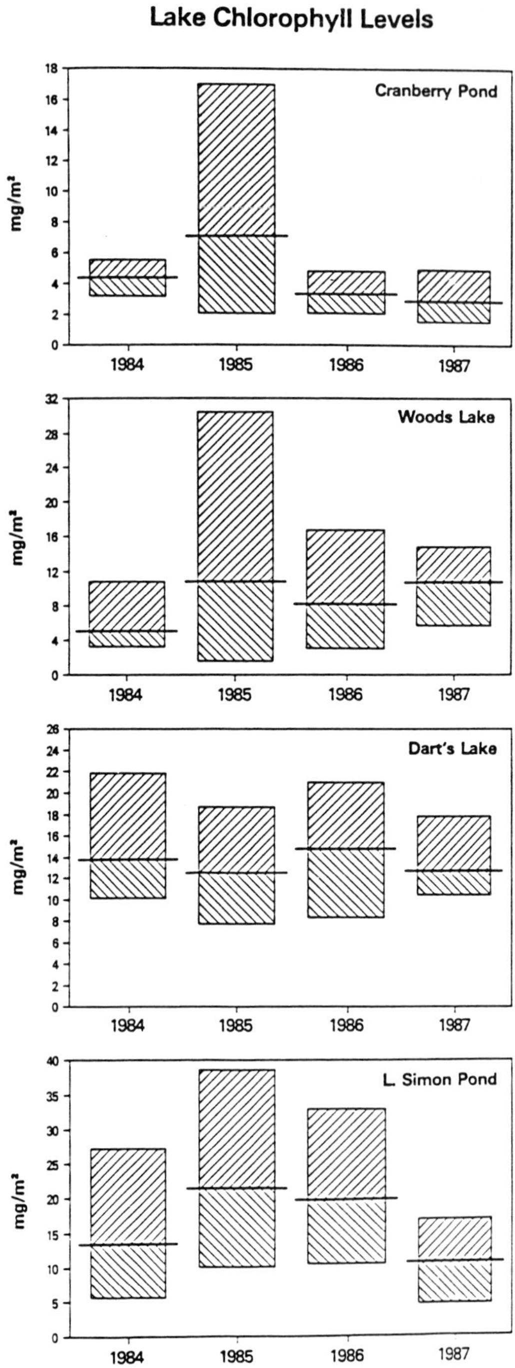

Figure 1: Whole-lake pigment concentrationsas mg chlorophyll per m^2 of lake surface area. Bars denote range of values observed for all sampling dates. Lines denote mean.

In some regions, acid-impacted systems are being treated by the addition of base materials such as $CaCO_3$ to neutralize acidity and increase buffering capacity. Restoration of water quality in acidified lakes has been the focus of research projects and operational programs conducted in Sweden, Norway, Canada, the United Kingdom and the United Staates (see review by Porcella et al., 1989). The Lake Acidification Mitigation Project (LAMP) was initiated to examine the effects of liming on clearwater acidic lakes located in the Adirondack Mountain region of New York state. The goal of this project was to evaluate the utility of whole-lake base addition in mitigating the effects of acidification on aquatic biota including plankton, macroinvertebrate and fish communities. In addition, data collected during the reacidification of treated lakes provides an opportunity to examine chemical and biotic responses to increasing lake acidity.

Liming effects on phytoplankton community composition have been well studied in both European and North American lakes. There is general concurrence that base treatment of acidic lakes results in increased phytoplankton diversity and a shift in species composition from dominance by acid-tolerant taxa to those characteristic of oligotrophic lakes (Eriksson et al., 1982, 1983; Hasselrot and Hultberg, 1984; Henrikson et al., 1984; Horstrom and Ekstrom, 1985). Typically, communities comprised dinoflagellates (e.g., *Peridinium* and *Gmodinium* and cryptophytes are succeeded after liming by chrysophytes and, in some cases, diatoms and chlorophytes. However, previous studies have not reported consistent effects on phytoplankton abundance. Liming of lakes in the Gjerstad region of Norway caused little change in algal biomass (Brettum and Hindar, 1985). Yan and Dillon (1984) reported no significant change from pre-treatment levels in algal biomass of three lakes in the Sudbury region of Canada In contrast, phytoplankton abundance in Lake Gardsjon, Sweden increased within a short time after liming (Lazarek, 1986). Similarly, Lakes Blanksjon and Iglafallsjon in Sweden exihibited enhanced phytoplankton growth within one year following treatment (Eriksson et al., 1983). In some cases, changes in phytoplankton biomass have been accompanied by changes in the abundance of aquatic marcophytes and attached algae (Hultberg and Andersson, 1982; Lazarek, 1986). Thus, interpretation of the effects of liming on lake primary production may be complicated by shifts in the relative importance of pelagic and littoral productivity.

Research on primary producers at the LAMP study sites has documented the response of phytoplankton, epiphyte and macrophyte communities following base treatment of three acidic Adirondack lakes. Changes in biomass, productivity, species composition and diversity were monitored during pre- and post-treatment periods to quantify the effects of base addition on lake productivity. The effects of liming on macrophyte communities at the LAMP sites are discussed in Bukaveckas (1988). Bukaveckas (1989) describes changes in phytoplankton abundance during the first summer following treatment of two highly acidic lakes (Cranberry Pond and Woods Lake).

This paper considers the long-term changes in phytoplankton abundance and productivity observed at these sites. Data were collected over a four year period during which one site (Woods Lake) was maintained at a circumneutral pH while a second site (Cranberrry Pond) was allowed to reacidify. Data describing short-term treatment effects on a less acidic lake (L. Simon Pond) are also presented. Similar data were collected from a control siter (Dart's Lake). The LAMP experimental design is presented in Porcella (1989) and summarized in Table I.

2. METHODS AND MATERIALS

2.1 Site and Treatment Description

Study sites are located in predominantly forested and undeveloped watersheds in the west-central (Cranberry Pond, Woods Lake and Dart's Lake) and central (L. Simon Pond) Adirondack region. Lake and watershed characteristics are summarized in Table II. A detailed description of watershed hydrology, geology and vegetation for Cranberry Pond and Woods Lake is presented in Staubitz and Zarriello (1988). Similar data for Dart's Lake and L. Simon Pond are available in McKinley (1985). Prior to $CaCO_3$ addition the chemical characteristics of these sites were typical of clearwater acidic lakes in the Adirondack region (Driscoll and Newton, 1985). Detailed descriptions of pre-treatment water chemistry are available for Cranberry Pond and Woods Lake (Fordham and Driscoll, 1989), Dart's Lake (Driscoll and Schafran, 1984; Schafran and Driscoll, 1987) and L. Simon Pond (McKinley,

Table I

LAMP DESIGN

	CRANB. P.	WOODS L.	DARTS L.	SIMON P.
1984	ACIDIC 4.7-4.9	ACIDIC 4.9-5.2	STABLE 5.2-5.6	ACIDIC 5.3-5.7
1985	NEUTRALIZED 6.8-8.2	NEUTRALIZED 6.7-7.9	STABLE 4.9-5.4	ACIDIC 5.3-5.6
1986	REACID. 4.7-5.2	TRANSITION 5.8-6.2	STABLE 5.0-5.5	ACIDIC 4.9-5.5
1987	ACIDIC 4.8-4.9	NEUTRALIZED 6.7-6.9	STABLE 5.0-5.3	NEUTRALIZED 6.4-6.6
1988	ACIDIC 4.6-5.0	TRANSITION 5.9-6.3	STABLE 5.1-5.3	STABLE 6.0-6.4

Table II

Morphometric and chemical characteristics of LAMP study sites. The pH and ANC data are for integrated water column samples collected during the period of summer stratification.

Site Characteristics	Cranberry Pond	Woods Lake	Dart's Lake	L. Simon Pond
Lake Area (m2 x 1000)	72	230	570	626
Lake Volume (m3 x 1000)	207	802	4150	6340
Maximum Depth (m)	7	12	15	33
Mean Depth(m)	2.9	3.5	7.3	10.1
Residence Time (days)	90	214	80	450
pH	4.7-4.9	4.9-5.2	5.0-5.5	5.4-5.7
ANC (*u*eq L^{-1})	-25 to -5	-15 to +5	-10 to +10	-10 to +10
Monomeric AL (*u*mol L^{-1})	5 to 20	2 to 12	5 to 10	.5 to 3.5
DOC (*u*mol C L^{-1})	250 to 450	150 to 300	240 to 40	200 to 300

1985). At all sites, sulphate was the dominant anion and the equivalence of acidic anions exceeded base cations, resulting in low pH, reduced acid neutralizing capacity (ANC) and generally high concentrations of aqueous Al.

Cranberry Pond and Woods Lake are relatively small drainage lakes generally exhibiting pH < 5 and ANC < 0. Neither site supported fish populations at the time the study was initiated, although brook trout (*Salvelinus fontinalis*) had been present historically (Gloss et. al., 1987). Owing to its smaller volume, Cranberry Pond has a shorter hydraulic retention time than Woods Lake and was therefore expected to reacidify at a faster rate. L. Simon Pond had higher concentrations of basic cations relative to Cranberry Pond and Woods Lake and, as a result, higher pH and ANC and lower Al concentrations. Populations of brook trout, lake trout (*S. namaycush*) and white sucker (*Catostomus commersonii*) was present at the time of treatment, although recruitment was low. The lake is larger and deeper and therefore has a longer retention time than the other study sites. Data were also collected from a control site (Dart's Lake) to assess annual variability in phytoplankton production. Dart's Lake exhibits chemical and morphometric features intermediate between those of the largest, least acidic and smallest, most acidic of the treated lakes..

The three experimental sites were treated by application of a highly pure calcite slurry (98% $CaCO_3$). The slurry was applied by helicopter on a volumetric basis to achieve a uniform target pH/ANC. Treatments were conducted on May 30-31, 1985 at Cranberry Pond (7 t d.w. $CaCO_3$) and Woods Lake (23 t) and on August 7-9, 1986 at L. Simon Pond (86 t). Woods Lake was retreated (35 t) during fall of 1986. An account of the treatment program is presented in Young et al. (1989). Treatment effects on lake chemistry are described in Fordham and Driscoll (1988) and Driscoll et al. (1988) and summarized in Table I. Four days after treatment, yearling and fingerling brook trout were introduced into Cranberry Pond and Woods Lake.

2.2 Methods

Samples were collected over the deepest portion of the lake basin at 5- to 10-day intervals during the ice-free period. Small diameter tubing was lowered to within 1 m of the lake bottom to obtain integrated water column samples. Discrete depth samples were collected at 1- to 2-m intervals from the surface to the bottom. Samples were preserved in the field and retained for phytoplankton identification and enumeration or stored in the dark for subsequent analyses of photosynthetic pigments.

Phytoplankton samples were preserved by the addition of 50/50 glutaraldehyde/ formaldehyde to a final concentration of 3% preservative. Samples were processed using a modification of the permanent slide mounting technique developed by Dozier and Richarson (1975). Slide mounts were prepared by drawing measured volumes of sample through Millipore HA filters (0.45 um) under low vacuum pressure. Filters were cleared in glutaraldehyde and phytoplankton were examined under a magnification of 1250 X using enhanced contrast optics. Two to four replicate slides were prepared from each phytoplankton sample and counted at 500X. Error associated with calculated cell concentrations was estimated from the variance between replicate slide counts.

Samples for pigment analyses were processed within 2 hr of collection by filtration through Gelman A/E glass fiber filters. Filters were macerated using a Teflon tissue grinder and extracted in aqueous 90% acetone for 12 hr in the dark at 4^0C. Extracts were analyzed using a Varian DMS 70 spectrophotometer equipped with long path length (4 cm) cells and narrow (1 nm) band width. Chlorophylls a, b and c were determined from optical density readings at 663, 645 and 630 nm using the trichromatic equations of Jeffrey and Humphrey (1975). Concentrations of chlorophyll degradation projects were determined by measuring the optical density of the extract before and after acidification with 0.1 M HCl. Estimates of chlorophyll a corrected for phaeophytin a were calculated using the Lorenzen equations as modified by Speziale et al. (1984). Whole-lake concentrations of chlorophyll a, phaeopigments and total pigments (chlorophyll a + b + c) were calculated for each sampling date from water column profiles and lake morphometric data.

Phytoplankton productivity was measured by the 14C isotope technique described by Vollenweider (1969). Radiocarbon uptake experiments were performed at all sites at 10- to 20-day intervals during 1985 through 1987. Samples were collected using a non-metallic Kemmerer sampler and incubated in 300 ml BOD bottles. Each bottle was injected with 5 uCi radiocarbon as sodium bicarbonate. Two light and two dark bottles were incubated for 4 hr at midday (1000 to 1400) at each of 3 depths (Cranberry Pond: 2, 4 and 6 m, Woods Lake and Dart's Lake: 3, 7 and 11 M, L. Simon Pond: 5, 9 and 13 m). Samples were placed on ice and kept dark during transport to the laboratory (< 1 hr). Labelled phytoplankton were collected on Millipore HA membrane filters (0.45 μm) under low vacuum pressure. Filters were rinsed with acidified water and placed in scintillation vials containing 0.1 ml of 1 M HCL to remove unincorporated isotope (Lean and Burnison, 1979). After 2 to 3 hr, fluor was added (Aquasol-2, New England Nuclear Products) and activity was measured on a Beckman LS-233 liquid scintillation counter. Quenching was corrected by the external standard channels ratio technique. Concentrations of dissolved inorganic C were measured by gas chromatographic as part of the water chemistry sampling program at the LAMP study sites (Fordham and Driscoll, 1989).

3. RESULTS

3.1 Pre-treatment Characterization

A description of phytoplankton species composition and growth patterns during the pre-treatment period is presented in Bukaveckas (1989) and will be briefly summarized. Phytoplankton floras were similar at the four study sites and characteristic of acidic lakes in Sweden (Almer et al., 1978), the Adirondacks (Hendrey, 1980; Charles, 1982), and the Sudbury region of Ontario (Kwiatkowski and Roff, 1976; Yan, 1979). Dominant phytoplankton genera included the chrysophytes *Dinobryon* and *Mallomonas*, the cryptophyte *Cryptomonas* and the dinoflagellates *Peridinium* and *Gymnodinium*. Patterns of phytoplankton seasonal succession and vertical distribution were similar at the two smallest, most acidic sites (Cranberry P. and Woods L.). During early- and mid-summer, cell densities within the epilimnion were extremely low, corresponding to chlorophyll concentrations of less than 1 mg m^{-3} and C uptake rates less than 1 mg $C \cdot m^{-3} \cdot hr^{-1}$. Highest estimates of phytoplankton abundance and productivity were measured in deep-water chlorophyll maxima. Carbon uptake rates at the chlorophyll peak generally ranged from 2 to 3 mg $C \cdot m^{-3} \cdot hr^{-1}$ and were consistently greater than rates measured in the upper water column.

At the two smallest, most acidic sites, the chlorophyll maxima typically occurred within 1 to 2 m of the sediment-water interface and did not vary in response to seasonal changes in thermocline depth. Although both sites exhibited only weak thermal stratification during late summer, the deep-water chlorophyll peak persisted throughout the sampling period. During early summer, the deep water assemblages were dominated by chrysophytes, particularly *Dinobryon bavaricum* and several species of *Mallomonas*. Phytoplankton biomass increased throughout the water column in late summer owing to higher densities of small chlorophytes, particularly *Kirchneriella obesa,* and small (<10 *u*m) cocciod unicells.

The larger, less acidic lakes (Dart's Lake and L. Simon Pond) differed from the smaller sites in that they exhibited a distinct spring maximum of phytoplankton growth. During the spring peak, chlorophyll levels increased throughout the water column but were highest in the metalimnion. These sites also differed in that blue-green algae comprised a greater proportion of the phytoplankton community. *Merismopedia* and *Chroococcus* increased in abundance in late summer and were associated with increasing densities of large filter feeding cladocerans. In contrast, *Daphnia* was not present prior to treatment of Cranberry Pond and Woods Lake. A description of the zooplankton communities at the LAMP study sites is presented in Schaffner (1989).

3.2 *Treatment Effects*

The mean and range of chlorophyll levels observed during 1984 through 1987 are presented in Figure 1 as mg chlorophyll m^{-2} of lake surface. Prior to treatment, differences in chlorophyll levels among the study sites followed a gradient of lake size and acidity. The summer mean and range of chlorophyll levels was lower at the smallest, most acidic sites (Cranberry Pond and Woods Lake) and higher at the larger, less acidic lakes (Dart's Lake and L. Simon Pond.). Base addition resultred in increased phytoplankton abundance at the two smallest, most acidic sites. At Cranberry Pond and Woods Lake, mean chlorophyll levels were 2X greater than pre-treatment levels during the summer following base addition. At both sites, maximal pigment concentrations were observed in July of 1985 and were 3X higher relative to peak concentrations in 1984.

Both sites experienced significant reacidification within 1 yr following the 1985 spring treatment. The pH of Cranberry Ponds decreased to pretreatment levels by summer of 1986. Following reacidification chlorophyll concentrations were reduced. The mean and range of pigment levels during 1986 and 1987 were comparable to those observed prior to treatment. In contrast, the pH of Woods Lake remained relatively high (pH > 6), despite significant loss of ANC. This site was retreated during the fall of 1986 to maintain a circumneutral pH. Mean summer chlorophyll levels for 1986 and 1987 remained elevated relative to the pretreatment period. The seasonal range of lake chlorophyll levels was smaller in 1986 and 1987 compared to 1985 since maximal concentrations did not approach those observed in mid-July of 1985. Furthermore, minimum concentrations were greater than those recorded immediately following calcite application in May 1985. At both Cranberry Pond and Woods Lake, chlorophyll levels were reduced by over 50% within three days following calcite application. A similar response was observed at L. Simon Pond after treatment in August 1986.

In comparison, mean annual chlorophyll levels did not vary significantly between years at the control site. The coefficient of variation (CV) calculated for the annual means was 10.4% and was lower than the seasonal variance observed during each year. At L. Simon Pond the mean and range of lake chlorophyll levels observed were similar during 2 of the 3 yr of pre-treatment data. Reduced chlorophyll concentrations in 1984 may be partially attributed to the shorter sampling period that year. Sampling was initiated in June of 1984 and therefore did not include the spring maximum which occurred during May of subsequent years. Pigment concentrations were reduced after treatment in August of 1986. Reduced chlorophyll levels during the first summer after treatment reflect the absence of the spring phytoplankton bloom observed during previous years (**Figure** 2). Lake pigment levels during May of 1987 were only 5 to 10% of those observed in 1985 and 1986.

Estimates of production based on radiocarbon uptake and lake morphometric data are presented in Table III as mg C fixed $m^{-2}.hr^{-1}$ of lake surface area. Interpretation of the effects of liming on phytoplankton productivity is limited by a lack of pretreatment data from the two shallow sites and by the high degree of annual variability observed at the control site (Dart's Lake). The coefficient of variation for 3 of the 4 mean and maximal levels of production were in 1986. Production was significantly reduced in the lower water column of

Table III

Phytoplankton production as mg C $m^{-2} \cdot hr^{-1}$ of lake surface area estimated from light, dark net and gross radiocarbon uptake.

Year	Variable	Minimum	Maximum	Mean
		Cranberry Pond		
1985	Light Uptake	2.02	13.36	5.59
	Net Uptake (L-D)	0.56	7.00	2.61
	Gross Uptake (L+D)	3.48	19.71	8.58
1985	Light Uptake	4.76	15.54	7.91
	Net Uptake (L-D)	3.64	11.05	6.21
	Gross Uptake (L+D)	5.11	20.04	9.62
1985	Light Uptake	1.52	10.58	4.41
	Net Uptake (L-D)	1.44	9.63	4.11
	Gross Uptake (L+D)	1.61	11.53	4.72
		Woods Lake		
1985	Light Uptake	1.19	14.98	6.94
	Net Uptake (L-D)	0.17	7.14	2.99
	Gross Uptake (L+D)	2.21	22.82	10.90
1985	Light Uptake	5.35	10.05	7.75
	Net Uptake (L-D)	4.76	9.52	7.11
	Gross Uptake (L+D)	5.94	10.59	8.38
1985	Light Uptake	6.58	12.87	9.21
	Net Uptake (L-D)	6.21	9.11	7.89
	Gross Uptake (L+D)	6.95	16.64	10.53

Table III (con't)

Year	Variable	Minimum	Maximum	Mean
		Dart's Lake		
1985	Light Uptake	2.80	6.32	4.63
	Net Uptake (L-D)	1.20	2.62	1.94
	Gross Uptake (L+D)	4.36	10.01	6.83
1985	Light Uptake	12.38	46.79	25.05
	Net Uptake (L-D)	10.80	44.62	23.23
	Gross Uptake (L+D)	13.57	48.97	26.87
1985	Light Uptake	5.87	19.33	12.58
	Net Uptake (L-D)	5.52	18.80	11.80
	Gross Uptake (L+D)	6.21	19.87	13.37
		L. Simon Pond		
1985	Light Uptake	2.64	4.86	3.70
	Net Uptake (L-D)	1.40	2.40	1.75
	Gross Uptake (L+D)	3.87	7.33	5.65
1985	Light Uptake	7.68	13.19	10.27
	Net Uptake (L-D)	6.71	11.06	8.81
	Gross Uptake (L+D)	8.65	16.19	11.73
1985	Light Uptake	3.68	11.45	6.41
	Net Uptake (L-D)	1.39	11.11	5.76
	Gross Uptake (L+D)	3.84	11.80	7.05

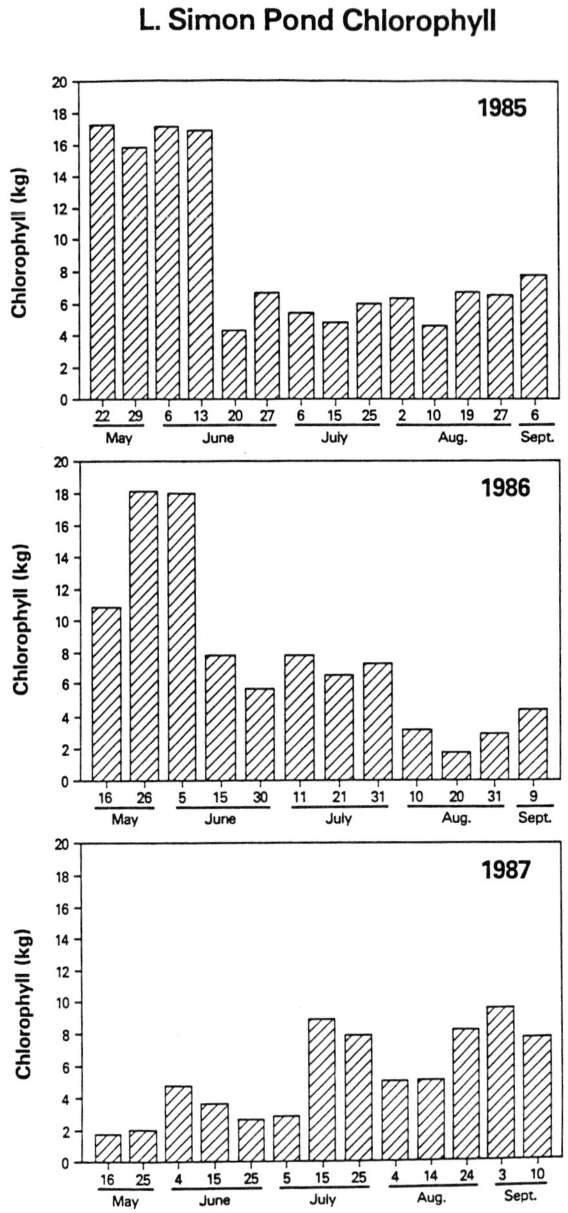

Figure 2: Seasonal patterns of lake pigment levels at L. Simon Pond during 1985-1987.

L. Simon Pond during the post-treatment period (**Figure** 3). The metalimnetic and hypolimnetic zone accounted for 20 to 30% of water column production prior to base addition but, accounted for less than 10% of production during the summer following treatment. Decreased growth in the lower water column may be attributed to reductions in lake transparency. Light attenuation as estimated by the attenuation coefficient for downwelling irradiance (Kd) was significantly higher in 1987 compared to previous years (Bukavekas and Driscoll, 1988). Percent of irradiance reaching the metalimnion (7 m) declined from 6 to 7% in 1985-6 to less than 2.5% in 1987.

4. DISCUSSION

Summer chlorophyll levels increased following the treatment of Cranberry Pond and Woods Lake. Increased phytoplankton biomass following base addition has also been reported for lakes in Sweden (Hornstrom and Ekstrom, 1985) and Norway (Brettum and Hindar, 1985). Changes in phytoplankton abundance following lake treatment have been attributed to increased nutrient availability. Boyd and Cuenco (1980) and Hindar and Nilssen (1984) have suggested that phytoplankton growth in acidic waters may be limited by low concentrations of dissolved inorganic carbon (DIC). Laboratory experiments have demonstrated that the levels of DIC which occur in acidic lakes are below the photosynthetic compensation point of some species (Williams and Turpin, 1987). If DIC availability limits phytoplankton growth, the addition of carbonte-based materials to softwater lakes may stimulate production.

Furthermore, base addition may increase nutrient availability by removing P complexing agents such as Al from the water column. At Woods Lake and Cranberry pond, a shift in the speciation of Al from acid-soluble to monomeric forms was observed following treatment (Driscoll et al., 1989). Monomeric Al concentrations were reduced by an order of magnitude within 6 to 8 weeks.

At Woods Lake, chlorophyll levels were higher following base addition and phytoplankton production exhibited a slight increasing trend throughout the post-treatment period. At Cranberry Pond, changes in phytoplankton abundance corresponded to changes in lake acidity, however, rates of production did not. Although base addition results in increased

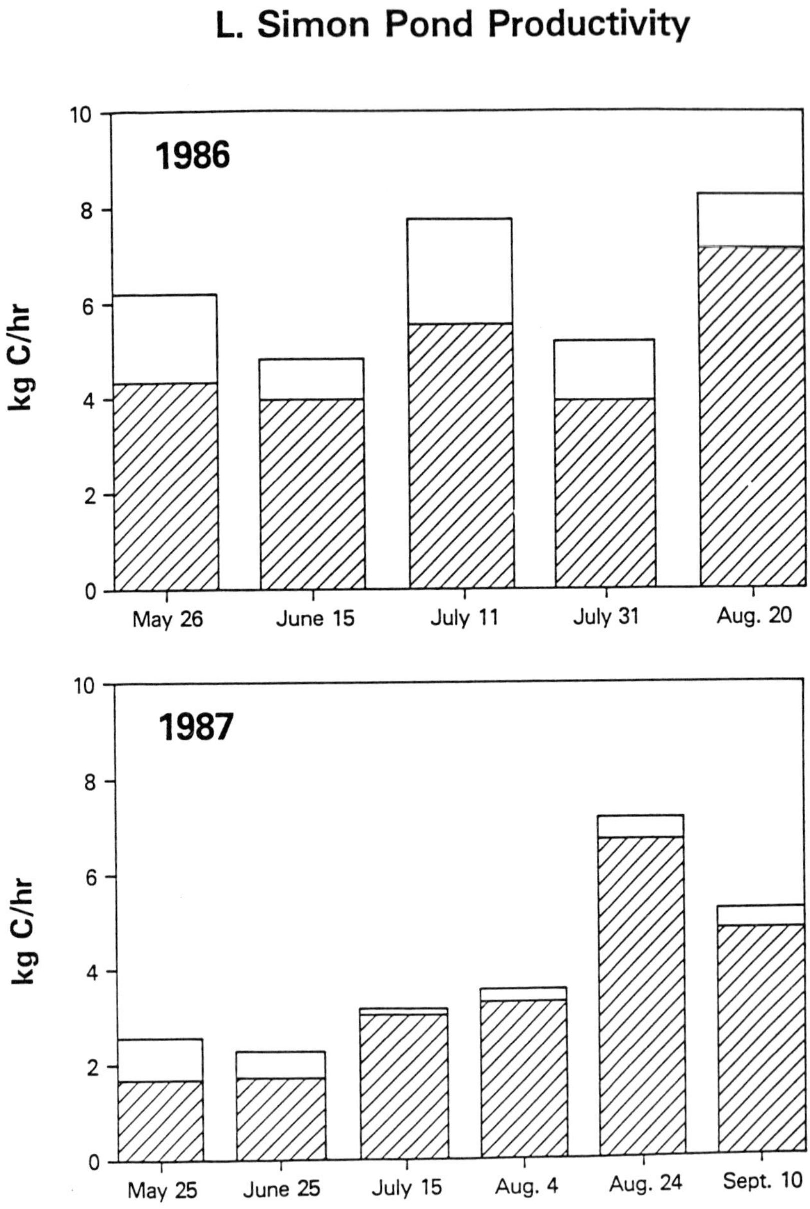

Figure 3: Seasonal pattern of phytoplankton production (as kg C hr-1) at L. Simon during 1986-1987.

phytoplankton abundance, it is unclear whether changes in phytoplankton abundance, it is unclear whether changes in phytoplankton standing crop reflect enhanced growth rates or reduced loss rates. Changes in loss rates are likely to accompany shifts in zooplankton species compositon following lake treatment. Densities of dominant zooplankton taxa decreased by 90% within one week after base application to Cranberry Pond and Woods Lake (Schaffner, 1989). The decline of the zooplankton community probably contributed significantly to the large increase in phytoplankton abundance observed at both sites in July of 1985 (Bukaveckas, 1989). Reductions in phytoplankton standing crop during late August corresponded to the recovery of the zooplankton assemblage and to the appearance of *Daphnia catawba*. *Daphnia Catawba* occurred at both sizes in 1985 and 1986, although it had not been present prior to treatment. In Cranberry Pond, densities of *Daphnia catawaba* were lower in 1986 compared to 1985; however, the abundance of *Bosmina longirostris* increased markedly during the summer of 1986. Greater densities of cladocerans would result in increased grazing pressure and could account for reduced chlorophyll levels in 1986, despite increased production rates.

At L. Simon Pond, chlorophyll levels were lower during the first summer after treatment. Reduced chlorophyll concentrations reflect the absence of the spring phytoplankton peak during 1987 and a decrease in phytoplankton growth in the lower water column. Eriksson et al. (1982; 1983) report that base addition in Swedish lakes had little effects on phytoplankton abundance until P concentrations in the water column had increased. In many lakes, no effect on biomass was observed during the first summer after treatment. Following spring and fall overturn, sediment-released P was mixed into the upper water column and phytoplankton abundance increased. Data collected at the LAMP sites is consistent with Eriksson's hypothesis that the rapidity of the response by the phytoplankton community is influenced by lake mixing patterns and hydraulic retention time. Cranberry Pond and Woods Lake have relatively small hypolimnetic zones (<25% of lake volume) and short hydraulic retention times. At these sites, significant changes in phytoplankton abundance, productivity and species composition were observed within 50 days following treatment. L. Simon Pond is a deeper lake with a longer residence time and a greater proportion of lake volume below the thermocline. Biomass was reduced during the first summer after treatment and little change in species composition was observed (P. Bukaveckas, unpub. data).

However, it is important to note that the treated lakes differed not only in their morphometry but also in their pretreatment chemical characteristics. The rapid response observed at the most acidic sites may reflect greater nutrient limitation prior to treatment. DIC concentration were lower at Cranberry Pond and Woods Lake relative to L. Simon Pond. Furthermore, levels of inorganic Al were higher at these sites and may have contributerd to reduced phosphorus availability. Thus, differences in response among lakes exhibiting a size gradient may reflect the degree to which the sites had been impacted by acidification. Different mechanisms may regulate the direction and rate of response at varying levels of acidification.

Furthermore, the relative importance of these mechanisms may change with time after based addition. The rapid increase in phytoplankton abundance observed within a few months following treatment of Cranberry Pond and Woods Lake likely reflects a release from zooplankton grazing. Chlorophyll levels were lower after the zooplankton community recovered but remained above pretreatment levels. At L. Simon Pond, the reduction in phytoplankton growth during the summer after base addition is attributed to decreased water column transparency. However, the long-term importance of these mechanisms is difficult to assess since liming simultaneously affects the physical, chemical and biotic environment of primary producers. Current research at the LAMP sites utilizes experimental enclosures in which light levels, nutrient availability, metal concentrations and zooplankton abundnace may be manipulated independently of whole- lake conditions. Results from these experiments may provide further insights on the mechanisms by which changes in lake acidity regulate the growth of photosynthetic organisms.

ACKNOWLEDGEMENTS

Research was funded by a grant from the Electric Power Research Institute (Donald Porcella, Project Manager) to the author and Donald R. Whitehead. Additional support was provided by the Empire State Electric Energy Research Corporation and Living Lakes, Inc. I am especially grateful to G. Adams, T. Heffernan, C. Rodriguez and M. Robbins for assistance during the field components of this study.

AN EVALUATION OF NEW YORK STATE LAKE LIMING DATA AND THE APPLICATION OF MODELS FROM SCANDINAVIAN LAKES TO ADIRONDACK LAKES

Steven P. Gloss
Carl L. Schofield
Ruth E. Sherman

ABSTRACT

Limestone dissolution efficiencies and reacidification rates observed in ten small Adirondack Mountain lakes, treated in 1983 to 1984 as part of the Extensive Liming Study (ELS), were compared with Scandinavian model predictions of dissolution (Sverdrup and Bjerle, 1983), and reacidification (Wright, 1985). The standard deviation of predicted initial dissolution was 15.4% of the observed fraction of limestone dissolved. Model predictions of dissolution for the Scandinavian lakes were similarly within 8 to 14% of observed values. Further analysis of the ELS data indicated that of the dissolution model parameters, dose rate alone was the best predictor of initial dissolution efficiency. Dissolution rates declined exponentially with time to undetectable levels within 2 to 3 yr following treatment. Total limestone dissolution efficiencies were in the range of 17 to 59% for the ELS lakes, which are comparable to levels observed in Scandinavian treatments with similar limestone materials (26 to 64%). Analysis of data from other Adirondack lakes limed by private groups and the New York State Department of Environmental Conservation for fisheries management programs, yielded similar estimates of dissolution efficiency for calcite based materials (average 36%). However, some of these lakes which were treated with slaked lime [$Ca(OH)_2$], exhibited initial dissolution efficiencies approaching 100%. The simple two box dilution model of reacidification, satisfactorily predicted Ca loss rates in the ELS lakes, indicating the importance of hydrology (water retention time) as a factor controlling reacidification rates in these small, limed lakes. For the ELS lakes, the ratio of watershed area/lake volume satisfactorily predicted Ca loss rates ($R^2 = 0.96$) and this simplified empirical model was applied to other Adirondack lakes where inadequate water chemistry and hydrologic data were available to utilize the dilution model. Limed Adirondack lakes with mean water retention times less than 4 mo reacidified within 1 yr after treatment. Given the preponderance of acidified lakes in the Adirondack region with retention times less than this threshold value of 4 mo (approximately 80% of lakes <10 ha surface area), simple whole lake liming practices would not be adequate for maintaining water quality suitable for the support of viable fish populations in these lakes.

Water, Air, and Soil Pollution **41** (1988) 241–278.

1. INTRODUCTION

The principal mitigation measure applied to acidified surface waters has been base addition (e.g., stream and lake liming). Numerous basic materials have been utilized to neutralize acidic lakes although major operational and research programs in Scandinavia, Canada, and the United States use $CaCO_3$. This calcite is frequently referred to as agricultural limestone in the United States although $CaCO_3$ materials are available in varying degrees of purity and particle size.

Sweden currently has the world's largest operational liming program, involving the application of approximately 800,000 t of calcite yr^{-1} (Sverdrup, 1984). Programs in Norway, Canada, and the United States tend to be more research oriented and involve fewer lakes and less material (Fraser and Britt, 1982). Within the United States it is generally recognized that the Adirondack Mountains region of New York is the area most severely impacted by acid deposition. Estimates from the late 1970's suggested that over 200 lakes or approximately 3000 ha of surface area were severely impacted by acid deposition. The New York State Department of Environmental Conservation (NYSDEC) has been using lake liming as a management technique since the early 1960's (Blake, 1981).

Early applications of neutralizing materials in New York State were made primarily on naturally acidic (dystrophic) bog ponds. This program has involved over 125 treatments on approximately 60 waters (Kretser and Colquhoun, 1984). However, operational liming since 1964 has focused on waters thought to be anthropogenically acidified. Most of the early treatments utilized hydrated lime [$Ca(OH)_2$] although $CaCO_3$ was used in later treatments and is now the material of choice for the NYSDEC program. In addition to the NYSDEC efforts, liming has been conducted during the past two decades on private lands in New York where waters have been too acidic to support healthy fish populations (Flick et al., 1982). State and private waters in New York represent the most substantial data base for investigating the effectiveness of liming in the United States.

The NYSDEC programs have been described by Blake (1981) and Kretser and Colquhoun (1984) and a data compilation was published by Britt and Fraser (1983). A review of some data from the privately owned waters was provided by Flick et al. (1982). However, no systematic analysis of the existing U.S. liming data has been undertaken with respect to dissolution (dose) efficiency or the reacidification process, i.e., predicted vs measured. The utility of these data bases may be limited because of the limited pre- and post-treatment sampling conducted. Our objectives were to assess the usefulness of these data for the application of existing dissolution and reacidification models and thereby characterize the effectiveness of previous liming efforts.

The primary objective of lake liming is to maintain or change water quality conditions in lakes affected by acidic deposition in order to support healthy fish populations. The most efficient means of liming (both in cost and calcite utilization) is to achieve maximum initial lime dissolution because neutralization of acidic waters and generation of bicarbonate alkalinity is dependent on the amount of calcite dissolved. Liming only has a temporary effect on lake chemistry; as acidic runoff from the terrestrial catchment uses up the acid neutralizing capacity (ANC) and flushes out the neutralized water, reacidification occurs. If maximum dissolution is achieved (i.e., greater ANC), the rate of reacidification is decreased, increasing the time before re-liming is necessary. Residual, or undissolved calcite which reaches lake sediments, may also play a role in slowing the reacidification process, dependent upon lake morphometry and hydrology. Dissolution efficiency is dependent on the type and particle size of limestone material used, method of application, and dosage (Sverdrup and Bjerle, 1983; Sverdrup, 1984). The rate of reacidification is dependent on the lake flushing rate, the mixing regime of the lake, and the additional long-term dissolution of calcite (Wright, 1985). We examined these factors as they affect the dissolution efficiencies and rates of reacidification of several limed Adirondack Lakes. A flushing model developed by Wright (1985) which predicts the rate of reacidification based on the dilution of limed water with acidic runoff was applied and compared with actual rates. Dissolution efficiencies were estimated using Sverdrup's (1984) sinking dissolution model and compared with observed values.

2. METHODS

Three separate data sets were evaluated with regard to dissolution efficiency and reacidification rates. The fist data were from an ongoing research project on lake liming

Table I

Morphometric characteristics of ELS and ALC lakes

Lake	Estimated Surface Area (m^2)	Volume (V) (m^3)	Watershed Area (W) (m^2)	W:V	Mean Depth (m)	Maxium Depth (m)	Retention time (months)	Actual dose ($g\ m^{-3}$)	Lime Applied (t)
ELS LAKES									
Group I									
Mountain	59,720	278,743	497,500	1.78	4.67	8.53	9.0	17.2	13.25
Big Chief	12,096	53,020	195,000	3.68	4.38	11.0	4.6	34.2	5.00
Little Rock	40,657	55,105	1,795,000	32.6	1.36	2.44	0.54	32.5	5.00
High Rock	39,984	140,163	167,500	1.195	3.51	8.23	12.3	20.7	8.00
Trout	36,578	46,016	675,000	14.67	1.26	7.26	1.2	43.4	5.50
Group II									
Barto	54,843	156,313	725,200	1.121	2.85	7.32	3.1	33.5	14.45
Jones	42,663	336,507	168,000	.499	6.39	14.63	27.9	21.0	19.50
Indigo	56,784	207,750	210,500	1.013	3.66	14.2	14.2	17.9	10.25
Pocket	11,817	34,718	236,500	6.812	2.94	11.58	2.5	73.1	7.00
Silver Dollar	5,023	20,596	97,500	4.734	4.1	8.53	3.6	52.8	3.00
ALC LAKES									
Chambers	112,330	384,846	742,500	1.929	3.63	9.15	8.1	24.3	
Deer	131,661	149,000	1,750,000	11.74	1.13	4.88	1.4	152.4	
4th Bisby	245,481	704,089	1,930,000	2.74	2.90	8.23	5.8	33.8	
Jones	202,727	421,851	705,000	1.671	1.83		8.5	19.71	
Mountain	42,595	67,000	672,500	10.04	2.01	6.10	1.8	13.0	

(Extensive Lake Study - ELS) wherein 10 small acidic lakes of variable hydrology and morphometry were neutralized with calcite (Schofield et al., 1986). These data represent the most extensive information on lake system response to liming in the U.S. and were used to validate the application of the models to Adirondack waters. We then extrapolated results of these model predictions for this comprehensive data set to the more limited data available from the other two sources. Pre- and post-liming lake chemistry data (Ca, pH, and ANC), dosages, application method, and lake morphometry were obtained from the NYSDEC studies (Krester and Colquhoun, 1984) and from five private waters located on the Adirondack League Club near Old Forge, New York (D. Webster - unpublished data).

Most of the lakes and ponds evaluated in this study are located in the west-central Adirondack Mountains, New York. The area is a forest preserve, underlain by granitic bedrock, and the lakes are located at elevations above 600 m (1900 feet). Average annual precipitation is 1.27 m (50 inches). Annual evapotranspiration was estimated to be 0.51 m (20 inches) or 40% of the total annual precipitation, and average annual runoff as 0.76 (30 inches) based on a 30-yr record (1931-1960) for the Black River Basin (Waller and Ayer, 1975).

The 10 ELS lakes selected to validate the models represent a wide range of hydrologic characteristics (**Table I**). Five of these lakes were limed in the fall of 1983 (Group I), and five were limed in the fall of 1984 (Group II). Dry agricultural limestone (Eastern Rock Products, 80% $CaCO_3$, mean particle diameter 45 μm) was applied over the lake surface by fixed wing aircraft and the lakes, their tributaries and outlets were sampled approximately monthly following liming (Schofield et al., 1986).

The five Adirondack League Club (ALC) lakes were limed during the period from 1973 to 1983 using both Eastern Rock Products limestone and unspecified sources (**Table I**). Information from these lakes comprise the most comprehensive data base available on privately owned limed waters in the U.S. Methods of lime application included aerial spraying, hand spreading, boat application, shoreline applications, and on the ice. Successive applications were made at different times of the year, at different stages and on an irregular basis (**Table II**). Routine annual water quality sampling occurred during the month of August for these five lakes and was used to compare liming efficiencies and reacidification rates.

Table II

Dates and amount of lime (t) applied to ALC lakes

Chambers Lake	Deer Lake	Fourth Bisby Lake	Jones Pond	Mountain Pond
				3/73 - 10
				11/74 - 10
	6/75 - 3	7/75 - 3	7/75 - 26	11/75 - 10
6/76 - 19	6/76 - 4	6/76 - 4		6/76 - 10
				3/77 - 10
	3/78 - 5	3/78 - 10*	3/78 - 15*	3/78 - 10
	2/79 - 4	3/79 - 8**	2/79 - 5*	1/79 - 10
	4/80 - 4	1/80 - 8	2/80 - 2*	2/80 - 10
	1/81 - 4			1/81 - 5
	1/82 - 4			1/82 - 5
	2/83 - 5			1/83 - 5

* along shoreline
** bay only

Lakes treated by the NYSDEC also represented a wide range of flushing rates (0.3 to 6.0 replacement vol/yr), however, data collected for these lakes were inconsistent (**Table III**). Water quality was monitored on an irregular basis for most lakes and the frequency of sampling varied considerably between lakes. Hence, our ability to assess liming efficiencies and reacidification rates varied for each lake according to the availability of data.

Dissolution efficiencies and lake reacidification rates were determined from the budget of each lake. The Ca budget of a lake provides the easiest measure for determining lime dissolution (Wright, 1985) and Ca concentrations can readily be translated into alkalinity and pH. ELS lakes showed virtually a one to one ratio of initial Ca dissolution to carbonate alkalinity generation (Schofield et al., 1986). The amount of Ca (g) dissolved during each hydrologic period for the lake was calculated as (Wright, 1985):

$$\text{Dissolution} = [\text{Ca-out}] - [\text{Ca-in}] + d[\text{Ca}] \qquad (2)$$

where, [Ca-out] = Ca concentration in outlet * (watershed + lake surface area) * runoff (Q)

[Ca-in] = Ca concentration in tributary * watershed area * Q

d[Ca] = change in Ca concentration in the lake from one hydrologic period to the next * lake volume.

Reacidification rates were calculated by the following equation (Wright, 1985): $C_t = [\exp(Q/VOL)]^*(C_o - C_b) + C_b$ (3)

where, C_t = theoretical Ca concentrations (g m^{-3})
C_b = background Ca concentrations (g m^{-3})
C_o = initial post-liming Ca concentrations (g m^{-3})
Q = total volume of runoff (m^3)
VOL = volume of lake mixing (m^3)
Q/VOL = flushing rate

This equation estimates Ca concentrations for one hydrologic period based on the dilution of limed water from the preceding period with low-Ca runoff water.

Each year was divided into five hydrologic periods for the ELS lakes: Winter (ice cover to snowmelt), early spring (snowmelt to ice thaw), late spring (ice thaw [spring turnover] to

Table III

Summary of data available for lakes limed by NYSDEC

	% of lakes
pH measured 1 mo prior to liming*:	13
pH measured 1 mo post liming:	37
pH measured 6 mo post liming:	48
pH measured 1 yr. post liming:	65
lakes with morphological data:	55
lakes with fish stocking data	61
lakes with qualitative data on fish survival	46
lakes with quantitative data on fish survival	0
lakes treated with $Ca(OH)_2$	76

*** pH was measured more frequently than any other water chemistry parameter**

summer stratification), summer (to fall circulation) and fall (to ice cover). For the ALC lakes, only winter (ice cover) and the rest of the year were separated due to limited availability of data. Treatment of the state data varied for each lake according to the availability of data. Runoff was estimated for each hydrologic period as:

$$Q = PPCT * ET \tag{4}$$

where, Q = runoff (m)
PPCT = precipitation (m)
ET = coefficient (runoff:precipitation ratio).

Thirty-year (1931-1960) average monthly runoff:precipitation ratios for the Black River Basin (Waller and Ayer, 1975) were used to estimate ET (Equation (4)) to determine runoff from precipitation data for each hydrologic season (**Table IV**). Precipitation data were used from the Old Forge, New York meteorologic station of the National Oceanic and Atmospheric Administration.

Runoff:precipitation ratios determined for Woods Lake (Peters 1984), located in the west-central Adirondacks and based on two years of runoff data (1980-1981), were similar in pattern to the Black River data (**Table IV**). Monthly Ca concentrations for the inlets and outlets of the ELS lakes were averaged over each hydrologic period.

Whole lake Ca concentrations were estimated from concentrations measured at surface, mid-depth, and 1 m above the bottom during each hydrologic period. Calcium concentrations were volume-weighted for each strata, based upon temperature profiles at each sampling period. Reacidification rates for the surface and bottom waters were modeled separately because of spatial variations in Ca concentrations observed during thermal stratification and under ice cover.

The Sverdrup model estimates lime dissolution efficiency based on the particle size of the liming material added, the mean lake depth, pH, the fraction of the lake covered by lime, and the dose rate. Sverdrup and Bjerle (1983) have shown that the instantaneous dissolution of a sinking calcite powder can be described by the following equation:

$$X_{inst} = A \cdot dH^{+} \cdot S \cdot \sum_{i=1}^{n} \frac{d0_i}{R_i^3} \quad \frac{1}{1 + \frac{A \cdot M}{M_E \cdot P} \sum_{i=1}^{n} \cdot \frac{d0_i}{R_i^3}} \tag{5}$$

Table IV

Precipitation and runoff data (cm) from the Black River Basin and Woods Lake

Month	BLACK RIVER BASIN 1931-1960			Season	WOODS LAKE 1980-1981		
	PPCT	Runoff	R:P		PPCT	Runoff	R:P
Dec	9.78	5.97	.61		7.55	4.95	.66
Jan	8.89	6.10	.69	Winter	5.36	2.79	.52
Feb	7.87	4.83	.61		9.51	10.07	1.06
Mar	8.51	8.89	1.04		9.49	7.95	.84
Apr	5.84	14.86	2.54	Spring	9.65	11.98	1.24
May	9.39	7.87	.84		5.51	2.39	.43
June	8.38	3.94	.47		10.66	4.04	.38
July	9.78	2.92	.30	Summer	15.23	3.75	.25
Aug	8.51	2.54	.30		9.98	5.18	.52
Sep	10.29	3.05	.30		14.00	4.49	.32
Oct	10.54	4.19	.40	Fall	15.64	10.21	.65
Nov	10.16	5.72	.56		7.56	7.91	1.05

where, x = fraction of calcite dissolved
dH^+ = H^+ concentration
S = sinking depth
A = integration constant
P = fraction of lake covered by lime
M_E = neutralization equivalent mass
M^* = dosage per lake surface area
$d0_i$ = fraction of particles in an amount with mean size i
R_i = initial average radius of fraction i.

The total amount of lime dissolved until time t after liming can be approximated by the following equations (Sverdrup and Bjerle, 1983):

$$x_{tot} = 3/4\,(X_{inst}) + 1/4(1-e^{-Kht}) \tag{6}$$

where, x_{tot} = total fraction of lime dissolved
x_{inst} = instantaneous fraction of lime dissolved
e^{-Kh} = humus inhibition rate constant

The fraction released from sediments approaches zero over time, and after a 5 yr period, lime dissolution can be approximated by

$$x_{tot} = 3/4(X_{inst}) + 1/4. \tag{7}$$

Equation 7 was utilized to calculate long-term dissolution rates.

The standard deviation of the predicted fraction of lime dissolved was calculated using the following equation (Sverdrup and Bjerle, 1983):

$$\sigma = \frac{1}{n-1}\left(\sum_{i=1}^{n}\left(1 - \frac{x_o}{x_t}\right)^2\right)$$

where, n = number of observations
x_o = observed fraction of lime dissolved
x_t = theoretical fraction of lime dissolved.

3. RESULTS AND DISCUSSION

3.1 Lime Dissolution

Estimates of initial lime dissolution efficiencies for the ELS lakes compared favorably with observed values with the exception of Little Rock Pond and Trout Pond (**Figure 1**). Both Little Rock Pond and Trout Pond are shallow with high flushing rates (**Table I**). Differences in actual (unmeasured) runoff from the long-term average utilized in the model may have contributed to deviations in predicated and observed dissolution. Excluding Little Rock Pond and Trout Pond, the standard deviation of the predicted fraction of lime dissolved was 15.4% of the observed fraction dissolved. Sverdrup's and Bjerle's (1983) model predicted dissolution efficiencies in Adirondack lakes similar to those for Scandinavian lakes where theoretical predictions were within 8% (n=43, Sverdrup, 1984) and 14% (N=8, Sverdrup and Bjerle, 1983) of observed values. The instantaneous fraction of lime dissolved in these Adirondack lakes (**Table V**) was low compared to dissolution efficiencies of Swedish and Norwegian lakes limed with $CaCO_3$ where initial dissolution ranged from 14 to 42.8% (Sverdrup, 1984; Sverdrup and Bjerle, 1983).

Lake depth and dose rate were the two variables in Sverdrup and Bjerle's (1983) model (Equation (5)) which differed between ELS lakes, therefore differences in liming efficiencies between the lakes were explained by these two variables alone. Dose rate was found to be the best predictor of lime dissolution (**Figure 2**); inclusion of lake depth in the model did not contribute to the regression at $P \leq 0.05$. The resulting empirical relationship:

$$\textbf{\% Lime Dissolution} = \mathbf{17.094 - (Dose \times 0.1165)} \quad \mathbf{R^2 = 0.66} \qquad \mathbf{(8)}$$

estimated lime dissolution efficiencies to within $\pm$14.3% of the observed values for the ELS lakes. The above equation can be used to approximate initial lime dissolution efficiencies of Adirondack waters treated with agricultural limestone.

Calcite utilization in the ALC lakes was approximated using equation 8 (**Table VI**). The actual amount of lime dissolved was only calculated for two lime applications as other treatments did not have both pre- and post-liming water chemistry data. The estimated values were in reasonable agreement with the observed fraction of lime dissolved in these lakes,

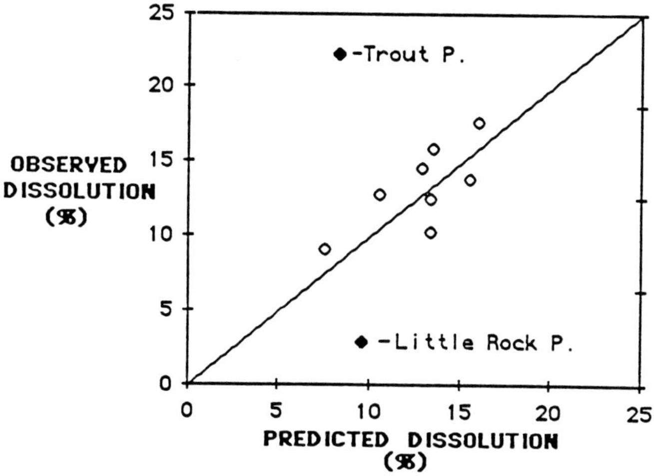

Figure 1: Comparison of observed initial dissolution rates in ELS lakes with theoretical model predictions.

Table V

Observed and predicted % lime dissolution efficiencies for ELS lakes.

	Observed initial lime dissolution	Predicted initial lime dissolution	Observed long term dissolution (1-2 yr)	Predicted long term dissolution (5 yr)
GROUP I				
Mountain	17.5	16.0	44.1	37.0
Big Chief	10.2	13.4	26.0	35.1
Little Rock	2.8	9.7	16.6	32.3
High Rock	14.5	12.9	32.6	34.7
Trout	22.1	8.3	58.9	31.2
GROUP II				
Barto	12.2	12.7	33.3	34.5
Jones	13.9	15.8	26.9	36.8
Indigo	16.3	13.7	25.7	35.3
Pocket	10.0	9.0	22.7	31.8
Silver Dollar	10.6	12.4	21.5	34.3

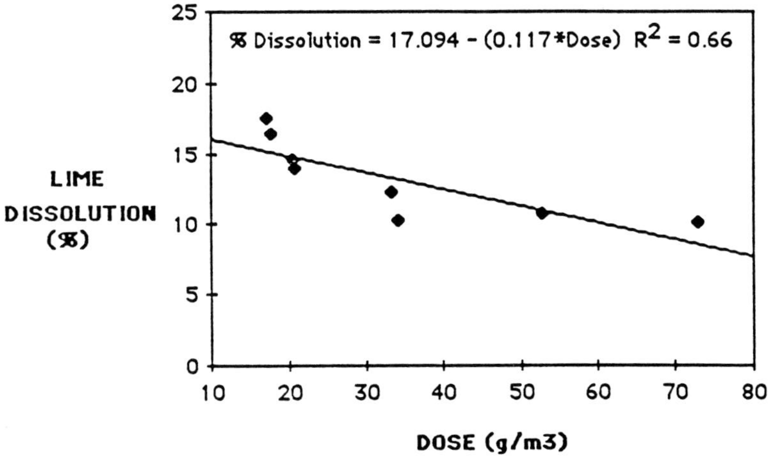

Figure 2: Relationship of initial lime dissolution efficiencies to dose rates in ELS lakes.

Table VI

Liming efficiencies of ALC lakes (%).

Lake	Observed initial dissolution	Predicted* initial dissolution	Predicted long term dissolution	Predicted % reacidif. after 1 year
Chambers	---	14.9	36.2	24.3
Deer 1	13.7	16.2	37.2	152.4
4th Bisby	---	16.6	37.5	33.8
Jones	---	14.3	35.7	19.7
Mountain	8.7	10.3	32.7	130.0

however, more data would be necessary before the accuracy of these predictions could be assessed. Lime application methods used to treat the ALC lakes differed from those used for the ELS lakes, and these differences may contribute some error to the values predicted by Equation 8 as liming efficiency depends in part on the application method (Sverdrup and Bjerle, 1983; Wright, 1985).

The NYSDEC treated lakes with hydrated lime up until the late 1970's. Due to differences in the physical and chemical properties of $Ca(OH)_2$ and $CaCO_3$, Equation (8) was applicable only to lakes treated with $CaCO_3$. Initial lime dissolution efficiencies were approximated for these NYSDEC lakes (**Table VII**); actual calcite utilization could not be calculated because of insufficient data. Dissolution was also calculated for those lakes treated with $Ca(OH)_2$ which had adequate data (**Table VIII**). The predicted percentage of lime dissolve initially, 26.9 to 107.8%, was much greater than that for lakes limed with $CaCO_3$. Hydrated lime has a smaller particle size and is more soluble than $CaCO_3$, yielding higher dissolution efficiency (Sverdrup, 1984). The long term effect of lake liming is somewhat related to continued dissolution of calcite from sediments; limestone not dissolved while sinking through the water column settles on the lake bottom forming a layer of calcite in the surficial sediments. The small fraction of lime dissolved initially in the ELS lakes treated with $CaCO_3$ indicated that the sediments were receiving a substantial dose. Sverdrup (1984) and Sverdrup and Berjle (1983) found that additional dissolution of residual limestone was minimal in most cases as calcite became buried in the sediments and was inactivated within a year or two. The Group I ELS ponds showed that residual limestone continued to dissolve over a 2 yr period (**Figures 3 to 7**) although the amount of lime dissolving during each subsequent hydrologic period decreased. The amount of residual lime dissolved averaged 64.4% and ranged from 55.5 to 83.1% of the total amount of lime dissolved over two years (**Table IX**). Dissolution of residual lime in Scandinavian lakes was much less, accounting for an average of 32.5% of the total lime dissolved over a 2 yr period (Sverdrup and Bjerle, 1983). However, the total amount of lime dissolved for the Adirondack lakes (**Table IX**) was similar to that dissolved in Scandinavian lakes which ranged from 25 to 63.5% (Sverdrup and Bjerle, 1983). Dissolution of residual limestone was also observed in the Group II ELS ponds over the first year since liming (**Figures 3 to 4**). Long-term dissolution of the lime was estimated for

Table VII

Lakes limed by the NYSDEC treated with $CaCO_3$

	Predicted Initial Dissolution %	Predicted Long Term Dissolution %
Boottree	15.4	36.6
Cooler	16.8	37.6
Deer	16.4	37.3
Deer	16.4	37.3
Townline	16.7	37.5
Brewer	14.9	36.2
Buck	16.1	37.1
Clear L	16.6	37.5
Clear P	14.9	36.2
Cleveland	13.7	35.3
Falls	15.1	36.3
Florence	15.3	36.4
Horn	14.2	35.7
Horn	14.5	35.9
Ice Mouse	15.6	36.7
Payne	14.8	36.1
Avalanche	11.4	33.6
Livingston	10.7	33.0
Tamarak	15.2	36.4
Nine Corner	15.5	36.6
Nine Corner	16.4	37.3

Table VIII

Lakes limed with $Ca(OH)_2$ by the NYSDEC

	Predicted Initial Lime Dissolution %				
	1st lime applic.	2nd lime applic.	3rd lime applic.	4th lime applic.	
W. Polliwog	96.1	85.2	77.2	40.3	
L. Echo	77.4*	54.4			
L. Egg	28.7*				
Black	59.3*				
L. Black	40.7	52.5	59.3	52.5	
NE Amphitheater		93.7	93.7	96.5	62.5
SW Amphitheater		71.7	96.4		
NW Amphitheater		39.3	51.4	62.5	
LE Copperas	39.0*				
E Copperas	47.7*				
N Whey	71.8*				
Turnoff	26.9*	16.8	62.1		
Sunday	107.8	54.1	19.7		

***based on 1st lime application**

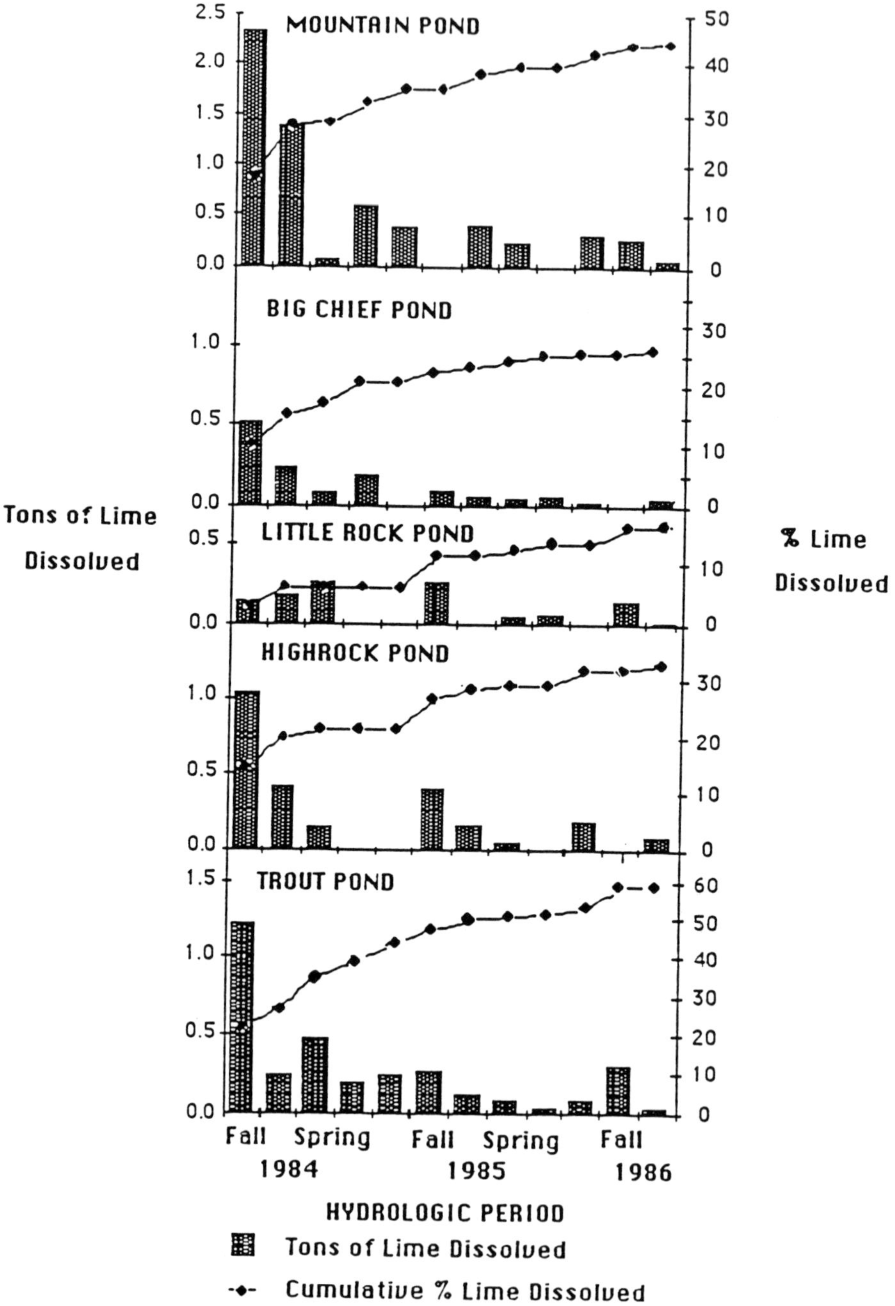

Figure 3: Cumulative % lime dissolution and t of lime dissolved.

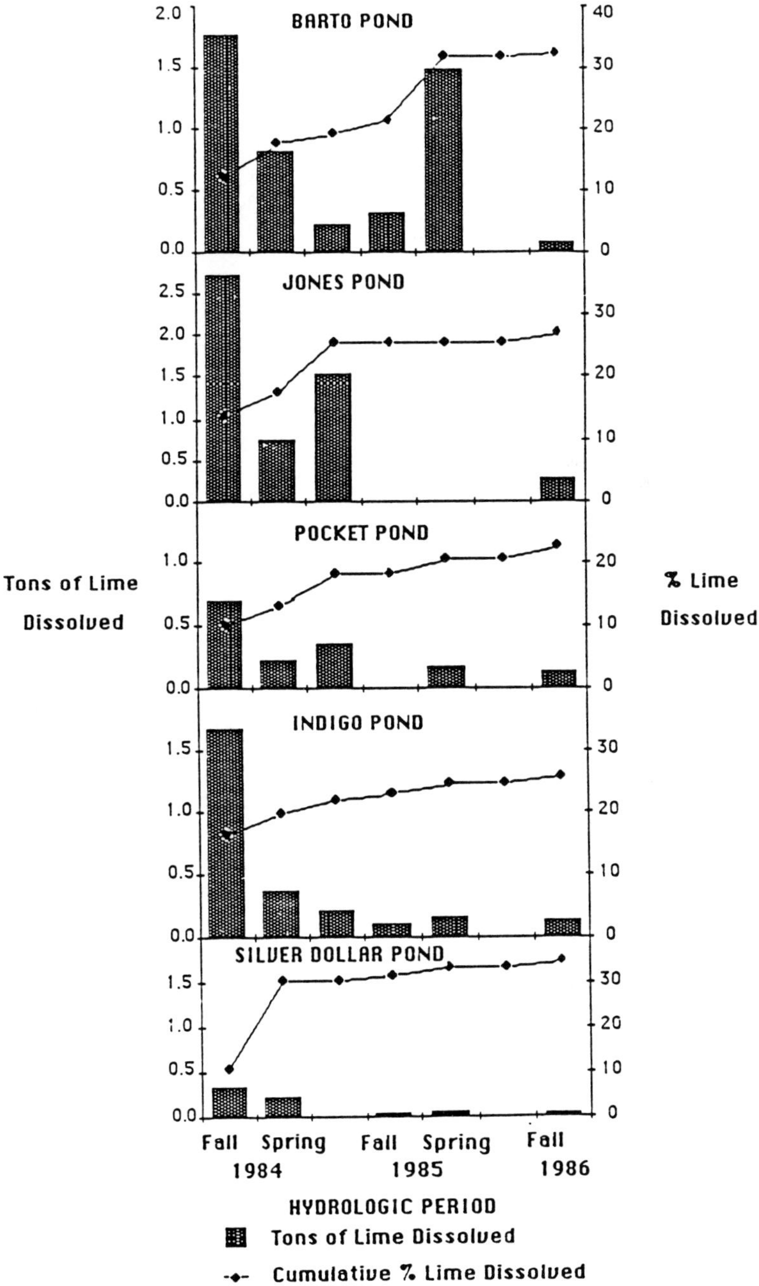

Figure 4: Cumulative % lime dissolution and t of lime dissolved.

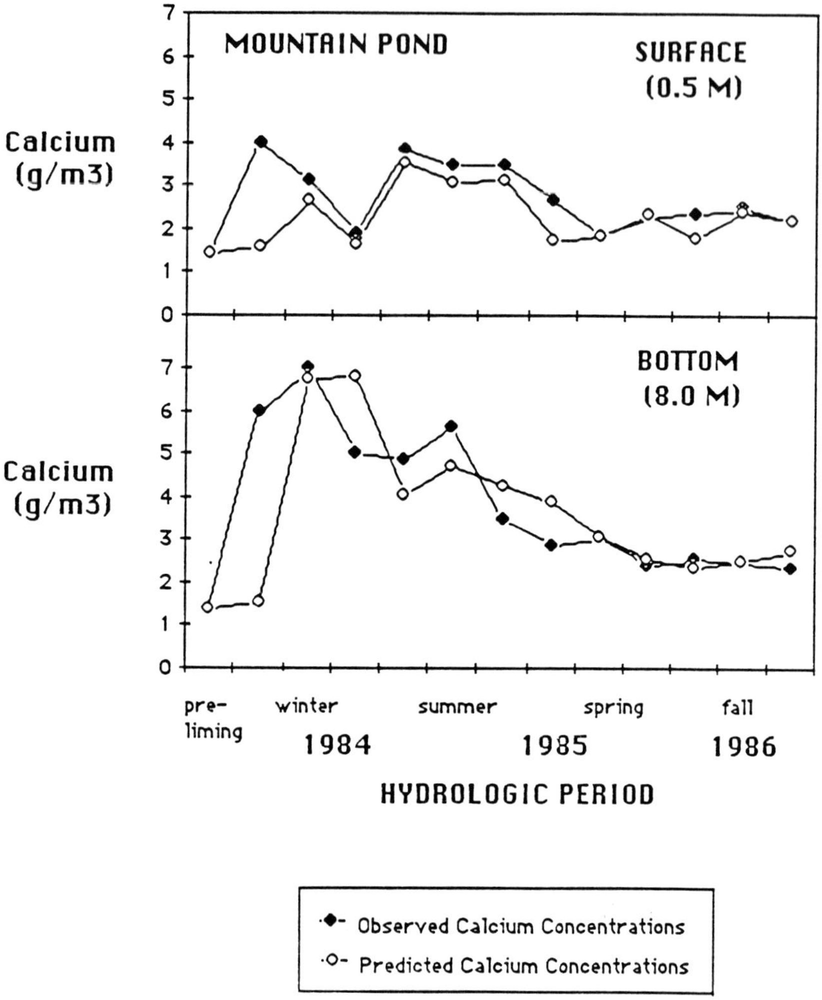

Figure 5: Observed and predicted Ca concentrations in Mountain Pond.

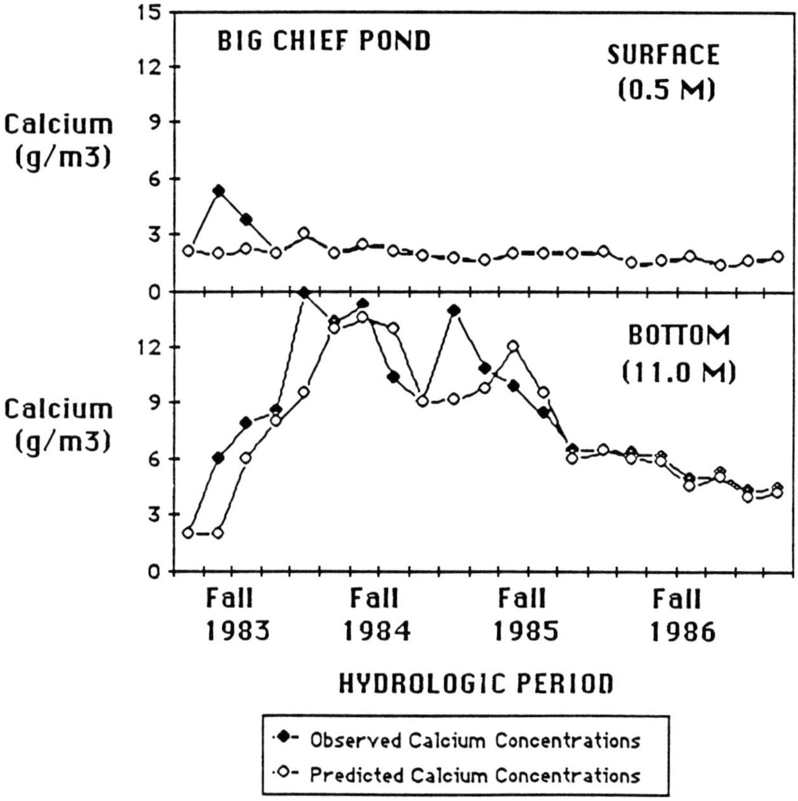

Figure 6: Observed and predicted Ca concentrations in Big Chief Pond.

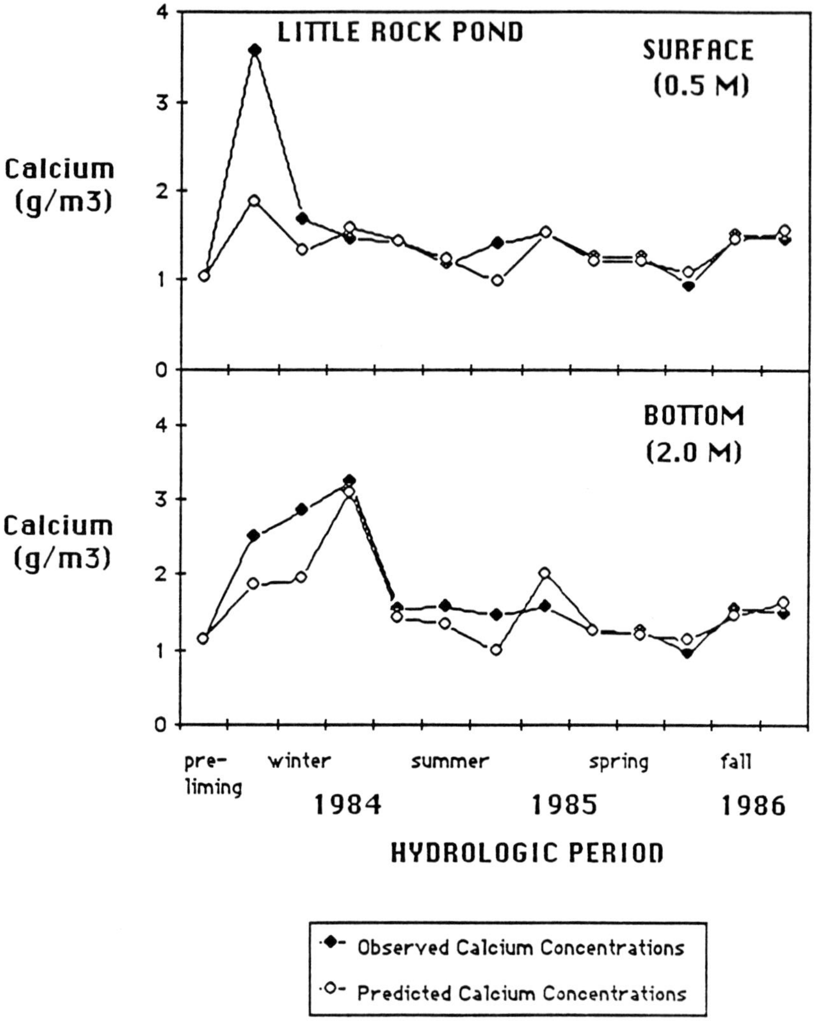

Figure 7: Observed and predicted Ca concentrations in Little Rock Pond.

Table IX

Calcite utilization efficiency for ELS lakes (%)

Lake	Dissolved Initially	Dissolved 1st year	Dissolved 2nd year	Dissolved Total	Part Dissolved from bottom
GROUP I					
Mountain	17.5	17.1	9.5	44.1	60.3
Big Chief	10.2	11.9	3.9	26.0	60.8
Little Rock	2.8	8.8	5.0	16.6	83.1
High Rock	14.5	12.0	6.1	32.6	55.5
Trout	22.1	25.5	11.3	58.9	62.5
GROUP II					
Barto	12.2	20.1	---	33.3	56.9
Jones	13.9	13.1	---	26.9	48.3
Indigo	16.3	9.4	---	25.7	36.6
Pocket	10.0	12.7	---	22.7	55.9
Silver Dollar	10.6	10.9	---	21.5	50.7

the ELS lakes using Equation (6) (**Table** V); the predicted values for the Group I ponds were similar to the observed values while the predicted values for the Group II ponds were greater than the observed values. Comparison of the observed and predicted values indicated that further lime dissolution in the Group I ponds will be minimal while lime should continue to dissolve in the Group II ponds.

The long-term dissolution of lime could not be determined for the private waters or for most of the NYSDEC lakes because of the insufficient data. Predictions of long-term lime dissolution were made using equation 6 for those lakes where the initial fraction of lime had been estimated (**Tables VI and VII**); the estimated values were similar to the long term dissolution observed in the ELS lakes. The bog ponds, limed by the NYSDEC from 1960-1963, were sampled seasonally and long-term lime dissolution could be calculated, however, there were no water chemistry data available for the tributaries and outlets for these lakes so the calculated values represent only approximations (see Equation (2)). Very little residual lime dissolved, due in part to high initial dissolution (**Table VIII**); this indicates that lime applications for these pond probably had a comparatively short-term effect on lake chemistry. Blake (1981) noted that $CaCO_3$ treated ponds remained neutralized longer than those treated with $Ca(OH)_2$.

3.2 Reacidification

The change in Ca concentrations predicted for each hydrologic period in the ELS ponds followed the same general pattern of change as the observed concentrations (**Figures 5 to 15**). The rate of reacidification varied among the lakes due to differences in the hydrologic and morphometric characteristics of each lake (**Table I**). A lake with a short retention time can *a priori* be expected to reacidify at a faster rate than a lake with a longer retention time, however, the mixing regime of a lake will also influence the rate of reacidification. The ELS lakes are primarily dimictic and are thermally stratified during the winter, early spring, and summer. During summer and winter stratification, flushing of a lake by acidic runoff only affects the epilimnion while the hypolimnetic waters remain isolated and unaffected by acidic inputs. The volume of the mixing layer varies seasonally, under ice cover and in very early

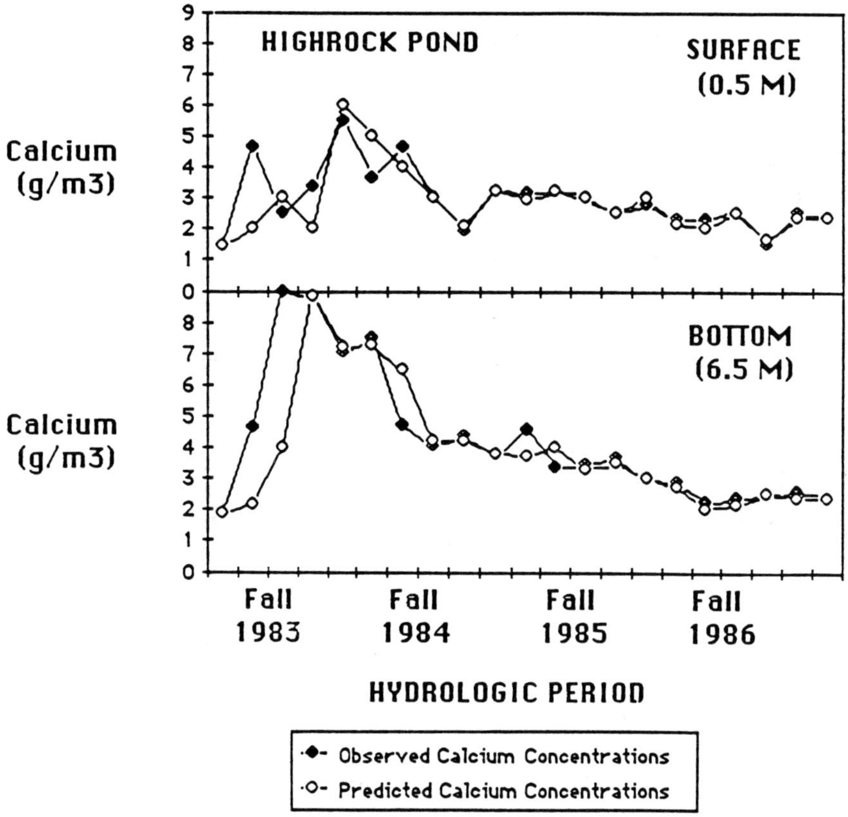

Figure 8: Observed and predicted Ca concentrations in Highrock Pond.

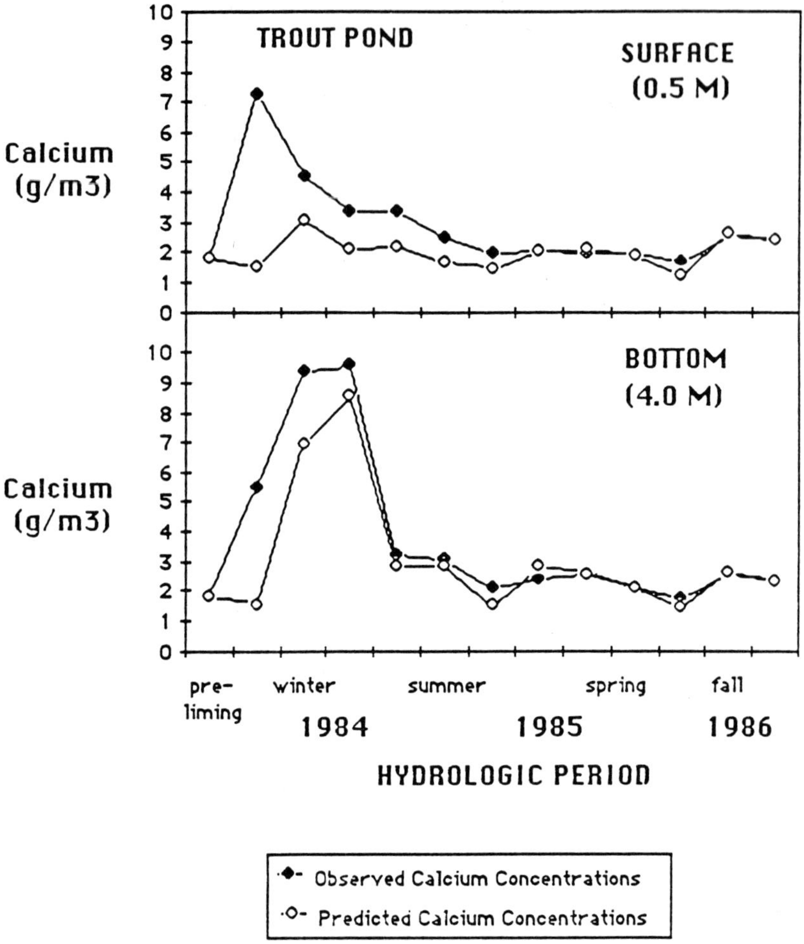

Figure 9: Observed and predicted Ca concentrations in Trout Pond.

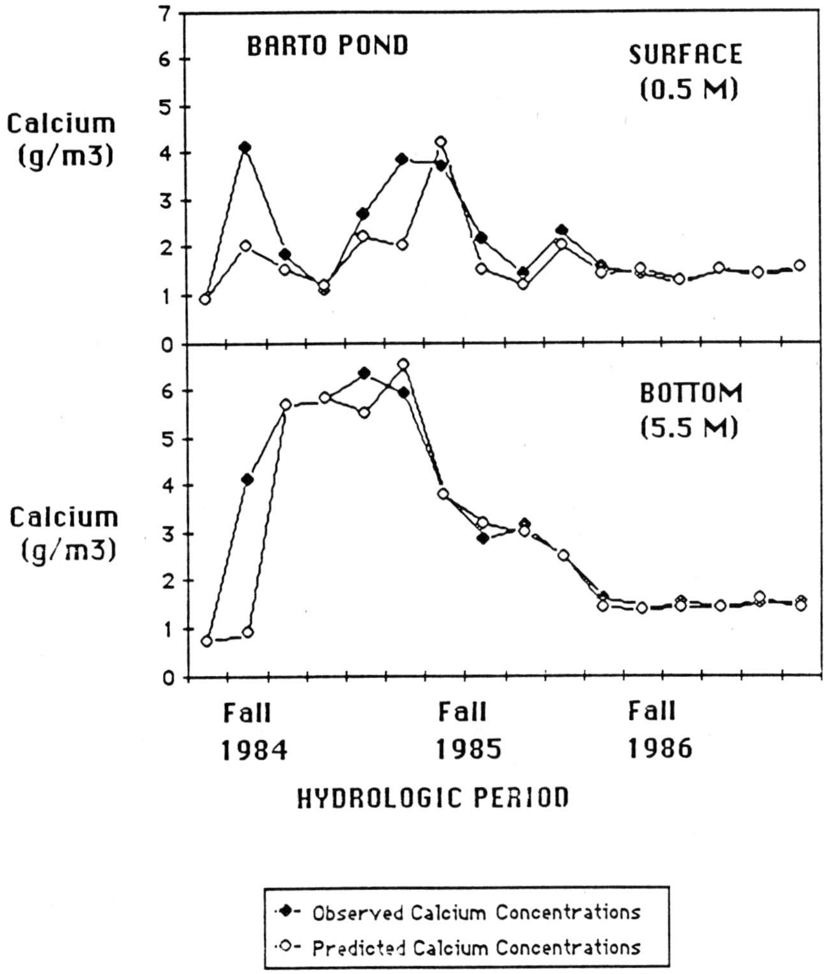

Figure 10: Observed and predicted Ca concentrations in Barto Pond.

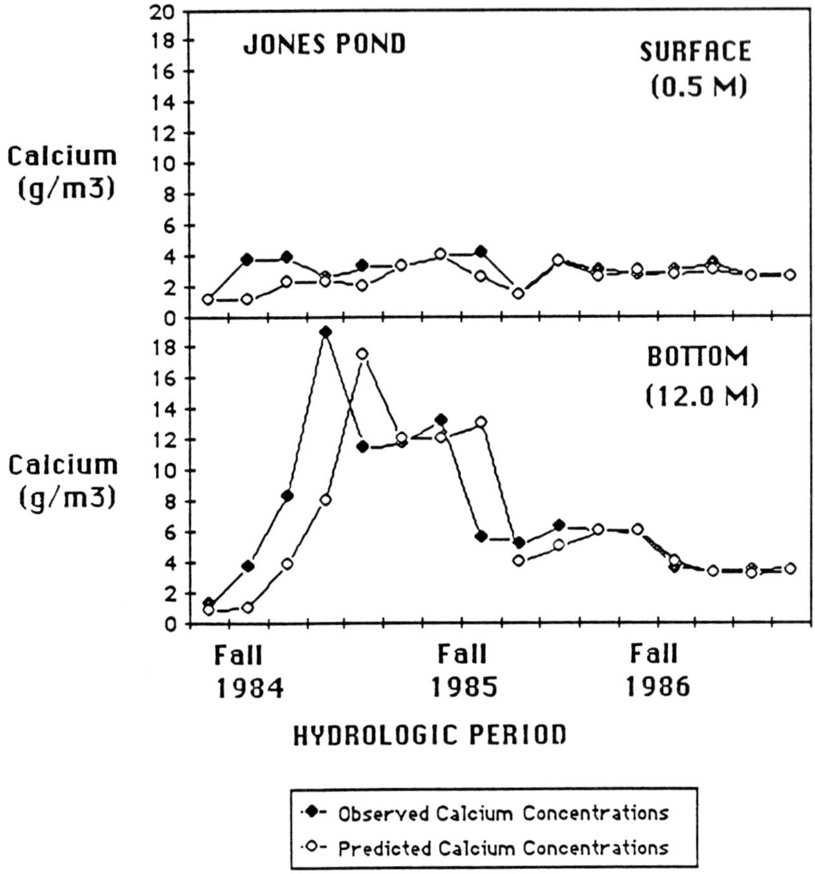

Figure 11: Observed and predicted Ca concentrations in Jones Pond.

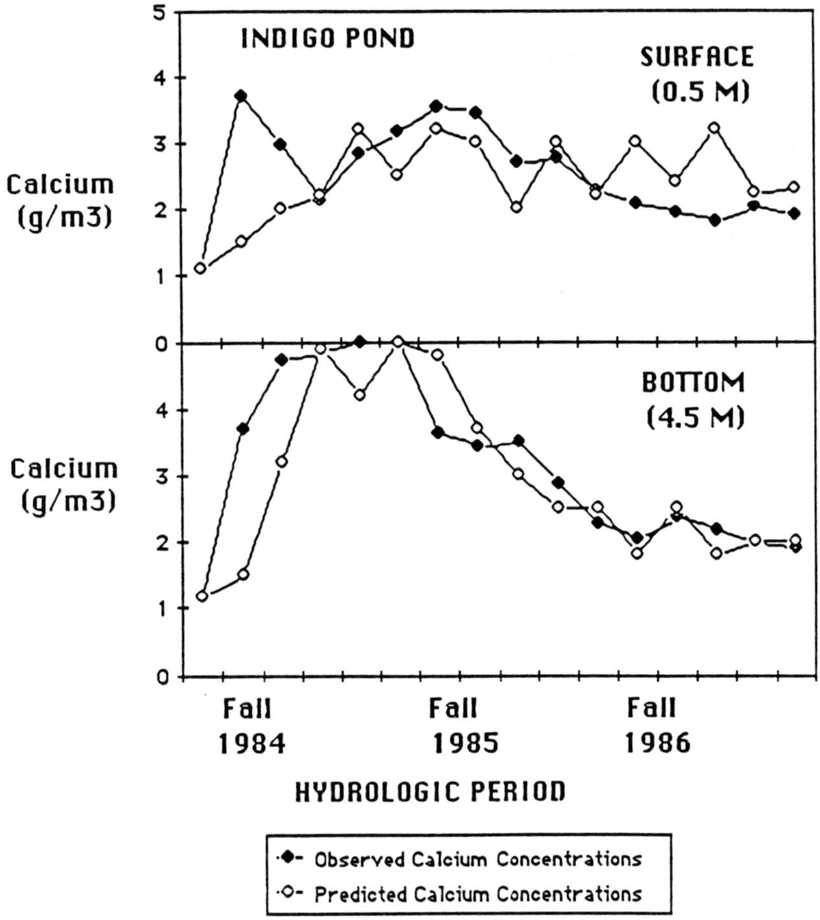

Figure 12: Observed and predicted Ca concentrations in Indigo Pond.

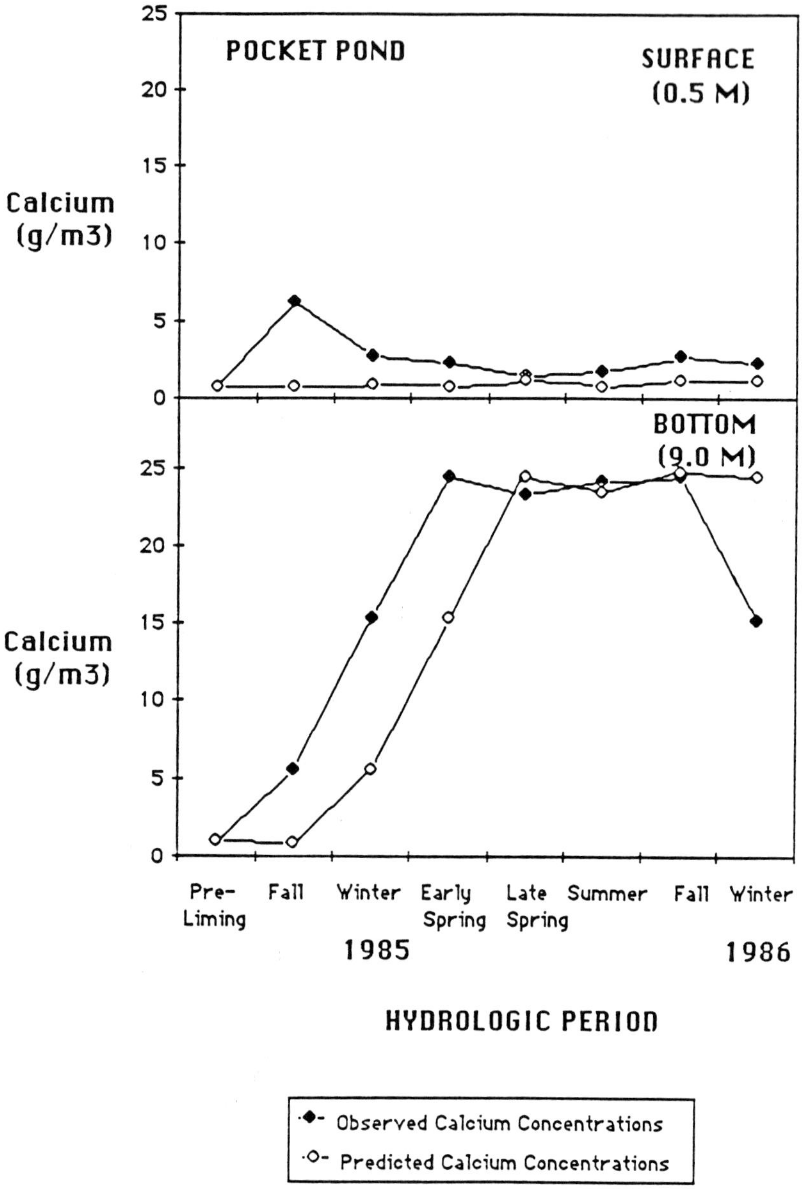

Figure 13: Observed and predicted Ca concentrations in Pocket Pond.

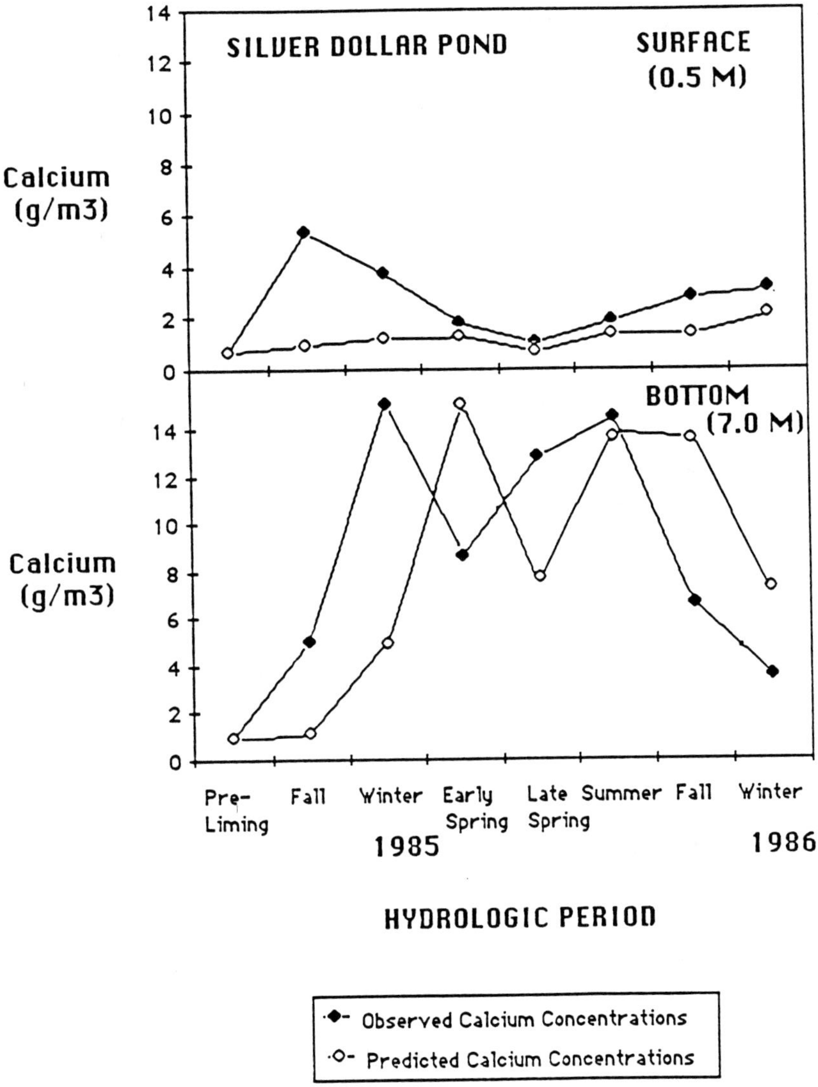

Figure 14: Observed and predicted Ca concentrations in Silver Dollar Pond.

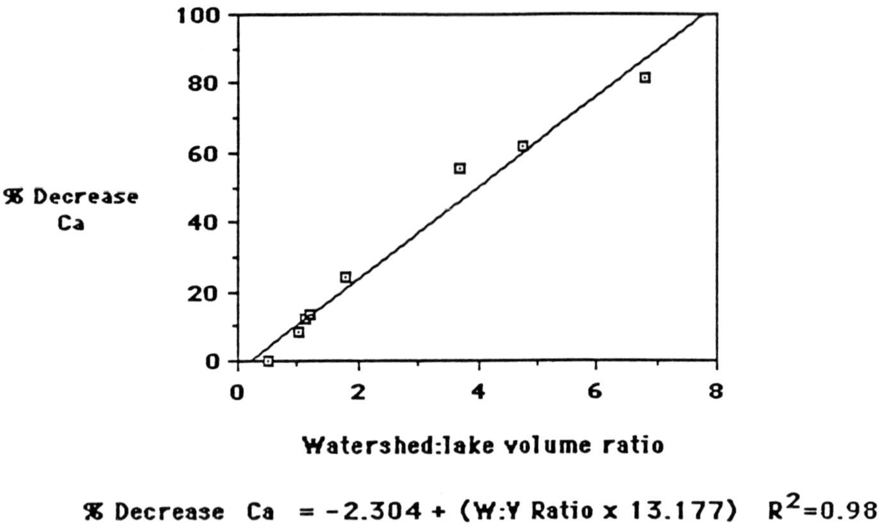

Figure 15: Relationship between % decrease in surface Ca concentrations 1 yr post liming and watershed area: lake volume ratios in ELS lakes.

spring, and only about the top meter of water mixes while during the summer the volume of the epilimnion increases (**Table X**). Differences in the volume of the mixed portion of the lake can affect the rate of reacidification. A lake with a large hypolimnetic volume can retard reacidification as the more Ca rich bottom waters mix with the Ca poor surface water (for example: Figure 5 and 8, late spring 1984) while a lake with a large epilimnion will dilute the Ca rich bottom water, enhancing reacidification (for example: Figure 7 and 9, later spring 1984; Figures 5 and 8, fall 1984). Lakes that do not always turnover completely can have acidic surface waters but alkaline bottom waters (for example: Figure 6, 11, 13, 14).

Available data were inadequate for application of Wright's (1985) dilution model to the ALC and NYSDEC lakes. A simplified dilution model, derived from the ELS data, was used to approximate reacidification in these lakes; a regression was fitted to the % decrease in surface Ca concentrations 1 yr post-liming vs the watershed area to lake volume ratio (**Figure 15**), resulting in the following equation:

$$\textbf{\% Decrease [Ca]} = -2.304 + (\textbf{W:V Ratio x 13.177}),\ \mathbf{R^2} = \mathbf{0.96} \qquad \textbf{(9)}$$

A 100% decrease in Ca concentrations indicates that a lake has reverted to pre-liming conditions. Application of this model (Equation (9)) to the ALC lakes shows that Deer Lake and Mountain Pond would reacidify within a year while the other three lakes would reacidify at a slower rate (**Table VI**). Inspection of the liming strategies employed for these lakes (**Table II**) shows the frequency of lime applications for each lake is in accordance with the rates of reacidification estimated for these lakes by the model, i.e., Deer Lake and Mountain Pond were limed on a yearly basis while the other lakes were limed less frequently. August pH levels for these ponds were, for the most part, above pH 5.5 indicating the treatment program for these lakes was effective.

Watershed areas were not readily available for many of the lakes treated by the NYSDEC nor was it certain that the above model (Equation (9)) would be applicable to lakes limed with $CA(OH)_2$; therefore, estimates of reacidification rates were not made for these lakes using

Table X

Seasonal epilimnetic and hypolimnetic volumes (m^3) of ELS lakes.

Lake	Epilim. Volume	Hypol. Volume	E:H	Epilim. Volume	Hypol. Volume	E:H
Mountain	59,720	219,023	.273	209,020	69,723	2.998
Big Chief	12,096	40,924	.296	24,192	12,096	2.000
Little Rock	40,657	14,448	2.835	40,657	14,448	2.814
High Rock	39,984	100,179	.399	119,952	20,211	5.935
Trout	36,578	9,438	3.879	36,578	9,438	3.876
Barto	54,843	101,470	.540	109,686	46,616	2.350
Jones	52,663	283,844	.186	184,320	152,196	1.210
Indigo	56,784	150,966	.376	113,568	94,261	1.210
Pocket	11,817	22,901	.516	17,725	17,016	1.040
Silver Dollar	5,023	15,573	.323	7,534	13,060	0.580

Equation (9). However, a different approach was taken to assess the effectiveness of the NYSDEC program. Fish stocking information, as available, was compiled and evaluated todetermine which treatments appeared to have been most successful in regard to fish population response. Data on fish survival were not usually complete, however, the available data suggested that survival was usually poor. Re-liming of the lakes was often done after the lake had already reacidified, thereby stressing the fish. If a limed lake is not carefully monitored and reacidification and pH decrease occur, a lake may cycle between reacidification and neutralization, with resulting repetitive pH shocks which might stress the fish population. The lack of consistent water quality monitoring data for these lakes and the poor survival of the stocked fish indicate that the liming program was not very effective.

4. CONCLUSIONS

Models developed to estimate lime dissolution efficiencies and reacidification rates in Scandinavian lakes generally were applicable to Adirondack waters. These models were validated using data from the Adirondack Extensive Liming Study and predictions of water quality changes induced by liming and reacidification were in close agreement with the observed changes. Modified versions of these models were recently applied to limed lakes in Canada with equally good results (Sverdrup, 1986). DePinto et al. (1987) applied a more sophisticated reacidification model, incorporating the interaction of sediments as sources and sinks for acid neutralizing capacity (ANC) to lakes limed as part of the LAMP study. Although sediment-water transfer of ANC was found to be of importance (particularly in shallow lakes with high ratios of sediment area/water volume), flushing of ANC from the lakes was considered to be the primary factor controlling reacidification rates. Given the importance of hydrology as a controlling factor in small Adirondack drainage lakes, it is not surprising that the relatively simple dilution model of reacidification applied in this study adequately predicted reacidification rates.

ACKNOWLEDGMENTS

Appreciation is expressed to the New York State Department of Environmental Conservation for providing data on limed state waters. Data for private waters in the Adirondacks were provided by the late Dwight A. Webster. Dan Josephson was primarily responsible for data collection and field operations in the Extensive Liming Study. Financial support for this study was provided by the U.S. Fish and Wildlife Service and the U.S. Environmental Protection Agency.

DISCLAIMER

Although the research described in this report has been funded by the U.S. Environmental Protection Agency through Interagency Agreement No. EPA-82-D-X0581 to the U.S. Fish and Wildlife Service, it has not been subjected to the Agency's peer and policy review and does not necessarily reflect the views of the Agency. Mention of trade names or commercial products does not constitute endorsement or recommendation for use by the federal government. This research was funded as part of the National Acid Precipitation Assessment Program by the U.S. Fish and Wildlife Service and the U.S. Environmental Protection Agency.

DIFFERENCES IN PRE- AND POST-TREATMENT WATER QUALITIES FOR TWENTY LIMED LAKES

Michael D. Marcus

ABSTRACT

Data analyses for pre- and post-treatment levels for 22 analytes from 20 lakes suggest that liming will, in general, increase water levels of ANC, Ca, pH, DIC, conductivity, as well as possibly NO_3 and TN, and will decrease concentrations of many metals, including AL, Zn and possibly Pb. ANC was the single analyte that best characterized pre- and post-treatment differences in these samples. The metal data from these lakes are equivocal due to the fact that most of the limed lakes in this study were only mildly acidic prior to liming and concentrations for many metals were near detection limits. Comparison of pre- and post-treatment water quality conditions in these lakes indicates that liming was successful in improving water quality conditions for fisheries.

1. INTRODUCTION

Additions of alkaline materials, especially limestone, to soft-water lakes and ponds having either circumnuetral or acidic pH levels have long been recommended to aid in establishing water qualities capable of supporting productive fisheries. During the 1940's, liming was used to enhance fisheries survival and production in acidic, dystrophic bog ponds in the midwest U.S. (Hasler et al., 1951, Waters and Ball, 1957). Also, during the 1940's and earlier, liming frequently was used to increase fish production in European and Asian culture ponds built on soils having low Ca content (Nesse, 1948). Similarly, liming has been recommended since the 1950's to increase fish production in circumneutral and acidic soft-water culture ponds in the Southeastern U.S. (Acre and Boyd, 1975). Limestone additions also have been used to improve water qualities and protect fisheries in mine drainage waters by reducing both acidities and concentrations of metals (Herricks and Cairns, 1977).

Increased interest in liming for fisheries management stems from the recent apparent spread of acidic lakes and streams as a result of acid deposition. Consequently, extensive liming research and application programs have been initiated in both northern Europe and

Water, Air, and Soil Pollution **41** (1988) 279–291.

North America (e.g., Bengtsson et al., 1980; Booth et al., 1986; Dickson 1985; Dillon et al., 1979, Porcella 1987; Schreiber et al., 1988; Watt 1986). The most extensive liming program in the United States is funded through Living Lakes, Inc. (LLI). The primary objective of this program is to demonstrate the capability of operational liming techniques in neutralizing acidic waters in lakes and streams, and in restoring and protecting ecologically, recreationally, and economically important fisheries (LLI 1987a).

Thus far, the LLI program includes nearly 30 lakes and streams in the Northeastern, mid-Atlantic, and upper mid-Western U.S. These include acidic waters treated to restore or renovate self-sustaining, put-and-take, or put-grow-and-take fisheries. This management strategy using chemical neutralization is termed "mitigative liming." Another management strategy used for some of these LLI treatment lakes and streams is "maintenance liming" where chemical treatments are used to protect unique or important biological populations (e.g., fisheries) in non-acidified surface waters considered especially susceptible to acidification.

The data analyses discussed here present pre- and post-treatment differences in water qualities for 20 limed lakes included in the LLI program. In total, statistical characteristics for 22 physical and chemical analytes are presented and discussed. Previously, relative differences for 9 analytes in 15 of these lakes were discussed by Brocksen et al. (1987). Table I lists the lakes included in the present analyses and briefly describes their general locations, lake types and treatment information. Additional detail on the specific treatments applied to each lake and on the relationships between pH, acid neutralization capacity (ANC), Ca and conductivity in these lakes are provided in a companion paper by Adams and Brocksen (1988).

2. METHODS

Analyte data discussed here were obtained via the LLI on-line database (LLI 1987b). Field collection and laboratory analysis procedures for these 22 analytes are presented (LLI 1987c) while the audit procedures used to assure and control the quality of these data are addressed in (LLI 1987d).

Table I

LLI treated surface waters included in the analyses discussed here.

State	Name	Water Type	Treatment Type	Date Treated
MA	Baker Pond	Seepage Lake	Maintenance	23 Jun 1986
MA	Curlew Pond	Seepage Lake	Maintenance	27 June 1986
MA	Lawrence Pond	Seepage Lake	Mitigative	25 June 1986
MA	Lout Pond	Seepage lake	Maintenance	27 June 1986
MA	North Pond	Drainage Lake	Maintenance	15 Sep 1986
MA	Schoolhouse Pond	Seepage Lake	Mitigative	23 June 1986
MA	Shubael Pond	Seepage lake	Maintenance	21 Nov 1986
MA	Snake Pond	Seepage lake	Maintenance	25 June 1986
MA	Wallum Lake	Drainage Lake	Maintenance	13-15 Nov 1987
MA	Round Pond	Seepage Lake	Mitigative	14 Nov 1987
MA	South Pond	Drainage Lake	Mitigative	19 Oct 1987
NY	Little Simon Pond	Drainage Lake	Mitigative	7 Aug 1986
NY	Lake Kanacto	Seepage Lake	Mitigative	9 Aug 1986
NY	Lake Tejekna	Seepage Lake	Mitigative	9 Aug 1986
NY	Woods Lake	Drainage Lake	Mitigative	16 Sep 1986
NY	Lake Madeleine	Drainage Lake	Mitigative	28 Oct 1986
NY	Barnes Lake	Seepage Lake	Mitigative	16 Oct 1987
NY	East Pond	Drainage Lake	Mitigative	31 Oct 1986
PA	White Deer Lake	Drainage Lake	Maintenance	26 Oct 1987
WV	Flat Top Reservoir	Reservoir	Maintenance	13-20 Jul 1987

Prior to data analysis, all values were screened to remove data marked in the database as failing to pass the acceptableness or reasonableness checks of the quality control program. Also, all data marked as failing to meet the instrument detection limits specified by the quality control program were converted to zero values. All other data were retained for further evaluation.

For many analytes, the database includes results produced using two or more alternate analytical methods on each individual sample, but one technique usually predominated across all lakes in the analyses for each analyte. Therefore, the following data analyses were completed using data primarily derived using the predominant analytical technique for each analyte; data derived using alternative techniques were included only to supplement data not available for samples using the predominant analytical technique. For example, Ca analyses were completed using atomic absorption (AA) spectroscopy on field filtered (FFA) and lab filtered aliquots (LFA), and using ion specific electrodes (ISE). The bulk of the Ca data used in the data analysis came from AA analysis of FFAs; to the extent possible missing FFA data were supplemented by LFA and ISE data. Such data "lumping" across analytical techniques was done for most analytes to increase the information available for analysis. This lumping includes the assumption that differences created by alternative analytical techniques are not biologically significant.

To provide an overview of pre- and post-treatment differences, the data from all 20 lakes were pooled for each analyte and analyzed using the Mann-Whitney test, but pooling pre- and post-treatment data from all lakes obscures potentially interesting differences for analytes within each lake. Therefore, the Mann-Whitney test was applied to examine pre- and post-treatment differences for each of the 22 analytes in each of the 20 lakes. This non-parametric procedure was used for these analyses, rather than its parametric equivalent t-test, to reduce computational problems posed by small sample sizes and/or by zero variances occurring for some pre- or post-treatment analyte sets for some lakes.

To further explore characteristics that define pre- and post-treatment differences in this set of study lakes, discriminant analysis was used to identify the minimum set of analytes necessary to best reclassify the 388 individually collected lake data sets correctly into pre- and post-treatment conditions. (For dichotomous classifications, each discriminant analysis produces only one discriminant function.) These analyses included both step-wise (PIN=0.050, POUT=0.051) and manual selection of analyte sets to be included in the unrotated discriminant functions computed using Mahalanobis distances. The prior probabilities of group membership were based on SIZE, i.e, the sample proportion of cases actually included in each group. Manual selection of analyte sets was undertaken to examine and to reduce the consequences of multi-collinearity in masking the possible influences among variables in deriving the discriminant reclassification equations. Standardized canonical discriminant function coefficients from three of the analyses are present; these coefficients are analogous to coefficients for independent variables in standardized regression equations.

All data analyses reported here were completed on a Compaq Deskpro 386 using SPSS/PC+ (Norusis 1986a,b). Determination of significant differences for all analyses was based on a $p \leq 0.05$ level.

3. RESULTS

The means, ranges and number of pre- and post-treatment samples collected for all 22 analytes from the 20 LLI study lakes included are shown in Table II. These results show that only levels of ANC, Ca, conductivity, dissolved inorganic carbon (DIC), pH, and total nitrogen (TN) were significantly greater, while Zn was significantly less in the post-treatment samples.

Results from the Mann-Whitney tests for pre- and post-treatment differences for each analyte in each lake are shown in Table III. These results are summarized in Table IV. With some exceptions (shown in the tables), levels for ANC, Ca, conductivity, DIC and pH were greater in the post-liming samples; while levels for Al, Cd, Cl, Fe, K, Mg, Mn, Na, NH_3, SiO_2, SO_4, total phosphate (TP) and Zn tended to be similar in both pre- and post-treatment samples for 11 of 20 lakes, less in one, and similar in eight lakes. For Pb, the concentration differences remained about equally distributed across the three categories, with six lakes showing greater, six showing lesser, and eight showing similar concentrations in post-versus pre-treatment samples.

Table II

Mean, ranges, and numbers (N) of samples for analytes before and after liming of the LLI study lakes.[1]

	Pre-treatment Analytes			Post-treatment Analytes		
	Means	Range	N	Means	Range	N
Al	0.04	(0.00 to 0.20)	112	0.04	(0.00 to 0.18	234
ANC*	25.38	(-35.40 to 90.40)	122	193.30	(1.20 to 638.90)	231
Ca*	1.78	(0.43 to 3.50)	115	5.09	(1.49 to 13.33)	236
Cd	0.00	(0.00 to 0.00)	112	0.00	(0.00 to 0.00)	117
Cl	5.98	(0.20 to 20.76)	15	21.02	(0.21 to 70.90)	36
COND*	39.49	(0.00 to 97.40)	130	64.05	(13.80 to 278.00)	222
DIC*	1.62	(0.00 to 9.11)	137	2.91	(0.00 to 13.30)	216
DOC	2.96	(0.91 to 11.86)	128	2.91	(0.00 to 13.43)	224
Fe	0.03	(0.00 to 0.83)	113	0.05	(0.00 to 1.23)	112
K	0.51	(0.24 to 1.05)	16	0.45	(0.00 to 0.96)	26
Mg	0.95	(0.09 to 2.40)	28	1.10	(0.14 to 2.22)	32
Mn	0.06	(0.00 to 2.70)	113	0.03	(0.00 to 1.45)	105
Na	3.38	(0.13 to 9.63)	15	12.33	(0.08 to 48.80)	34
NH_4	0.01	(0.00 to 0.11)	15	0.04	(0.00 to 0.32)	27
NO_3	0.33	(0.00 to 1.71)	112	0.35	(0.00 to 1.90)	117
Pb	0.01	(0.00 to 0.03)	113	0.01	(0.00 to 0.03)	117
pH*	5.51	(4.44 to 6.99)	128	6.62	(5.40 to 8.63)	226
SiO_2	1.60	(0.00 to 6.65)	14	1.82	(0.00 to 6.49)	24
SO_4	5.67	(0.00 to 8.29)	112	5.58	(0.62 to 7.61)	117
TN*	0.11	(0.00 to 1.50)	112	0.61	(0.00 to 82.14)	215
TP	0.00	(0.00 to 0.05)	112	0.00	(0.00 to 006)	217
Zn*	0.03	(0.00 to 0.21)	113	0.00	(0.00 to 0.09)	118

1All units in mg/l except COND (uS/cm) and pH.

*Significant difference ($p \leq 0.05$) between pre- and post-liming means for the analyte based on the Mann-Whitney U test.

Table III.

Mann-Whitney test results showing post-liming analysts that were significantly greater (G), lesser (L), and similar (0) relative to pre-liming samples.

Analyte	Baker Pond	Curlew Pond	Lawrence Pond	Lout Pond	North Pond	Schoolhouse Pond	Shubaul Pond	Snake Pond	Wallum Pond	Round Pond
Al	O	G	O	G	O	O	O	O	O	O
ANC	G	G	G	G	G	G	G	G	G	O
Ca	G	G	G	G	G	G	G	G	G	O
Cd	O	O	O	O	O	O	O	O	O	O
Cl		O	O	O	O	O	O	O	O	
Cond	G	G	G	G	G	G	G	G	G	O
DIC	G	G	G	G	G	G	G	G	O	O
DOC	O	G	O	O	O	O	O	O	G	O
Fe	O	O	O	O	O	O	O	O	O	O
K		O	O	O	O	O	O	O	O	
Mg		O	O	O	O	O	O	O	O	
Mn	O	O	O	O	O	O	O	O	O	O
Na		O	O	O	O	O	O	O	O	
NH_4		O	O	O			O	O	O	
NO_3	O	G	O	O	O	O	O	O	O	O
Pb	G	G	G	G	O	G	O	G	O	O
pH	G	G	G	G	G	G	O	G	G	O
SiO_2		O	O	O	O		O		O	
SO_4	L	L	O	O	L	O	O	L	O	O
TN	G	G	O	O	G	G	G	G	G	O
TP	O	O	O	O	O	O	O	O	G	O
Zn	O	O	O	G	O	O	L	O	O	O

Table III. (con't)

Analyte	South Pond	Little Simon Pond	Lake Kanacto	Lake Tejekna	Woods Lake	Lake Madeleine	Barnes Lake	East Pond	White Dear Lake	Flat Top Reservoir
Al	O	L	L	L	O	O	O	O	L	O
ANC	O	G	G	G	G	G	O	G		
Ca	O	G	G	G	G	G	O	G		G
Cd	O	O	O	O	O	O	O	O	O	O
Cl	O	O					O	O	G	
Cond	O	G	G	L	G	G	O	G	L	
DIC	O	G	O	O	O	O	O	G	O	L
DOC	O	G	G	G	O	O	O	L	O	G
Fe	O	O	O	O	O	O	O	G	O	O
K	O	O					O	O	O	O
Mg	O	O					O	O	O	O
Mn	O	O	O	O	O	O	O	O	O	O
Na	O	O					O	O	O	
NH_4	O	O					O	O	O	O
NO_3	O	O	O	O	O	O	O	O	G	G
Pb	O	L	L	L	L	L	O	O	O	L
pH	O	G	G	G	G	G	O	G		O
SiO_2	O	O					O	O	O	O
SO_4	O	L	L	O	O	G	O	O	O	O
TN	O	G	O	G	G	L	O	O	G	O
TP	O	O	O	O	O	O	O	O	G	O
Zn	O	L	L	L	O	O	O	O	O	O

Table IV

Summary of Mann-Whitney results from Table III.

ANALYTE	TOTAL COMPLETED TESTS	TOTAL NUMBER GREATER	TOTAL NUMBER LESSER	TOTAL NUMBER SIMILAR
Al	20	2	4	14
ANC	19	15	0	4
Ca	20	16	0	4
Cd	20	0	0	20
Cl	14	1	0	13
Cond	19	14	2	3
DIC	20	10	1	9
DOC	20	6	1	13
Fe	20	1	0	19
K	15	0	0	15
Mg	16	0	0	16
Mn	20	0	0	20
Na	14	0	0	14
NH_4	14	0	0	14
NO_3	20	3	0	17
Pb	20	6	6	8
pH	19	14	0	5
SiO_2	12	0	0	12
SO_4	20	1	6	13
TN	20	11	1	8
TP	20	2	0	18
Zn	20	1	4	15

Including all analytes shown in Table II in step-wise discriminant analyses resulted in selection of ANC, Pb, DOC, Zn and NO_3 as the variables producing the one function discriminant equation, one which yielded a 80.7% correct reclassification of the lake data into pre- and post-treatment groups (Analysis 1). But because this analysis included even those analytes having relatively few measurements (see Table II), the fit of this function to the overall data set was limited; 356 of the 388 cases were excluded during analysis because one or more discriminating variables were missing. Therefore, for the next step-wise analyses, Cl, K, Mg, Na, NH_4 and SiO_2 were excluded. In this analysis, where 211 of the 388 cases were excluded because of missing discriminating variables, step-wise discrimination selected ANC, DOC, NO_3, pH, SO_4, and TP as the analytes producing the function having the best fit; one producing 88.1% correct reclassification of all 388 cases. Including only these six analytes in the next analyses allowed an additional 26 cases to be included in developing the fit to the function, which produced 89.2% correct reclassification (Table V, Analysis 2).

Repeated discriminate analyses using all combinations of these six variables showed that the greatest correct reclassification of cases (90.5%) was produced when using ANC, pH, TP and NO_3 (Table V, Analysis 3). Interestingly, using ANC, Ca, DIC, pH and conductivity each individually in discriminant analysis resulted in functions producing 89.4%, 85.3%, 84.8%, 80.4%, and 64.7% correct reclassifications, respectively.

4. DISCUSSION

The preceding results must be considered with the understanding that the majority of the liming projects under LLI sponsorship are conducted as applied management activities (LLI 1987a). Consequently, data are collected during this program with the intent to monitor project effectiveness and to avoid creating deleterious conditions; they are not collected with the intent to test hypotheses or to evaluate scientific and technical issues. Despite this, these data can help in understanding the overall consequences of liming.

The data analyses indicate clearly that ANC, Ca, conductivity, DIC and pH are greater in the post-liming samples, agreeing with the conclusions of Adams and Brocksen (1988). Increases in these analytes should be expected with liming.

Middleton and Rovers (1976) have argued that averaging pH measurements themselves is mathematically equivalent to calculating the geometric mean for the corresponding hydrogen ion concentrations; they suggested that such values better reflect the typical ("average") hydrogen ion concentrations to which resident aquatic organisms are exposed. In contrast,

Table V

Standardized canonical discriminant function (DF) coefficients for the first DF from the analyses, with percentages of correct reclassification.

ANALYTE	DISCRIMINANT ANALYSIS		
	1	2	3
ANC	1.463	0.941	0.952
DOC	-0.983	-0.317	
NO^3	0.691	0.611	0.270
Pb	0.676		
pH		0.429	0.274
SO_4		-0.428	
TP		-0.317	-0.305
Zn	-0.700		
% Correct	80.7	89.1	90.5

they suggest that computing average pH based on H^+ concentrations that correspond to measured pH values, yields a value that reflects more the acidity extremes to which the organisms are exposed. In analyzing the LLI data, the principal concern was whether the treatments were adequate to protect against the more extreme acidic conditions, thus calculations showing "average pH" were based actually on mean hydrogen ion concentrations. Those limed lakes not showing a significant difference in pH following liming (1) were treated under a maintenance liming strategy, so that their treatment pH changes were relatively slight (Shubael Pond, Flat Top Reservoir); and/or (2) had small sample sizes, limiting the power of the pre- to post-treatment comparisons (Round Pond, Wallum Lake, Barnes Lake).

The data analyses also indicated that other chemical changes can accompany liming. Specifically, NO_3, TN and DOC tended to be greater, while SO_4, Zn and other metals, including Al, tended to be lesser in the post-liming samples from some lakes. Some of the apparent change in N concentration in some LLI lakes may relate to poor analytical precision during the initial periods of this project. But, other researchers have similarly found concentrations for some forms of N to be greater in some lakes and lesser in others following liming (Dillon et al., 1979). These researchers also indicated that relative proportions of total inorganic N decreased in the total N concentrations following liming, probably a result of the increased biological productivity in those lakes. Increased biological productivity may also account for the greater post-treatment DOC found in six of the LLI lakes.

The solubilities of most metals decrease rapidly with increasing pH (Drever 1982). Therefore, decreases in dissolved metals concentrations should be expected to follow lake liming, as has been observed in earlier studies (Dillon et al., 1979). The non-significant pre- and post-treatment differences in metal concentrations for most LLI lakes may be a consequence of the fact that pH in most of these lakes was only moderately acid prior to liming, and the metals present were at concentrations near their analytical detection limits. Also, a problem with Zn-contaminated filters affected results for some of the pre-liming samples and excluded these results from these data analyses.

As with metals, SO_4 solubilities also decrease with increasing pH, but all concentrations reported from the LLI studies were well within the solubility limits. Therefore, the slight differences occurring between pre- and post-treatment levels may be due solely to temporal variation.

Discriminant analyses suggested that pre- and post-treatment differences in ANC, pH, NO_3 and TP contribute slightly to improving the differentiating ability between pre- and post-treatment samples from these lakes, which further suggests that changes in these analytes may relate to liming. Differences in the first three of these analytes have already been discussed, and the Mann-Whitney test indicated that possible increases in TP occurred in only two of the 20 lakes. The slight increases noted in TP for these two sample sets, however, may have been only a result of low analytical precision with the pre-treatment samples from these waters.

The discriminant analyses also suggested that of the five analytes most correlated with liming, ANC is best able to distinguish between pre- and post-treatment samples, while conductivity is least able to discriminate. In fact, ANC alone appears approximately equal to any combination of analytes in discriminating between pre- and post-treatment conditions for samples from lakes and ponds in this data set. This can be at least partly explained by the significant correlation of ANC with Ca ($r = 0.92$), DIC ($r = 0.71$), pH ($r = 0.68$), Zn ($r = -0.34$), conductivity ($r = 0.30$), TP ($r = 0.20$), and DOC ($r = 0.14$).

In conclusion, the analyses discussed here suggest that liming will, in general, increase water levels of ANC, Ca, pH, DIC, conductivity, as well as possibly NO_3 and TN, while decreasing concentration of metals, as suggested in these data by Al, ZN and possibly Pb. ANC was the single analyte apparently best characterizing pre- and post-treatment differences in these samples. The metal data from these lakes are equivocal due to the fact that most of the limed lakes in this study were only mildly acidic prior to liming, and concentrations for many metals were near detection limits. Questions remain, however, on the potential fates and effects of these metals once they are removed from the water column.

Finally, comparison of pre- and post-treatment water quality conditions known or suspected to be detrimental to fisheries (Marcus et al., 1986) indicated that this liming program has been successful in improving water quality conditions for fisheries.

ACKNOWLEDGMENTS

This work has been financially supported by Living Lakes, Inc. I thank R. Brocksen, T. Adams and D. Gulley for their continued contributions in helping me to understand the intricacies in these data.

EVALUATION OF LIMING OPERATIONS THROUGH BENEFIT-COST ANALYSIS

Patricia Lavin Riely
David B. Rockland

ABSTRACT

This paper presents a methodology for assessing the benefits and costs of liming acidified waters. Public decision makers generally have to choose among a variety of projects for public investment. Benefit-cost analysis offers one set of tools to assist the decision maker. This paper provides the benefit-cost tool to the decision maker to allow him or her to understand whether a liming project is justified from an economics perspective. The rudiments of benefit-cost analysis are first derived with particular attention paid to issues such as accounting stance, discounting, non-market benefit estimation and computation of applicable economic costs. The paper also provides detailed information on the costs associated with liming acidified waters and the appropriate methods by which benefits are to be computed. A hypothetical examples is provided to assist the reader in understanding the application of benefit-cost analysis to liming projects. Available data sources for future analyses are described for seven states and the federal government.

1. INTRODUCTION

Acidified waters can be neutralized by the application of base materials (a procedure called liming). Liming has been successfully used in Europe and North America to reduce the acid level of lakes and streams and enable water systems to support aquatic populations. This study will provide analysts and decision makers with a method of evaluating a stream or a lake liming project from an economic perspective.

Several basic economic principles are presented in Section 1 and the theory behind benefit-cost analysis is explained. Section 2 provides an overview of identification and valuation of economic benefits associated with a liming project. The costs of such a project are identified in Section 3. Hypothetical benefit-costs analyses are presented in Section 4. A step-by-step format is used in order to clearly present the economic procedure. Section 5

Water, Air, and Soil Pollution **41** (1988) 293–328.

identifies available state and federal databases and provides information on the collection o' primary data when sufficient databases are not available.

?. THE THEORY BEHIND THE BENEFIT-COST ANALYSIS

Economics is defined as the study of the allocation of scarce resources among competing users. The branch of economics known as welfare economics is concerned with the allocation of those scarce resources so as to maximize the welfare of society. The term "welfare" does not refer to income support programs, but is a collection of measures economists use to evaluate benefits or costs to society. Benefit-cost analysis is a tool used by economists to assist decision makers in distinguishing between the projects, programs, or activities which will make society better off and those that will make it worse off. Benefit-cost analysis is applied welfare economics (Anderson and Settle, 1979).

An overview of the theory behind the benefit-cost analysis can be best provided through a simple example. Suppose you are a fisheries professional and are considering recommending that a public body of water be limed to create a new sport fishery. Before undertaking the project, decision makers would want to compare the expected benefits of the project to the costs, to determine if it would be a worthwhile endeavor. The cost of the liming project would be financed by general funds which are derived mainly from tax revenues. The economic benefits stemming from the liming project would include income to local businesses, tax revenues and the value of the sport fishery that resulted from the improved water quality.

The difference between the cost of the liming operation and the tax revenues returned to the government coffers due to the economic activity generated by the sport fishery, is the net cost to the government. However, sport fishing, itself, has an economic value that is distinct from how much income or tax revenues the sport fishery generates. Thus, the difference between this economic value (i.e., benefits) and the expenditure of general funds (i.e., cost) is a more appropriate net benefit since it compares the cost of producing the product (i.e., sport fishing) to the value of that product. The decision makers should then use this information to determine whether the society is made better off by the project.

2.1 When is Society Made Better Off?

Economists use two distinct criteria to determine whether a society is better off from a project. These criteria are: 1) economic efficiency; and (2) income-distributional equity. Economists look to increases or decreases in the value of goods and services in the economy, as a whole, when evaluating the economic efficiency of a project, program, or activity. For example, economists look at the economic value of sport fishing relative to the cost of liming when economic efficiency criteria are used. The value of goods and services is defined in terms of society's "willingness to pay." Willingness to pay for goods and services is revealed by the demand for these items. The economic cost of a project is defined as value of the goods and services that could be rendered if the resources to be employed by the project were instead used in their next highest valued use (Ibid.) This cost is also termed by economists as the "opportunity cost."

The income-distributed equity criterion is often less clear-cut than economic efficiency. A project that appears to benefit society in terms of economic efficiency may result in a decrease in the welfare of a certain sector of society. For example, the damming of a stretch of the Rogue River in Oregon may increase the welfare of society, as a whole, through the generation of electrical power to nearby cities, however, such construction would eliminate white-water-rafting activity in the area and would also have a negative impact on the local salmon and steelhead recreational fisheries. The welfare of those individuals who take part in these two activities (rafting and fishing) would be decreased.

In order to apply the income-distributional equity criterion it is necessary to weigh the welfare of the different sectors of society. It has been commonly held, however, that economists should point out the income-distributional equity issues to policy makers who must then assign the weights to the different sectors of society before deciding on the resultant equity of a proposed project.

2.2 Selecting the Accounting Stance for a Benefit-Cost Analysis.

It should be clear by now that the purpose of a benefit-cost analysis is to ascertain the effect a project will have on the welfare of society. We have not, however, defined "society." The term society can be viewed from a global accounting stance to a national, regional, state, or local accounting stance.

To determine the appropriate accounting stance, the analysts should look at the overall project and determine where the bulk of the benefits and costs occur. If the project is, for example, the construction of a major interstate highway which would impact the entire northeast section of the U.S., then a regional accounting stance would be most appropriate. If, however, the project is, for example, the construction of a water treatment plant in a small Colorado town, then a local accounting stance would be indicated. It also should be kept in mind that (1) the broader the accounting stance, the more costly the study, and (2) the underlying purpose of the analysis is to provide decision makers with useful information and not to necessarily account for absolutely every effect the project may have.

2.3 Selecting a Methodology.

In addition to selecting an accounting stance, the analyst must also select a methodology. The two basic methodologies which are used in benefit-cost are the general equilibrium approach, and partial equilibrium approach.

In the general equilibrium approach the analyst seeks to identify and assure all gains and losses resulting from a proposed project whether the effects are primary or secondary. The partial equilibrium approach concentrates only on the primary effects of a project. The economic value of sport fishing created by liming acidified waters is a primary effect. A secondary effect would be the purchases at a restaurant by sport fishermen enroute to the newly created fishing site. Under most circumstances those two approaches will render the same substantive information (since the secondary effects often cancel out), however, the partial equilibrium approach involves substantially less work (and, hence, cost) and is less subject to misinterpretation (Ibid.).

Anderson and Settle (1979) point out that as "a practical guide to obtaining a measure of social benefit and costs, it is recommended that the partial equilibrium approach be adopted as the basic methodology because it will generally achieve the desired results and yet is easier to perform than the general equilibrium analysis." If the analysts have good reason to believe the secondary effects of a project will not be self-cancelling the analysis can then be expanded to the general equilibrium approach.

The basic theoretical approach is estimating the benefits and costs of a liming operation when sport fishing is the project resulting from the liming is:

$$\text{Benefits} = \text{ABT} * \text{EAT} \quad (1)$$
$$\text{Costs} = \text{AD} + \text{PC} + \text{AC} + \text{MC} \quad (2)$$

where:
ABT = Average angler benefit per trip
EAT = Expected angler trips resulting from the restored fishery
AD = Administration costs
PC = Purchase costs
AC = Application costs
MC = Monitoring costs

The benefits and costs of a project are then compared through either a ratio of benefits over costs or the difference between benefits and costs. If a ratio is used, a value greater than 1 would indicate a favorable result while with the difference method a positive value indicates a favorable project. In some cases, there may be costs other than those shown in Equation (2) that need to be included in the benefit-cost difference or ratio. An example would be opportunity costs of public agency resources.

2.4 Accounting for Benefits and Costs Over Time

Although the cost of liming a lake or a stream may be experienced today, the benefits of such a project may not accrue immediately but instead over a number of years as sport fishermen and other recreationalists use the restored lake. Since a dollar today is worth more than a dollar in the future (due to interest growth) how then are the benefits and costs of a project compared when they do not occur simultaneously? The answer lies in both discounting and accounting for inflation, which are discussed here.

In order to compare present dollar amounts with future amounts the latter must be converted into present value equivalent terms. This process is called discounting. Once all benefits and cost are discounted (expressed in present values) they can then be compared with one another.

The present value (PV) of a future value (FV) is defined as the amount of money that would make the recipient indifferent between receiving the present value now and receiving the future value at a specified time in the future. The present value can be calculated by:

$$PV = FVn/(1 + r)n \tag{3}$$

where r is the discount rate and n is the number of units of time (e.g., week, mon, yr) for which r is expressed (Anderson and Settle, 1979). For example, r could be an interest rate of 10% per yr^{-1} and n could refer to the relevant year, and as year s.

When more than one time period is involved Equation (3) can be generalized for a stream of benefits (B) as:

$$Bpv - BO + B1/(1+r) + B2/(1+r)^2 + \ldots + BN/(1+r)^N. \tag{4}$$

To illustrate the concept of discounting let us turn to an example of liming a lake used for recreational fishing. Suppose the cost of liming the lake is \$173/ha and the lake is comprised of 4.05 ha for total cost of \$700. The liming cost must be paid upon treatment (i.e., at the beginning of the first yr). The benefits of the liming accrue over a 3 yr period (received at the end of each yr) from the restored recreational fishery and amount to \$1,000 per yr^{-1}. Using a discount rate of 10%, Equation (4) can be applied to calculate the present value of the benefits of the liming as:

$$\$1,000/(1+.10) + \$1,000/(1+.10)^2 + \$1,000/(1+.10)^3 = \$2,486.85. \tag{5}$$

Once the annual future benefits have been transformed into commensurable present values they are added together (as shown above) to obtain a measure of the total present value of all benefits.

The costs for the liming example are all faced in the beginning of the first year and hence are not discounted. Had the costs of the liming operation accrued over time, the present value of all costs (C) would have been obtained through:

$$Cpv = CO + Cl/(1+r) + C2/(1+r)^2 + \ldots + CN/(1+r)^N. \tag{6}$$

When the present value of the benefits ($2,486.85) is compared with the cost of liming ($700) it is apparent that liming is an economically justifiable decision since the benefits exceed the costs.

The discounting for this liming example is done using a discount rate of 10%. The selection of this rate is arbitrary for the example. A conventional practice for selecting a discount rate for public sector projects is to use a measure of the opportunity cost of capital in the private sector (Ibid.). This opportunity cost is the rate of return to investment of the private sector funds or resources when they are placed in their highest value use. (For further discussion of the determination of this opportunity cost, see Anderson and Settle, 1979.)

A second approach for selecting a discount rate public sector project is to use the interest rate at which the governmental entity can borrow funds. This interest rate represents the opportunity cost of capital in the public sector.

Criticism has been leveled at both approaches to discount rate selection since some economists feel that neither approach adequately values future consumption and savings in lieu of such actions in the present. A social discount rate, estimated at less than the private and public rates of return, would more equitably value future economic activity. However, since no conventional method exists for estimating the social discount rate, selection of a discount rate is often based on the rate of return on investment in the public or private sectors.

Although there is some controversy over the selection of an appropriate discount rate, in many cases the value selected will not significantly affect the outcome of a benefit-cost analysis. Some projects offer such a high or low rate of return to investment that the discount rate selected does not affect the apparent desirability of the project.

Anderson and Settle (1979) indicate that the conventional method for identifying those situations in which the exact value of the discount rate is crucial is to preform a sensitivity analysis. That is, the analyst would discount the benefits and cost of a project using first a low discount rate (e.g., 5%) and then a high discount rate (e.g., 15%). The discounted benefits and costs would then be compared to judge the sensitivity of the estimates to the selected discount rate. Should this exercise indicate that the discounted values are sensitive to the discount rate chosen, such information is essential for policy makers so that they can judge for themselves the appropriate magnitude of the discount rate.

When the economy is experiencing inflationary times, the benefit-cost analyst must avoid allowing purely inflationary effects to influence the outcome of his or her study. Anderson and Settle (1979) suggest that inflation can be properly handled by either of these methods:

(1) Estimate all future benefits and costs in constant (non-inflated) dollars and use an estimate of private opportunity cost of capital in the absence of inflation as the discount rate; or

(2) Estimate all future effects current (inflated) dollars and use as the discount rate an estimate of the private opportunity cost of capital plus the expected inflation rate.

With either approach the analyst is left to make an educated choice between the private opportunity cost of capital with the effects of inflation removed or the private opportunity cost of capital plus the expected inflation.

3. IDENTIFYING THE ECONOMIC BENEFITS OF LIMING ACIDIFIED WATERS

Recreational users of lakes and streams substitute different sites and/or select different types of outdoor activities as a consequence of the increased acid level and subsequent reduction in indigenous fish, animal, and avian populations. The liming of acidified lakes and streams benefits all users of these water resources whether they be sport fishermen, waterfowl observers, hunters, or general naturalists. Restoration of water bodies has both an economic value and an economic impact.

3.1 Estimation of the Economic Benefits

Economic benefits fall into two general categories -- economic value and economic impact. Restoration of water bodies has both an economic value and an economic impact. Economic value, in the case of sport fishing, is the value anglers place on the opportunity to go fishing. If sport fishing was bought and sold in a market, this value would be expressed as the most the price of sport fishing could be before anglers would choose to do something else. This section focuses on the derivation of economic values, given the complexity inherent in the calculation of non-market economic values. Economic impact measurement is much more straightforward.

Economic impacts are the effects on an economy, as measured in jobs, income, taxes, sales, etc. In the case of sport fishing, these impacts result from sport fishermen's expenditures. If policy makers are also interested in the economic impact of a liming project the total expenditures by anglers, hunters, and naturalists should be estimated before and after the project. The difference between these two values (appropriately discounted) represents the changes in total expenditures. The relationships between expenditures and jobs, salaries, taxes, sales, and/or wages can be used to calculate the economic impacts.

The marginal change in the economic value accruing to recreational anglers, waterfowl hunters or observers, and naturalists by liming acidified water bodies can be estimated using one of the four valuation methods described in this chapter. The total change in economic value can be estimated by taking the difference between the total estimated economic value before the liming project and the total estimated economic value after the project. The appropriate discounting should also be conducted.

3.2 Valuation Methods for Non-Market Goods

Since lakes and streams are often non-priced public goods, that is that access to the resource is not bought and sold, the increase in the economic value of the resource brought about by the liming operation is not readily observable. Economists have developed several techniques for estimating the value of non-market goods. Among these are: (1) the travel cost method; (2) the contingent value method; (3) the hedonic method; and (4) the unit day method.

The travel cost method (TCM) for estimating the economic value of an environmental resource was introduced by Notelling in 1949 and is based on the inverse relationship between the price and quantity demanded of a good (see Appendix A for a thorough explanation of this concept). The TCM uses the travel costs incurred by the recreationalist as a proxy for the price paid to take part in the recreational activity. The relationship between the price proxy and the recreational trips demanded is then used to trace out a demand curve for the non-market good. The area under the demand curve is measured to derive an estimate of economic value. Early applications of the TCM to the valuation of outdoor recreation can be found in Trice and Wood (1958), Clawson (1959), Knetsch (1963), and Clawson and Knetsch (1966). This technique has evolved significantly since those works, as reflected in Sorg and Loomis (1986), Violette (1985), and Meyer, Brown, and Hsiao (1983).

Economists using the contingent value method (CVM) directly ask individuals to give a measure of the economic value they place on the resource. A cross section of the population is asked to reveal the maximum they would be willing to pay for a good in addition to what they may already be paying. Direct personal interview methods allow the use of the "bidding game" technique for obtaining the respondent's maximum willingness to pay. This commonly used technique requires the respondent to reply to a series of dollar values posed by the interviewer. The respondent answers "yes" or "no" to whether he or she is willing to pay the stated amount of money to use the resource. The interviewer increases or decreases the dollar value until the highest amount the respondent is willing to pay is determined. The intention is to prompt the respondent to think at length about his or her willingness to pay, and encourage the respondent to disclose the maximum value he or she would pay before foregoing the good. This maximum value is the individual's economic value for the fishery. (Willingness to pay is equivalent to net economic value when the price charged for a good is zero. If a price is charged for the good then willingness to pay equals the net economic value plus the total expenditure.)

The CVM is a powerful tool if you can elicit meaningful responses from individuals in the sample. The fact that it is hypothetical in nature can hinder the ability to elicit valid responses. It is called the "contingent value method" because the valuation is often contingent on a proposed change in reaction opportunities or resources.

The hedonic method measures values by comparing and contrasting known values or prices. It assumes that a good is a composite of characteristics with each characteristic representing a (lesser) good itself. For example, a fishing trip (a good) may be thought of as being made up of a number of characteristics including preparation for the trip, the fellowship on the water, the fish caught, and the return trip. If one could know the value of two trips that were identical except for the number of fish caught, comparing these two values would automatically identify the value of difference in fish caught. In this case, the "fish caught" represents a non-market good that is valued (implicitly) as a characteristic of a fishing trip.

The hedonic method relies upon actual market prices or values as a basis for estimates. It is, therefore, seldom available for use in the context of recreational fishing because it is rare to have data for values of goods that only differ by the single characteristic of fishing opportunity.

The unit day value method sets a dollar value for a day of a particular activity or group of activities in a particular place or region based on values from studies completed for similar activities in similar settings. Initially unit day values were determined about the year 1964 by experts who had access to very limited information and therefore the accuracy of these values was suspect. The unit day values have been revised since that time. Many of these values are derived from CVM and TCM studies.

The unit day method has been instrumental in gaining a general acceptance for the basic notion that there is a consumer surplus that needs to be recognized in land management and other environmental decisions. Users of this approach find it far more convenient than any other method for valuing fish and wildlife recreation. Since the number of recreation days can be known or assumed in many situations, net economic value is calculated by multiplying the unit day value times the number of days involved. There is no need for statistical work, no judgments must be made (once the unit day value becomes available, usually through a government agency), and there are no data to collect, except for the actual or anticipated days of use.

4. IDENTIFYING ECONOMIC COSTS OF LIMING ACIDIFIED WATER

The costs of liming a lake or a stream vary with the type and amount of neutralizing agent used, the application method, and the remoteness of the site. "Liming materials can be applied from planes, boats, and land vehicles, spread on the ice of frozen lakes and streams, or held in porous containers through which water flows" (Shepard, 1986). This section will provide some insight into the costs of liming by using examples from actual liming projects. The examples are limited to the liming of lakes since more research has been done with lakes than with the liming of streams.

4.1 Lake (lentic) Liming Methodologies and Costs

Living Lakes, Inc.

Living Lakes, Inc. provided average cost information for lake liming projects. The primary objective of Living lakes is to demonstrate the effectiveness of using liming techniques to neutralize acidic waters and not necessarily to find the most cost-effective method of liming. For this reason, the Living Lakes data may not always represent average costs and may indeed exceed state agency costs.

Living lakes, Inc. generally uses the slurry box liming methodology for lakes less than 16 to 20 ha. On average, 1 t of limestone will be sufficient for lakes of 3 acres (1.21 ha). The length of time over which this treatment remains effective is a function of several of the properties of the individual lakes, (e.g., the amount of flushing occurring in the water system). For this reason Living Lakes, Inc. did not provide an average treatment length figure.

Average costs for liming a 12-ha lake using the slurry box application can be itemized as follows:

1. Design of appropriate dose is a fixed cost based on a rate of pay of \$60 hr^{-1} for 10 hr and including an overhead rate of 2.25:

   \$60 hr^{-1} x hr x 2.25 = \$1,350

2. Material costs are estimated at \$40 t^{-1}. For a 12-ha lake 10 t would be required for a total material cost of \$400. The limestone used by Living Lakes, Inc. is food grade and thus more expensive than lower grades. Living Lakes chooses to use the food grade limestone in order to avoid the heavy metal problems which often result from the application of lower grades.

3. Transportation costs are estimated at \$20 to \$30 t^{-1} plus 15% if less than a full truck load is transported (this is always the case with the slurry box technique, but not with the barge application). Using \$25 t, a cost of approximately \$288 is realized for a 12-ha lake requiring 10 t of limestone.

4. Labor/application cost is approximately \$160 t^{-1} which converts to \$1,600 for the 12-ha lakes.

5. Miscellaneous costs are calculated at 10% of the labor costs or \$160 for this example. Miscellaneous costs include such things as gas, trash removal, etc.

The total costs for liming a 12-ha lake using the slurry box is approximately \$3,800 or \$317/ha. It should be noted that state/local permit costs are not included and may be substantial. The costs reported above also do not include site characterization studies or pre- and post-application monitoring.

Living Lakes, Inc. has used the barge application method for lakes exceeding 16 to 20 ha. Barge application is more efficient than slurry box application with an average application of 1.25 t required for 1 ha of lake. The barge cost is approximately \$198/ha - \$296/ha which covers all costs itemized above for the slurry box method.

New York State

Cost breakdowns are available for selected lake liming projects completed in New York State. Costs are described for the following application methods: helicopter, boats/bucket, and on ice.

The helicopter method, useful for remote lakes, is the most expensive procedure with costs totalling in excess of $8,000 for treatment of a 15 ha lake. The cost found in the New York projects is $562.22 per ha.

Liming of the selected New York State lakes by boats carrying buckets of the liming materials was the least expensive application process. Costs as low as $8.00 per ha-m were achieved through the deployment of unpaid volunteer personnel.

The cost of applying liming materials to ice either on foot or with the use of snowmobiles ranged from a low of $59.79 a ha-m to a high of $817.78 a ha-m.

Cost breakdowns for selected lake liming projects are provided by Fraser and Gremillion of International Science and Technology, Inc. (IS&T). the application methods are by slurry box and helicopter. These costs generally exceed the costs in the New York data. Reasons for this could be (1) the use of volunteer labor by New York which will lower application costs or, (2) the remoteness of some of the sites treated by IS&T which may increase their application and mobilization costs. The term mobilization refers to the costs of getting the needed equipment to the site and setting it up. Examples of the costs for the IS&T projects are:

- Flat Top Reservoir, WV/Slurry Box -- 25 ha @ $11,470;
- 3 sites in Massachusetts/Slurry Box -- 33.6 ha @ $12,585;
- 2 sites in New York/Helicopter -- 33 ha @ $40,867;
- 3 sites in New York/Helicopter -- 69 ha a $75,031; and
- 6 sites on Cape Cod/Helicopter -- 132.5 ha @ $76,880.

4.2 Restocking Costs

Often restocking the waters of a lake or stream is performed shortly after the liming operation to accelerate the return of the water body to its pre-acidified state. The cost of the restocking effort is dependent on the fish species, size, and number along with the remoteness of the stream or lake location. A source for the costs of purchasing hatchery raising fish is *Monetary Values of Fish*, American Fisheries Society, Bethesda, Maryland. The costs of transportation and labor related to the stocking must also be included.

5. CONDUCTING THE BENEFIT-COST ANALYSIS

This section conducts a step-by-step benefit-cost analysis of a liming project using the economic information provided in the previous chapters. For ease of explanation, the benefits of the liming operation are limited to the benefits accruing to recreational fishermen. The reader should note, however, that benefits from liming may also accrue to other water resource users (e.g., waterfowl observers and hunters, commercial fisherman, and naturalists). Also, some people may view liming as a cost, in that it may change water quality characteristics that some people find desirable.

5.1 Estimation of Hypothetical Benefits and Costs of Lake Liming

The benefit-cost analysis is explained through the following hypothetical example. The biological characteristics of this example have been "glossed over," for the purposes of showing, as clearly as possible, the workings of a benefit-cost analysis. Assume that a 40 ha cold water lake has shown signs of increased acidity. Fishery biologists have made the determination that without treatment the trout population will suffer reduced reproduction and will go into decline in the near future. The recreational fishery operating on this lake is also predicted to reproduction and will go into decline in direct response to the decline in the trout population. The regional planners are faced with the question: is it economically prudent to lime the lake and thus maintain the natural productivity of the trout fishery? To answer that question a benefit-cost analysis is performed.

The economic benefits resulting from liming the lake result from the recreational fishery and the number of catchable fish available. Both the local economic impacts and the user values will be considered. The economic benefits are dependent on the number of fishing trips taken to the restored lake and are defined as:

Total Benefits = Average Benefit/Trip * /Angler Trips Taken (7)

Average benefits per trip can be defined either with the local economy in mind, as the economic impact created per trip or, with the individual fishermen in mind, as the economic value per trip (see Appendix A for an explanation of these economic concepts).

The number of fishing trips expected to be taken is assumed to be a function of the number of fish available to be caught, the natural and fishing mortalities, and the catch rate needed to attract fishermen:

Angler Trips Taken = **(8)**

(Initial Fish Available * Total Morality Rate * Fishing Mortality Rate)/ Acceptable Catch Rate.

The regional planners also have the following facts at their disposal:

1. The initial number of trout available for harvest in the base year is 2,000.

2. The total mortality rate is 70%; of that 75% is from fishing mortality and 25% is from natural mortality. When the fish are able to reproduce the reproduction rate equals the morality rate and the fishery is in equilibrium (i.e., the population remains constant).

3. The average acceptable catch rate for the lake is 0.60 fish per trip.

4. The discount rate is established by the decision makers at 10%. Inflation is assumed to be constant throughout the time period.

5. The cost of liming is $173 per ha for the entire lake with the treatment remaining effective for 5 yr.

6. The average consumer surplus per trip is $10. This is a hypothetical value. In a real study, values derived from one of the valuation methods described in Section 2.2 would be used.

In order to determine the economic prudence of liming the lake, the regional planners must compare the discounted stream of benefits with the discounted stream of costs. A partial equilibrium approach is selected along with a local accounting stance.

Looking at benefits first in terms of user value, the following calculations are made:

Expected Angler Trips

2,000 * 70 * .75/.60 = 1,750; **(9)**

Total Consumer Surplus Per Year

1,750 * $10 = $17,500. **(10)**

The stream of discounted total consumer benefits are calculated based on Equation (8) and are presented below. The simplifying assumption is made that the benefits are received at the end of the year.

The total administrative, purchase, application, and monitoring costs are:

$173 * 40.5 ha = $7,000 **(11)**

This cost is incurred at the beginning of the first year and thus is not discounted. The following is a comparison of the benefits and costs of the hypothetical liming operation from the perspective of the economic value of the resultant sport fishing based on a 10% discount rate:

	Year	Benefits	Costs
	0	$ 0	$7,000
	1	15,909	0
	2	14,463	0
	3	13,148	0
	4	11,953	0
	5	10,866	0
Total		$66,389	$7,000

The resultant benefit-cost ratio is 9.48 and the difference between benefits and costs is $59,389.

Economic benefits are next evaluated in terms of their effect on the local economy. This effect is viewed in terms of the local tax revenue generated. In addition to the five facts listed above, the regional planners in this hypothetical example also have the following information:

1. The average expenditure per trip is $15.00.
2. For every dollar spent on fishing $0.50 in direct income is generated.
3. Sales and income taxes are both 5%.

Determining the discounted stream of benefits in terms of local economic impact involves the following calculations:

<u>Total Expenditures</u>

1,750 * $15.00 = $26,250; (12)

<u>Direct Income</u>

$26,250 * $.50 = $13,125; (13)

Total Tax Revenues

($1,312.50 * 0.05) + ($656.25 * 0.05) = $1,968.75. (14)

Here again the assumption is made that the total tax income is received at the end of each year while the costs are faced up front and are not discounted (see Equation (9) for cost calculation). The stream of discounted local tax revenues are calculated based on Equation (4) and a discounted rate of 10%. The resultant benefits and costs are:

	Year	Benefits	Costs
	0	$ 0	$7,000
	1	1,790	0
	2	1,627	0
	3	1,479	0
	4	1,345	0
	5	1,222	0
Total		$7,463	$7,000

The resultant benefit-cost ratio on the basis of local tax impact is 1.07 with a net gain to the government treasury of $463.

5.2 Analysis of Benefit-Cost Ratios.

A liming project is considered to be economically viable if the ratio of the discounted benefits to the discounted costs exceeds 1.0, or if the difference between the discounted benefits and discounted costs is positive.

Turning to the estimates above it is apparent that the lake liming project analysis is considered economically prudent since the ratio of discounted benefits (in terms of economic value) to discounted costs exceeds 1.0 and the difference between the benefits and the costs is positive. When benefits are defined in terms of local tax impact as presented above it is apparent that the project also returns a positive contribution to the government coffers.

6. DATABASES

This section identifies data available to conduct benefit-cost analyses of liming acidified waters. In addition, methods by which to gather primary data are described. As is shown below, while most states collect much of the biological data needed, economic data at the state level is generally absent.

6.1 Identification of Available Databases

This section describes available databases. As most liming projects are local, most available data are at the state level.

The major National database to calculate economic benefits is the National Survey of Fishing, Hunting, and Wildlife-Associated Recreation which has been conducted every 5 yr by the U.S. Fish and Wildlife Service since 1955. The Survey is designed to gather information about Americans who engage in the recreational activities of hunting, fishing, and non-consumptive activities such as bird watching and wildlife photography. A principal purpose of the Survey is to gather information on the numbers of fishermen, hunters, and outdoorsman and women and their frequency of participation in various defined activities.

A secondary purpose for which the Survey has been used in recent times is to conduct economic analysis of various types. The economic analyses are possible because within the Survey are a host of questions relating to expenditures by the participants on the various goods and services they consume to pursue their sport. Specific categories of information include the socio-demographic profiles of the users such as the age group, participants by geographic region, race, sex, and income levels.

Participation also includes a breakdown of the number of days the sportsmen pursue the activity, the dollars they spend on various items like food and lodging, and the dollars they spend on major and minor equipment products like boats and fishing tackle.

This expenditure information may be used to conduct empirical analyses of the data. The technique of benefit-cost analysis (BCA) may be applied to the database to rank order projects in a logical manner. For example, when faced with the choice between two competing projects, or user groups, for a particular category of information derived from the Survey, the relative benefits of the activity may be assessed. The cost information must be obtained from local or state sources. With these two sets of information, a BCA may be conducted rank ordering projects according to their potential benefits.

Relevant also to the Survey is the potential to apply another type of marginal analysis to the database. Here, the motion of producer and consumer surplus may be incorporated into the analysis. Producer surplus is basically the additional (surplus) valued derived by the merchants (producers) of products for which their exists a demand, but whose price is in excess of market equilibrium. Consumer surplus, is the additional value associated with the market equilibrium price, but below the demand for which a market exists but has not been exploited. This additional market (surplus) of the consumer is of interest to policy formation as policies designed to tap into the surplus regions are considered, mostly non-distortionary. If, for example, I am willing to pay $100 for a day of recreational fishing but I only pay $95, then my surplus value of that day's worth of activity is $5. If a process is put in place to take away, usually by taxation, this $5, then I lose nothing, but society gains in the form of a new revenue base of $5. This consumer surplus region is especially relevant to the "user pay - user benefit" approach to fisheries programs sponsored by such bills as the Wallop-Breaux Amendments.

Information on available databases was requested from the fisheries chiefs of the following states: Alabama, Massachusetts, New Hampshire, Maine, Maryland, Michigan, Minnesota, New York, Pennsylvania, West Virginia, Wisconsin, and Rhode Island. The information available that was received is summarized below:

Maine

The Maine Department of Inland Fisheries and Wildlife (1986-1991) publishes a document entitled: *Planning for Maine's Inland Fish and Wildlife* which is updated every 5 yr. The document contains strategic management plans for Maine's individual fisheries. Each strategic plan is divided into:

History;
Past Management Goals;
Habitat and Abundance;
Supply and Demand;
Use Opportunity;
Management Goals and Objectives; and
Management Problems, Strategies, and Priorities.

In addition, some studies are currently underway which address the perceived effects of acid deposition in Maine waters.

Michigan

Complete inventories of Michigan lakes are available for about 1,000 lakes from 1930 and 1964, and partial inventories are available on many others. Few lakes have been completely inventoried by fisheries personnel since 1960 due to a de-emphasis of the program. Monitoring by the Water Quality Control Division of the Michigan Department of Natural Resources has recently increased (Schneider, 1975).

The Michigan Department of Natural Resources conducted creel surveys for 1975-82 to determine catch, catch rate, and fishing pressure for inland lakes, rivers, and specific Great Lakes areas. A report entitled: *On-Site Creel Surveys in Michigan 1975-82* by Ryckman and Lockwood, compile most of the on-site creel census data for Michigan waters.

A study by Schneider (1986) entitled: *Acid Rain: Significance to Michigan Lakes and Their Fisheries* provides a table of the methyl orange alkalinity of lakes sampled in the upper penninsula of Michigan along with a report of the status of Michigan lakes.

Minnesota

Approximately $515 x 10^6 is spent annually by Minnesota's sport fishery. This is one reason why Minnesota has been, through both research and legislative action, aggressive in addressing the issue of acid deposition. The Acid Precipitation Act of 1980 initiated investigations by the Minnesota Department of Natural Resources, Pollution Control Agency (MPCA), and the Department of Health into resource susceptibility and potential impacts from acid deposition. The results of these studies prompted the Acid Deposition Control Act of 1982, the first governmental legislation of its kind. (Minnesota Department of Natural Resources, 1985).

As a result of these legislative actions, Minnesota conducts annual surveys on the large walleye lakes (>10,000 ha) in the state. These annual surveys include:

1. Sampling the lakes by gillnet, seine, and/or trawl to ascertain population characteristics of the indigenous fish species.

2. Determining the water quality of the individual lakes.

3. Performing a creel census to help characterize the sport fishery on the individual lakes.

In addition, the Minnesota Department of Natural Resources has put together a special publication entitled: *Indexing Minnesota Fish Lakes Relative to Potential Susceptibility to Acidic Deposition.* This report contains data on the ecological classification and fish species composition for each lake deemed sensitive to acidic deposition. Limited physical and chemical characteristics are also provided for the sensitive lakes.

New Hampshire

The state of New Hampshire monitors the alkalinity of its ponds and lakes. Recent data indicates that 85% of the state's lakes and ponds are sensitive to acid deposition. Nearly 50% of the state's lakes are endangered or critical (PH < or = to 6) using the winter data, this figure drops to 20% using the summer data. Using the spring-time data, nearly 75% of the remote ponds are critical or endangered.

New York

"New York has had an ongoing program of lake neutralization using both hydrated lime and agricultural limestone since 1959. . . . The program continued at a low level through the late 1960's and early 1970's. In the mid-1970's, as problems associated with acidic deposition became well known, New York State Department of Environmental Conservation (NYSDEC) initiated a program of lake neutralization in lakes likely to have been acidified by atmospheric inputs. This program has continued through the late 1970's and early 1980's" (Kretser and Colquheun, 1984).

The NYSDEC published the *Acidity Status Update of Lakes and Streams in New York State* in April 1984. This publication is a summary of current data on water chemistry, fisheries surveys, and stocking policy for some areas of New York considered to be at risk of damage due to acidification.

The *1976-1977 New York Angler Survey* provides "in-depth information on fishery resource users, their preferences, opinions, backgrounds and distribution as well as information on harvest, waters fished and expenditures incurred by those fishing during the 1976-77 season" (Kretser and Klatt, 1981).

Voluminous information has been published by New York state on the effects of liming on water chemistry and fish populations. A few examples are: Gloss et al. (1980) and Schofield et al. (1988, 1987).

Information is also available on the estimation of cost functions for liming acidified waters. Examples include: Dutkowsky and Menz (1988) and Menz and Driscoll (1983).

Pennsylvania

The Pennsylvania Fish Commission has conducted a number of use and harvest studies in recent years, some of which included questions characterizing anglers. Although most studies were conducted on larger streams, lakes, and reservoirs that are less likely to become acidified in the near term, this information may be able to be applied to acidified waters to a limited extent. Only a limited amount of information is available on the economic benefits related to the fishery resource.

Biological information on fishing and natural resource mortality figures, catch rates, and average weight of captured fish are available for several of the larger waterways. Here again, this information may be applied to a limited extent to waters likely to become acidified.

The state of Pennsylvania has the following items available detailing the impacts of acid deposition and/or the costs and benefits of liming projects: Pennsylvania Fish Commission (1984), Lorson and Marcinko (1986), Marcinko (1986) and Marcinko, et al. (1986)

West Virginia

Creel censuses for the Ohio River and some other major rivers are available. No recent creel census data are available for acid impacted trout streams. Expenditure data by county are available. Some estimates of the value of liming a few of the larger trout streams have been made.

Some data are available on the effects of acidification on fish populations. Most data are based on the delays in spring trout stocking due to acid water. Catch rates by species are available for some waters where creel census are conducted. All others have catch rates lumped over species. Creel census data report catch by length class, not weight. Some estimates have been made. Most data are from reservoir studies. Fishing participation estimates are available from state and national surveys. Five years of data are available on the impact of acid deposition on trout streams. West Virginia has initiated treatment of a major trout stream with proposals for treatment of two others. The impacts of acidification and benefit-costs of liming have been made for these streams.

6.2 Collection of Primary Data When Necessary

Data on the biological, chemical, and economic parameters of a lake or stream considered for liming are necessary before a benefit-cost analysis can be undertaken. When such data on the specific site is not available to the analyst and data collected on other similar sites are not sufficient, the collection of primary data is necessitated.

What follows are "wish lists" for biological and economic data. Often, the collection of these data will be expensive. Therefore, a combination of critical on-site data and economic value information from the literature will be the requisite approach to the analysis.

Data on the biological parameters will help the analyst develop an ecological profile of the lake or stream in its current state. Historical biological information will assist in identifying changes in the lake or stream ecology induced by the acidification. Information on the following topics is desirable:

1. Overview of the aquatic, avian, and animal populations dependent on the waterbody, including:

    a. population biomass;
    b. mortality figures (natural and hunting or fishing);
    c. reproduction figures;
    d. reaction to increased acidity;

2. Physical biological information on flow and flushing rates for the river or lake, respectively.

A thorough knowledge of the chemical composition of the lake or stream is needed before a liming operation should be undertaken. Information on the following is needed for the lake or river:

1. Temperature and oxygen profiles;

2. Total alkalinity and total dissolved solids;

3. Transparency data (from secchi disc readings).

Historical and present data are needed on the recreational usage of the lake or stream to assist the analyst in ascertaining the economic importance of the water body. The following information is needed:

1. Fishery information
    a. average catch per trip (species specific)
    b. average size of fish caught
    c. frequency of trips
    d. economic value information (see appendix A)
    e. expenditure information

2. Waterfowl information
    a. average bag per trip or average sightings per trip
    b. average size of bagged waterfowl
    c. frequency of trips
    d. economic value information (see appendix A)
    e. expenditure information

3. Other use information (e.g., naturalists etc.)
    a. description of use
    b. frequency of use
    c. economic value information (see appendix A)
    d. expenditure information

APPENDIX A

INTRODUCTION TO ECONOMIC CONCEPTS USED IN BENEFIT-COST ANALYSIS

Benefit-cost analysis is a tool used by economists to evaluate the projected economic outcome of a proposed project. The economic benefits of a proposed project are estimated and are compared with the estimated economic costs of a project to determine if society is better off with or without the project. Before this analytical tool is applied to a liming project, an overview of some economic principles is needed. This appendix describes supply and demand in section A.1; user and non-user values in A.2; and economic impact concepts in A.3.

Demand and Supply

The laws of demand and supply lay the foundation for much economic theory and for the benefit-cost analysis. It is important that the reader understand each law individually as well as how they jointly interact in the market place.

The Demand Curve

An individual consumer's demand curve is a schedule of the various amounts the consumer is willing and able to pay for specific quantities of a good during some specific time period. To determine the demand curve for all the consumers in a given market, the individual consumer demand curves for all individuals are horizontally summed.

The consumer demand curve illustrates, at a price of $2.00, the consumer will demand 3 units of the good. Should the price increase to $4.00, the consumer will decrease his or her purchase to 2 units. This example reveals the inverse relationship between the price of a good and the quantity demanded and illustrates the "law of demand."

Although economists assume that the price of the good is the most important determinant of the quantity of that good demanded by the consumer, other non-price factors also play a role. These non-price factors are assumed to be held constant when drawing a demand curve and include: (1) tastes or preferences of consumers, (2) the population size of the consumers, (3) the income level of the consumer, (4) the prices of related goods, and (5) consumer expectations with respect to future prices and incomes. Changes in any or all of these factors will alter the slope of the demand curve.

Another economic concept illustrated by the downward sloping demand curve is that of diminishing marginal utility. The driving force behind the purchase of a good is the utility or benefits received by the consumers of that good. For example, consider the purchase of a fishing rod. The consumer is willing to pay \$5.00 for the purchase of one rod. The second rod is worth only \$4.00 to the consumer while the third rod decreases in value to \$2.00 for that same consumer. Thus, as the number of units purchased of a good increased, it generally follows, that the utility received from each unit (i.e., marginal utility) decreases.

The related concept of consumer surplus can be briefly described. Assume the price of the good is set at \$2.00. At that price the consumer will purchase 3 units of this good. The second unit, however, is actually valued by the consumer at \$4.00. By purchasing the third unit the consumer receives a benefit of area C equal to \$.50. For the units purchased the consumer pays out area D+E (\$6.00) and receives benefits (the area under the demand curve) of area A+B+C+D+E (\$10.50). The surplus benefits or consumer surplus is area A+B+C (\$4.50). Consumer surplus is defined as the benefits received from the purchase of a good above and beyond the actual purchase costs.

The Supply Curve

An individual producer's supply curve is a schedule of the various amounts of a given good the producer is willing and able to produce for the market at each specific price over a specified time period. To determine the supply curve for all producers in a given markets, the individual producer supply curves are horizontally summed.

The supply curve reveals the positive relationship between price and quantity supplied. As the price of a good rises so does the quantity of the good producers are willing to make available for sale. The relationship between price and quantity illustrates the economic concept of the "law of supply".

As with demand, economists assume that the price of the good is the most important determinant of the quantity of that good made available for sale by the producer. Other non-price factors also play a role. These other factors, assumed to be held constant when drawing a supply curve, include: (1) the production method, (2) price of inputs to production, (3) the prices of related goods, (4) price expectations, (5) the number of sellers in the market, and (6) taxes and/or subsidies. Each factor can affect the shape of the supply curve.

Another economic concept associated with supply, is the concept of producer surplus. Assume the price of the good is set at $4.00. At that price the producer will supply 4 units of the good. The first two units of the good only cost $2.00 to produce. By selling all four units for $4.00, each, the producer receives income of area A+B+C+D while only spending area A+B in producing the good. The producer, receives surplus income (producer surplus) of area C+D. Producer surplus can thus be defined as the benefits received from the sale of a good above and beyond the actual production cost.

Equilibrium

The equilibrium supply curve represents the market supply curve; similarly, the demand curve is market demand curve. Assume for the moment that price is arbitrarily set at $2.00. Could this be the equilibrium price for the good in question? The answer is "no." At a price of $2.00 consumers demand 5 units of the good while producers are only willing to supply 2 units. The result is a shortage in the market place of 3 units. The shortage encourages consumers to bid the price up so that the suppliers are willing to supply more good. This process continues until the quantity supplied by the producers equals the quantity demanded by the consumers and the market "clears". This occurs at a price of $4.00. At the market clearing price, 3 units of the good are simultaneously supplied by the producers and demanded by the consumers and the market is said to be in equilibrium. Conversely, had we started with an arbitrary price about the equilibrium price a surplus supply of the good in questions would be present in the market and the price of the good would be bid down to the point where quantity demanded once again equaled quantity supplied.

Economic Value

Economists commonly divide goods and services into private goods and public goods. Private goods (e.g., automobiles and houses) are generally traded in the market and priced in accordance to the laws of supply and demand as explained in section A.1. This price gives an estimate of the economic value of the good.

Most environmental resources have historically fallen under the public good category and, as part of the public domain, are available to all. The "common property" status removes these environmental resources from inclusion in the market place, so that most environmental resources and experiences are not priced automatically as are market goods. While publicly provided recreation in general is not priced and paid for by the direct beneficiaries of the service, this non-market nature of the service does not make the economic value less real. Non-market economic values occur to both users, non-users and indirect users of the environmental resource. Economic techniques used to estimate the value of non-market goods are presented in section 4.1.

User Value

User values take three forms: (1) consumptive use, (2) non-consumptive use, and (3) indirect use.

Consumptive Use

Consumptive use values for a natural resource can be most clearly explained with an example such as that of a recreational fishery. Anglers or divers who physically remove fish from the fishery are classified as consumptive users. Although these consumers may not be charged a fee for their use of the resource, the very fact that these individuals engage in fishing in preference to another activity and consequently spend money, time, and effort doing so, indicates a value exists for the opportunity to go fishing and for the use of the fish stock. In addition, if faced with the choice, many consumers would be willing to pay additional money over their costs of fishing before being excluded from using the resource. "Willingness to pay" over and above expenditures of money, time and effort is a further indication of consumptive use value. The entire "willingness to pay," which includes expenditures, is the total economic value of the resource to the consumptive user (Riely and Rockland, 1988).

Economists use the term "willingness to pay" and also "willingness to accept compensation (or sell)" as dollar measures of economic benefits. These terms stem from different assumptions about ownership of the resource. The choice of which concept is relevant to a particular resource depends on whether someone else (e.g., the state) owns the resource and hence whether individuals would be required to pay for the use of the resource, or the users own the rights to the resource and would thus have to be compensated to give up these rights. In this latter case, willingness to sell is the appropriate dollar measure of benefits. Generally, however, willingness to pay is the assumed choice because the resource most often is owned by the public or the government and not by the resource user (Ibid.).

Non-Consumptive Use

Just as anglers place a value on fish and fishing, so do photographers, snorkelers, aquarium visitors, and others who receive value from directly viewing the resource though not removing it from its environment. While this concept may have greater application to wildlife (i.e., birdwatching), it is also clearly relevant to fisheries. The popularity of non-consumptive snorkeling and scuba-diving attests to the validity of the concept of non-consumptive use value as does the prevalence of people who watch salmon on their annual spawning run. A willingness to pay additional monies over the existing costs of participation before foregoing the activity, is an indication of the economic value of the fishery resource to non-consumptive users as are the expenditures, time, and effort made to undertaken the activity. (Ibid.)

Indirect Use

An indirect resource use results when people derive personal satisfaction from a resource while not coming into direct contact with it. Examples of indirect use of water resources include reading about fish in textbooks or outdoor magazines or viewing televised waterfowl programs. People send money (e.g., on books, magazines, television) to use the resource indirectly which is, in itself, an expression of economic value. There may be additional monies that an indirect user would be willing to pay over existing costs before foregoing the opportunity for indirect use. (Ibid.)

Consumer Surplus

User values can be desegregated into expenditures and consumer surplus. The user (consumptive, non-consumptive or indirect) will value an environmental resource at least by as much as he or she spends to use it. Expenditures could include purchases of fishing tackle by consumptive users, snorkeling equipment by non-consumptive users, and outdoor magazines by indirect users. These expenditures are expressions of economic value associated with the resource only if they are specifically for the use of the resource and are not for another activity (Riely and Rockland, 1988).

In addition to the expenditures to use the resource, there is usually an additional value that the individual places on the opportunity to use the resource. The value is referred to as consumer surplus. Consumer surplus is defined as net value in benefit-cost analysis and represents an increase in society's well being as a result of, for example, fishing or hunting opportunities (Ibid.).

The concept of consumer surplus can be further clarified through the demand curve for recreational fishing examples. Imagine a fishing trip which costs \$5.00. For that fee an angler would go on one trip. If the cost were \$4.00, instead, two trips would be made; at a \$3.00 cost, three trips wold be made; at a \$2.00 cost four trips would be made; and at a zero price, six trips would be made. Now, if the prevailing fishing trip cost is \$2.00, the consumer would spend \$8.00 to go on four trips, and would receive a consumer surplus on the first three trips. For the first trip the consumer is willing to pay $5.00 but only has to pay $2.00, the surplus is $3.00 ($5.00 - $2.00). Surplus for the second trip is $2.00 ($4.00 - $2.00); and $1.00 for the third trip ($3.00 - $2.00). The fourth trip will not generate a surplus since the price paid is equal to the trip's value. The total value to the individual for fishing will be $14.00, total expenditures will be $8.00 resulting in a consumer surplus of $6.00 (Ibid.).

There has been an historical controversy in the economics literature surrounding the use of consumer surplus as a valid economic tool. (Hicks, 1956, Little, 1957, and Samuelson, 1947). The controversy stems from the effect of price changes on income. An increase in the price of a good means a reduction in real income for consumers of that good. This results in fewer purchases of the good whose price had risen, as well as fewer purchases of other goods. The idea of consumer surplus does not take into account these other goods. Therefore, economic theorists have argued that consumer surplus does not reflect the value the consumer please on the opportunity to purchase a good at a particular price (Riely and Rockland, 1988).

Two other measures of the surplus consumers receive suggested by economic theorists are compensating and equivalent valuation (CV and EV). CV is the most an individual would be willing to pay to use the resource at a lower price. It represents the recision in income that makes the individual equally as well off or equally satisfied, with a lower price (Ibid.). EV is the least amount the individual would accept to forego the price decrease. The individual would achieve a higher level of satisfaction due to an increase in income but this increased income is the result of a price decrease. With EV, the assumption is that the individual has some right to the lower price. With CV the assumption is that he does not have that right and therefore would be willing to pay for the price reduction (Ibid.).

Willig (1976) argued that consumer surplus approximates the CV and EV if the price changes and the income effect are small. The margin of error will be over shadowed by errors in the demand curve estimation. McKenzie (1979) commented that Willig's approach would only work if the price and income changes were minute. The reader should be aware that consumer surplus is a theoretically controversial subject and view the discussions in other sections of this report in that light (Ibid.). However, consumer surplus is the standard economic measure of value that resource economists use.

Non-user Value

Non-users of the resource also may place an economic value on a natural resource. These values are referred to as intrinsic values and result from sentiments about the resource not related to current-use. Non-use values are categories as option and existence values. Option values reflect uncertainty about future resource used and represent the value the individual places on the availability of resource use in the future. Existence value is the economic value a non-user places on the knowledge of the mere existence of a resource (Rockland, 1987).

Economic Impact

The expenditures made for private goods in the market place or for the consumptive, non-consumptive, or indirect use of a public good such as a natural resource have an economic impact on the local, state, and/or national economies. This impact may be expressed in terms of jobs, salaries, taxes, sales, or wages. The reasoning that the total economic impact is greater than the total expenditures results from the premise that sales in one industry not only impacts that industry but also impacts the industries that supply goods and services to the initially affected industry. Three levels of economic impact exist, namely: (1) direct impact (2) indirect impact, and (3) induced impact.

Direct Impact

The three levels of economic impact can be explained using an example of a recreational fisherman. The direct impact of the recreational fisherman is the initial purchases made to go fishing. Such purchases could include the rod and bait needed for a day of fishing.

Indirect Impact

The indirect impact of the recreational fishery would be the revenue from purchases made by the businesses from which the fisherman purchased the rod and the bait. For example, the rod manufacturer may purchase paints, metals, etc. to manufacture the rod. The revenue from these purchases by the rod manufacturer would be indirect impacts of the recreational fishing activity. Likewise, if the bait was purchased at the retail level, the retailer would have purchased the fish from a fish processor. The revenue from this purchase by the retailer is again an indirect impact of the recreational fishery. It is easy to see how the initial round of indirect purchases has further indirect impacts as the suppliers to the direct and indirect businesses make further purchases to meet that demand (Rockland, 1985).

Induced Impact

The retail, wholesale, and manufacturing businesses directly and indirectly impacted by the recreational fisherman, in our example, pay wages to their employees. These wages then are used to purchase goods and services which represent the induced impacts of a recreational fishing day. This pattern continues so that the initial induced impacts have additional indirect and induced impacts of their own.

SECTION III: STREAMS

Lakes and ponds are considered by most resource managers to be less difficult to neutralize than are streams. The basic flow characteristics make streams more difficult to treat as well as more expensive because of the continuous requirement for replenishing the liming material. In the following papers, several different methods for applying neutralizing materials to streams are described. The papers presented in this section are not as extensive as in the Lakes section because of the difficulty and cost-effectiveness of treatment, but they are no less interesting and informative.

EFFECTS OF LIMING ON WATER QUALITY OF TWO STREAMS IN WEST VIRGINIA

Tamara I. Ivahnenko
John J. Renton
Henry W. Rauch

ABSTRACT

In an effort to improve the fisheries potential of many West Virginia mountain streams by increasing the alkalinity and pH, the West Virginia Department of Natural Resources (DNR) is conducting research into the feasibility of a point source limestone application technique. This research was to provide data that would assist in the design of future projects and the selection of appropriate methodology. Thirteen metric t and 33.3 t of 0.64 cm crusher run ground limestone aggregate were applied as a single deposition to Crouch and Yokum Runs, respectively. Limestone was applied to Crouch Run a year prior to treatment of Yokum Run. The point source treatment produced discharge dependent changes in the downstream water chemistry. The pH increased from an average of 4.73 above the application point to 5.70 below. At high stream discharges, the pH dropped, but never to the pretreatment background level. Increases in Ca concentrations and ANC were also observed downstream of the limestone application site. Sediment analysis was performed to determine the calcite concentration within the stream bed by both an instrumental technique (x-ray fluorescence (XRF)) and an HC1 leach technique to indicate the extent of downstream movement and distribution of limestone. The analysis showed both a limited downstream transport and a variable spread of limestone across the stream channel. The XRF and wet laboratory methods of quantifying the calcite content of the sediments were compared. Although statistical analysis indicated a difference in precision between the x-ray fluorescence and HC1 leach methods, the acid leach method can be considered an economical alternative to x-ray fluorescence. Recommendations are made on methods to increase the effectiveness of the point source treatment and the design of future research.

1. INTRODUCTION

Many of the pristine mountain streams in West Virginia lack established reproducing trout populations because of low pH waters. The lowest tolerable reproduction limit for brook trout is pH 5.5, while for rainbow trout the lower limit is pH 6.0 (Sedgwick, 1973). The naturally

Water, Air, and Soil Pollution **41** (1988) 331–357.

occurring low buffered, moderately acid waters of these streams and their fisheries have in recent decades been further damaged by the effects of acid rain. In areas of active coal mining, the already acidic streams may be further polluted with mine discharges, thereby creating waters devoid of fish.

In an effort to neutralize the combined natural acidity and that introduced by acid rain and restore and create streams with reproducing trout populations, the West Virginia Department of Natural Resources (DNR) is conducting research to determine the feasibility and effectiveness of using limestone as an acid neutralizing agent. Experiments conducted by the DNR employing various methods of limestone introduction have indicated that the use of ground limestone aggregate improves stream water quality. Ground limestone aggregate for this project is defined as 0.64 cm crusher run limestone, the waste product of limestone gravel crushing operations. The limestone aggregate is typically composed of particles in the size fractions: 39% silt, 42% sand and 19k% gravel (WV DNR, 1986).

Other methods of using limestone to neutralize acid water have been tried both domestically and internationally. The procedures vary from water powered rotating drums (Zurbuch, 1984) to the dusting of watersheds by helicopter or plane (Fraser and Britt, 1982).

Most of the neutralizing treatment schemes used to date require substantial funding and personnel to maintain the systems. To minimize the technological and economic constraints of maintaining and monitoring a limestone neutralizing scheme, the DNR has suggested the "point source" application. A point source application involves the introduction of ground limestone aggregate at a convenient place in the stream and allowing the flowing water to disperse the limestone aggregate downstream. A pilot project has been designed and implemented to study the effect of point source application of ground limestone aggregate on two nearly identical juxtaposed watersheds on Cheat Mountain, Randolph County, West Virginia.

The purpose of the project was to: 1) determine the chemical effectiveness of ground limestone aggregate introduced at a single source to favorably alter water quality for the two study streams; 2) determine the extent of downstream transport and distribution of ground limestone aggregate from the point source introduction site, and 3) suggest a limestone aggregate introduction scheme, and the appropriate methodology to evaluate the effects of the treatment scheme on a third stream, Red Run, based upon the results from objectives one and two.

2. METHODS

2.1 Introduction

The two streams chosen for the study, Crouch and Yokum Runs, are similar in most physical and chemical characteristics. The streams are adjacent tributaries of Shavers Fork, located on Cheat Mountain, Randolph County, West Virginia and are easily accessible from Elkins, West Virginia and from US Forest Service roads. Crouch Run had been previously treated with ground limestone aggregate prior to the initiation of this study, therefore only monitoring of the stream was required.

The watersheds of the study stream are located on the High Appalachian Plateau physiographic province approximately 15 km west of the Appalachian Structural Front. The general geology of the area is lower Pennsylvanian and upper Mississippian strata, predominantly sandstone, shales and a few thin coal beds (Reger, 1931). The bedrock geology offers little in buffering capacity to the stream waters. A summary of watershed area, length and gradient are reported in Table I.

Table I

Stream Characteristics

Stream	Watershed Area (km^2)	Length (km)	Gradient ($m\ km^{-1}$)
Crouch Run	7.20	4.5	29.2
Yokum Run	7.12	4.5	25.8

Four water chemistry monitoring stations, evenly spaced along the stream length, were established for each of the two streams evenly spaced along the stream length as shown in Figure 1. One station was located upstream of the limestone application site to serve as the background control for the stream study. The second station on Yokum Run was located approximately 90 m downstream of the application site, while on Crouch Run the second station was 122 m from the application site. The second station on both streams corresponds to the Living Lakes, Inc. (1987) "mixing zone." The third and fourth sites were in the "target zone;" the fourth site being at the mouth of the stream.

Natural sediment trap sampling stations were located on both streams. Five natural sediment traps were located for sampling in each stream to monitor the movement of the limestone particles downstream from the application point. Four of these sites coincide with the stream water sampling stations. The fifth sediment sampling site was located as shown in Figure 1.

Four cross-section sampling sites, also corresponding to each water sampling station, were selected to determine the cross-channel limestone distribution for sediment channel sampling. One of each type of sediment sample collecting site was located upstream of the limestone point source to provide control sediment samples. The other stations were distributed downstream, with the last being at the mouth of the stream (**Figure** 1). All the sites were chosen to follow the protocols established by Living Lakes Inc. (1987) and for the convenience of sampling. The number of water and sediment sampling sites was deemed sufficient to observe changes brought about by the limestone addition in the stream environment.

Crouch Run was treated on December 11, 1985 with 13.3 t of Ordovician age, Trenton ground limestone aggregate obtained from the Germany Valley plant of the Greer Limestone Co. Yokum Run was treated with 33.4 t of the same material on December 3, 1986.

The treatment schedule used in this study provided the opportunity to ascertain whether or not any significant beneficial effects of liming could be observed 1 yr after application in the case of Crouch Run and immediately after application in the case of Yokum Run.

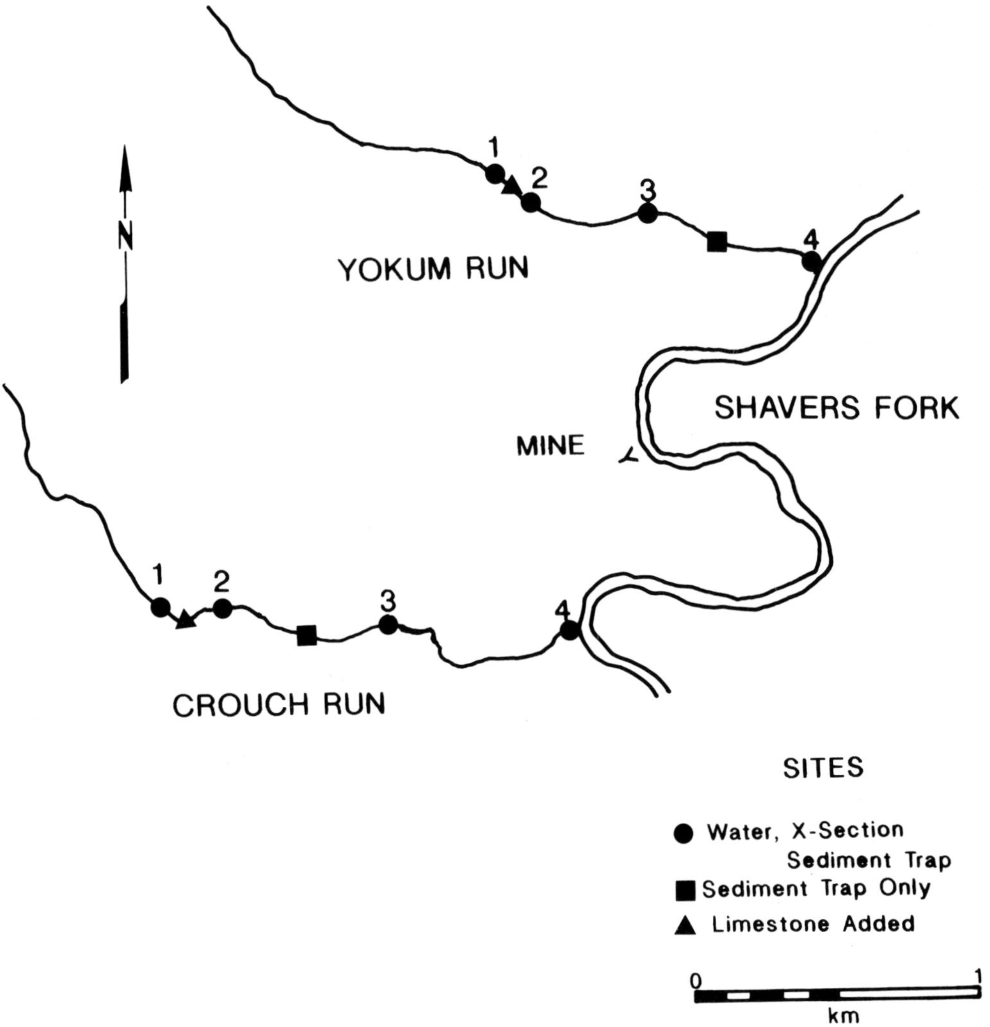

Figure 1. Location of water and sediment sampling sites on Crouch and Yokum Runs.

Both water and sediment sample collections for this project were designed to cover a 9 mo period starting from the time of limestone introduction to Yokum Run and ending in August 1987. Water samples were taken twice a month from the two streams. Sediment samples were collected from natural sediment traps and from cross-section sites at 0.61 m intervals across the stream channel width. Sediment trap samples were collected once a month, for a total of 9 sets. Cross-section samples were taken every 3 mo for a total of 3 sets.

2.2 Field Methods

The water samples from each water chemistry station were collected twice a month. The samples were taken in 1 L cubitainers leaving as little air space as possible remaining inside the container. One sample was filtered and acidified with 5 ml of 0.5M HNO_3. The samples were placed in a cooler for transport to the laboratory. During the winter months, sampling was terminated when heavy snowfall made the roads impassable to the water sampling stations. As the water samples were being collected, field pH and water temperature were measured at the water station.

Flow measurements were taken on Yokum Run station 4 by a Marsh-McBirney model 201 flow meter. A Stevens Model 71 continuous flow recorder was installed in Yokum Run. A gauge station discharge relationship was established, allowing gauge level readings to be converted to discharge values on Yokum run.

Natural sediment trap samples were collected by removing the upper centimeter of loose material into a sampling jar. Excess water and any floating organic matter were decanted off and the jars labeled.

Cross-section channel samples were obtained after first securing a wire marked in 0.61 m sampling intervals to a stake located on the bank or waters edge and stretching the wire across the channel. Sediment was not collected from some sites within the channel where rocks obstructed the collection location as determined by the marked wire. If the sampling location, as indicated by the wire, fell between the rocks, a sediment sample was taken.

Two core samples were taken in August 1987. One was taken in sediment trap 2 of Crouch Run and the second was taken in the corresponding sediment trap of Yokum Run. A 5.08 cm diameter hollow core tube was inserted into the unconsolidated sediment until bedrock was encountered. The tube was capped at the top, tilted horizontally and carefully withdrawn. The other end was then capped and the core was labeled.

2.3 Laboratory Methods

2.3.1 Water analysis. Within 2 hr of collection, specific conductivity, alkalinity and laboratory air equilibrated pH (Standard Methods, 1978) were determined on the unfiltered, unacidified sample at the DNR water laboratory in Elkins, West Virginia. The samples were then repacked in a cooler, with ice for transport to West Virginia University.

The unacidified samples were split for use in analyzing various parameters. One split was brought to room temperature and the acid neutralizing capacity (ANC) was measured using an endpoint titration method following the protocol established by Living Lakes, Inc. (1987). The other split was analyzed for chloride, nitrate, sulfate and fluoride ions using a Dionex Model 2000 Ion Chromatograph.

The acid stabilized samples were analyzed for Ca, Mg, Na, and K ions using a Perkin-Elmer 403 atomic absorption spectrophotometer (AA). Aluminum analyses were obtained by use of a graphite furnace installed in a Perkin-Elmer Model 5000 automatic atomic absorption spectrophotometer.

2.3.2 Sediment analysis. One of the purposes of the study was to determine the downstream distribution and transport of the limestone aggregate. In order to ascertain the calcite content of the sediment samples, two analytical techniques were developed and compared; an instrumental method, x-ray fluorescence (XRF) and a simple HCl leaching technique. Both procedures require the use of working curves which relate to magnitude of some measured parameter or parameters and the concentration of calcite within a series of standard samples.

In order to prepare working curves for the two techniques, a series of standard samples were prepared. The matrix for the samples was prepared from the typical bedload materials found in Crouch and Yokum Runs. The materials were first crushed in an iron mortar and pestle, then ground under alcohol in a Spex Mixer-Mill for 10 min, reducing the material to -325 mesh. Subsequent XRD analysis showed the material to be devoid of calcite. X-ray fluorescence analysis indicated a background CaO content of 0.23 wt %. The 0.23 wt % CaO content became the "zero point" from which the sediment calcite content was determined.

Pure calcite was similarly ground to -325 mesh and admixed to the ground bedload materials in the wt% indicated in Table II. The mixtures were thoroughly blended by tumbling for 1 hr. When mixed, representative portions of each sample were taken to be used for the individual analysis.

Table II

Calcite Standards Series
(wt% $CaCO_3$)

0.05	1.50
0.10	2.00
0.25	2.50
0.50	5.00
0.75	7.50
1.00	10.00

Quantification of a mineral component in an unknown by XRF is based upon the comparison of the measured intensity of a selected spectral line of an element contained within the mineral (in this case, Ca in calcite) reported as the oxide to the intensity of the same spectral line of the element reported as the oxide from a series of standard samples containing the mineral of interest in known concentrations.

HCl leaching technique is based upon the assumption that the major source of Ca within the unknowns is contained within acid soluble calcite. Previous analysis of bedload materials from Crouch and Yokum Runs confirmed this assumption. The procedure involved subjecting known weights of unknown samples to controlled leaching with fixed volumes of HCl of known molarity. The concentration of Ca in mg L^{-1} in the subsequent acid leachate is compared to the Ca in mg L^{-1} leachates similarly prepared from a series of standard samples containing calcite in known concentrations.

The working curves prepared for the XRF and HCl leaching techniques are illustrated in Figures 2 and 3, respectively. Comparison of the XRF instrumental and HCl leach procedures showed the fluorescence analysis technique to be the most accurate. The XRF calcite standards as illustrated in Figure 2, produced a regression coefficient of 0.9996. The probability of error associated with the regression line was not greater than 0.1%. The HCl technique, on the other hand, had a regression coefficient of 0.9939. The probability of error for the regression line in Figure 3 was 5%. Because of the higher precision, x-ray fluorescence was the primary analytical method used in this study to determine the calcite concentration of the sediment. Although the HCl leach technique provided the lower correlation coefficient for the two sediment analysis methods, because of its simplicity and low cost, it can be considered as an adequate substitute for the more costly x-ray technique.

The sediment samples were initially dried. A Jones splitter was used to obtain a representative subsample. In the event that particles too large to be split efficiently were present in the sample, the large particles were first reduced to sand-sized fraction using an iron mortar and pestle. The separated subsample was ground in a Spex-Mixer Mill under alcohol to -325 mesh. Samples were ground under methanol to minimize heating which may alter the crystallinity of the calcite lattice and subsequently affect the results of any x-ray diffraction analysis that may be conducted. The powdered samples were then subjected to various analyses.

2.3.3 X-Ray Fluorescence. To prepare a sample for the x-ray fluorescence, approximately 0.5 g of powdered sample was weighed out. Chemplex, an inert binding agent, was added to the sample equal to 10% of the sample weight. The samples were mixed by tumbling for 1 hr on a

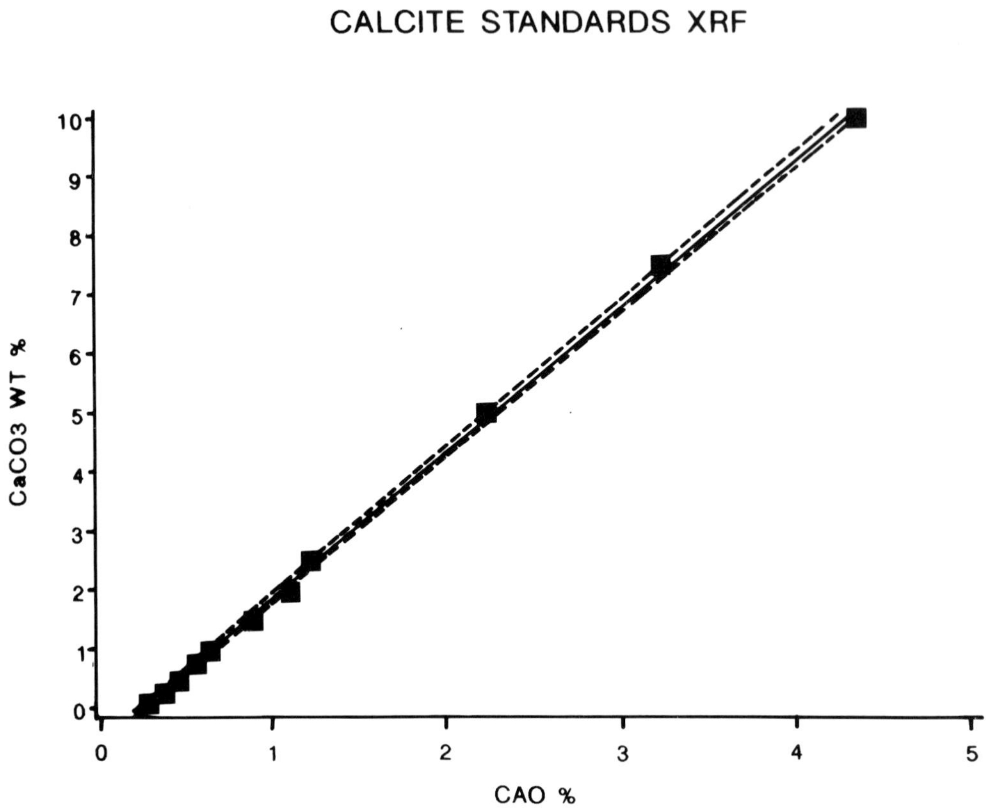

Figure 2. Regression line with 95% confidence interval limits for weight % calcite analysis vs standard values as determined by x-ray fluoresence.

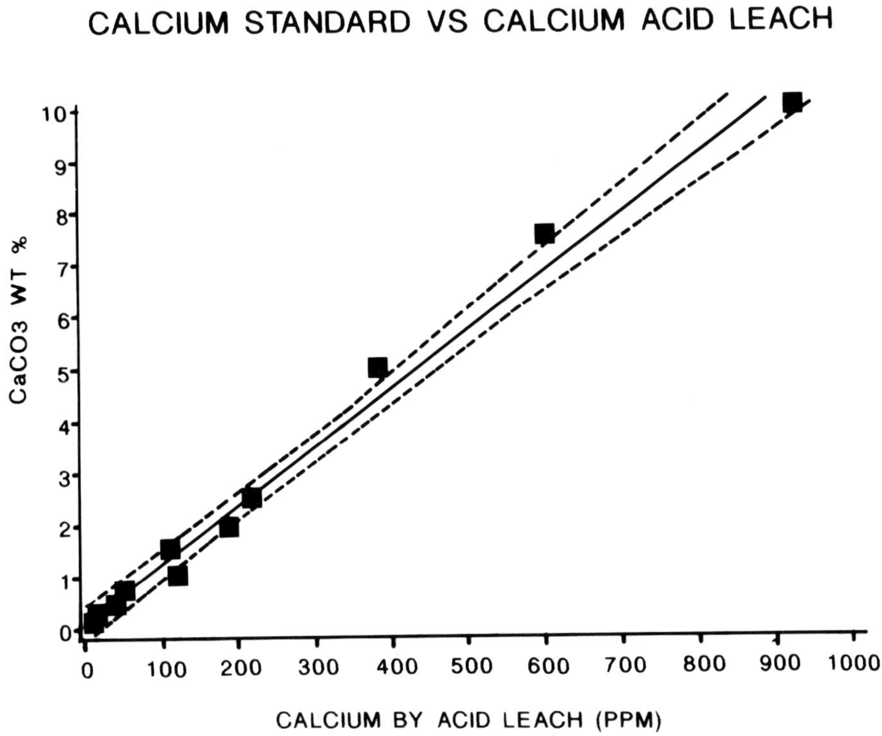

Figure 3. Regression line with 95% confidence interval limits for weight % calcite analysis vs standard values as determined by HCl leach.

Table III

Yokum Run pH, Ca and ANC Statistics

Station	pH			Ca Concentration ($mg\ L^{-1}$)			ANC ($\mu eq\ L^{-1}$)		
	Mean	Maximum	Minimum	Mean	Maximum	Minimum	Mean	Maximum	Minimum
1	4.73	4.90	4.40	1.67	2.10	1.30	-10.80	0.05	-39.50
2	5.47	6.51	4.52	2.87	6.60	1.60	45.20	289.50	-19.50
3	5.94	6.79	4.65	3.31	6.40	2.10	74.10	367.50	-3.50
4	5.85	6.64	4.68	3.08	6.50	2.10	42.30	208.50	-7.50

rotating wheel. The sample was then pressed into a 33 mm diameter pellet under 15.3 t total load.

2.3.4. Hydrochloric Acid Leach. One gram of sample was weighed out and placed into a ceramic evaporation dish. Fifty ml of 0.5M HCl were added to the sample and the mixture was heated for 1 hr to ensure total dissolution of the $CaCO_3$ as suggested by Gross (1971) and Molina (1979). The leachate was filtered through a glass filter and analyzed by atomic absorption for Ca content.

3. RESULTS AND DISCUSSION

3.1 Water Analysis

The chemical parameters which directly indicate the effects of limestone fines on water quality are pH, Ca concentration and acid neutralizing capacity (ANC).

Table III shows the maximum, minimum and mean pH values for each water chemistry monitoring station on Yokum Run. The general distribution of pH values show increasing values at all stations downstream of the limestone application site. However, a slight reduction in pH was observed in the case of both streams between the third and fourth stations. A student's t-test performed on the data from stations 3 and 4 showed that the pH values were not statistically different at an 0.05 alpha level of significance.

Figure 4 represents a plot of Yokum Run pH vs ANC. The plots show the ANC is a direct function of pH. A significant relationship exists between pH and ANC within the range of -20 to 30 $u$eq L^{-1} which can be represented by the predictive equation:

$$\text{ANC} = \text{pH} - 4.50/0.0212$$

allowing for an estimate of the ANC value from a pH reading within the limits of pH from 4.60 to 5.90.

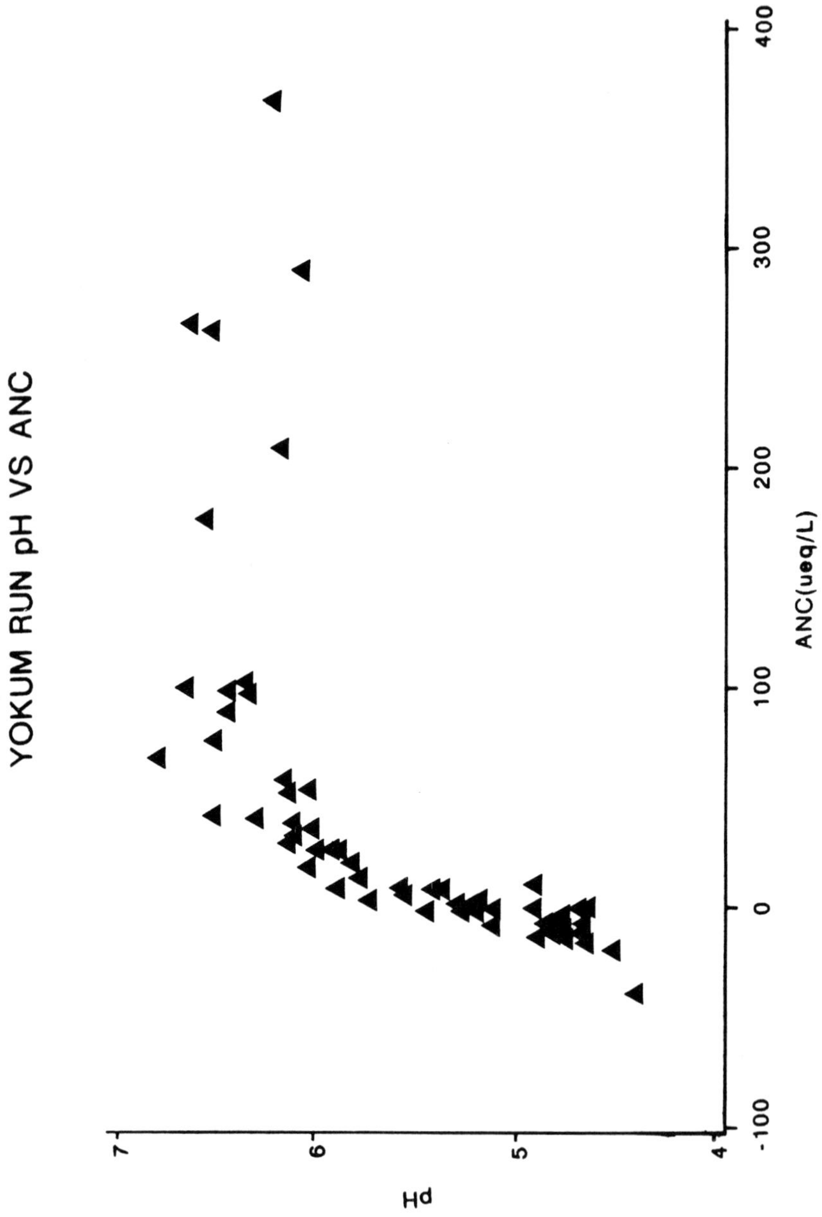

Figure 4. Illustration of the relationship between Yokum Run pH and ANC.

The Ca and ANC maximum, minimum and mean values for Yokum Run are listed in Table III. The distribution pattern for these parameters parallels that shown for the pH trends. The difference in the downstream decrease of these parameters from stations 3 to 4 was also found by a student's t-test not to be statistically significant for either ANC or Ca concentration.

An important relationship between pH and stream discharge is illustrated for Yokum Run (**Figure 5**). The data indicate a non-linear decrease in pH with increasing stream discharge. Similar results were observed by Gunn and Keller (1984). Increased stream discharge, resulting in increased acid loads, reduced the buffering capacity of the waters, decreased the contact time at the limestone particle-water interface and thereby reduced the effectiveness of the limestone particles as acid neutralizers. The plot does not contain sufficient data above 0.7 $m^3 s^{-1}$ to draw a conclusion as to the predictive activity of the H^+ in relation to discharges in excess of 0.7 $m^3 s^{-1}$. Higher discharge data are required to fully determine the effect of limestone aggregate on water quality. However, throughout the study, the pH downstream of the point of limestone introduction was never equal to that of the background level (water sampling site 1).

A complementary graph to Figure 5 is the illustration of pH and discharge for the duration of the study (Figure 6). The dashed lines on the discharge graph represent interpolations for missing data. The plot for Yokum Run clearly indicates the effect of discharge on pH. Within a month of limestone application, the pH began to decrease, and the fluctuations in water chemistry throughout the remainder of the study depended upon the amount of water in the stream system. The relationship can be best illustrated on Yokum Run by the marked increase in pH beginning in May 1987. At this time of the year, a summer drought occurred and the discharge of the stream by August 1987 had dropped to nearly zero, where only isolated pools of standing water were observed. The subsequent pH of the stream waters was greater than 6.0.

The maximum, minimum and mean pH, Ca concentration and ANC values for Crouch Run are listed in Table IV. The water quality data with respect to pH, Ca concentration and ANC for Crouch Run are identical to that of Yokum Run. Results of students' t-tests indicated no statistical significance to exist between stations 3 and 4, in terms of pH, Ca or ANC.

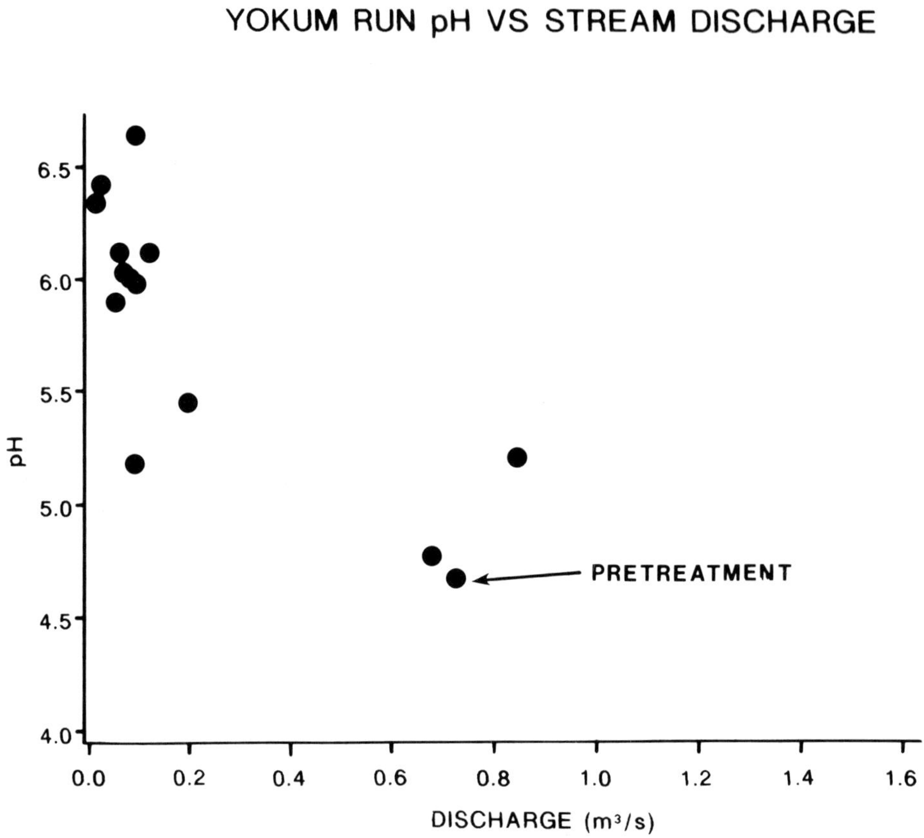

Figure 5. Illustration of the relationship between stream discharge and pH at station four.

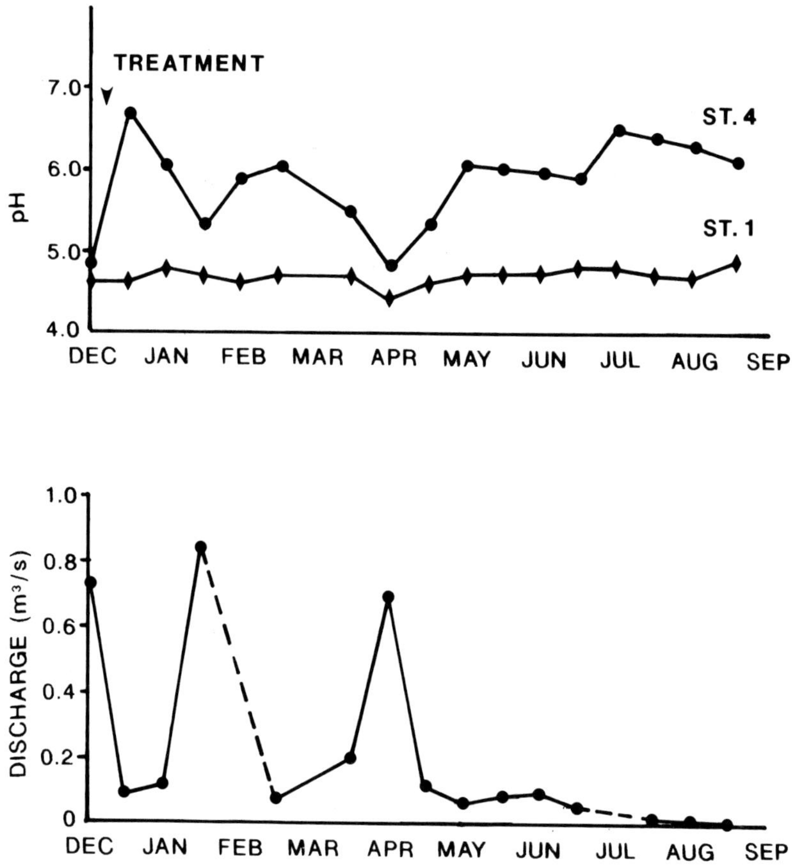

Figure 6. Illustration of pH and discharge of Yokum Run stations four and one over the study period.

Table IV

Yokum Run pH, Ca and ANC Statistics

Station	pH			Ca Concentration (mg L^{-1})			ANC (μeq L^{-1})		
	Mean	Maximum	Minimum	Mean	Maximum	Minimum	Mean	Maximum	Minimum
1	4.66	4.83	4.34	1.69	2.40	1.30	-17.40	4.50	-35.50
2	5.01	6.25	4.45	2.27	3.90	1.40	-4.61	48.50	-37.50
3	5.08	5.76	4.52	1.96	2.60	1.60	-1.62	32.50	-25.50
4	5.03	5.48	4.43	2.01	3.90	1.60	-5.23	9.50	-24.50

At the initiation of this study, the limestone had been present in the Crouch Run stream system for a year. One year after the introduction of limestone aggregate, the pH values below the application point on Crouch Run had not returned to that of the control station background levels. However, in terms of an established fishery, the water quality below the Crouch Run point source limestone introduction site was unsatisfactory for the survival of rainbow and possibly brook trout (pH < 5.5). It was clear that in order to maintain a suitable fishery water quality, the stream would require reapplication of the limestone or a continuous dosing system to treat the stream (Rosseland and Skogheim, 1984).

3.2 Sediment Analysis

Natural sediment trap calcite values for Yokum and Crouch Runs are plotted against downstream stations in Figures 7 and 8. The recognizable trend in both figures is the concentration maxima at station 2. The limestone application site is located between stations 1 and 2. Prior to the project, it was anticipated that the limestone particles would be distributed within the entire downstream system by water action. The data showed, however, that the downstream transport of the ground limestone aggregate only achieved a maximum distance of 562 m from the application site on Yokum Run. Most of the limestone remained within 90 m of the application site. Nonetheless, it is significant to note that the water quality along with the entire length of the stream downstream of the application site improved and remained good (pH > 5.5) for the length of the study.

The maximum downstream transport of limestone aggregate for Crouch run was 122 m from the application site. The shorter distribution distance may be attributed to the smaller amount of material applied to Crouch Run (only 13.3 t). Also, the sampling of Crouch Run for this project began a year after application, allowing ample time for the transported particles to dissolve.

Plots of the calcite content of the cross-section samples from station 2 of Yokum and Crouch Runs, showed variability across the channel (**Figures 9 and 10**). Field observations and study of the limestone content of the station 2 cross-section indicate that the limestone fines are randomly distributed from one stream bank to the other.

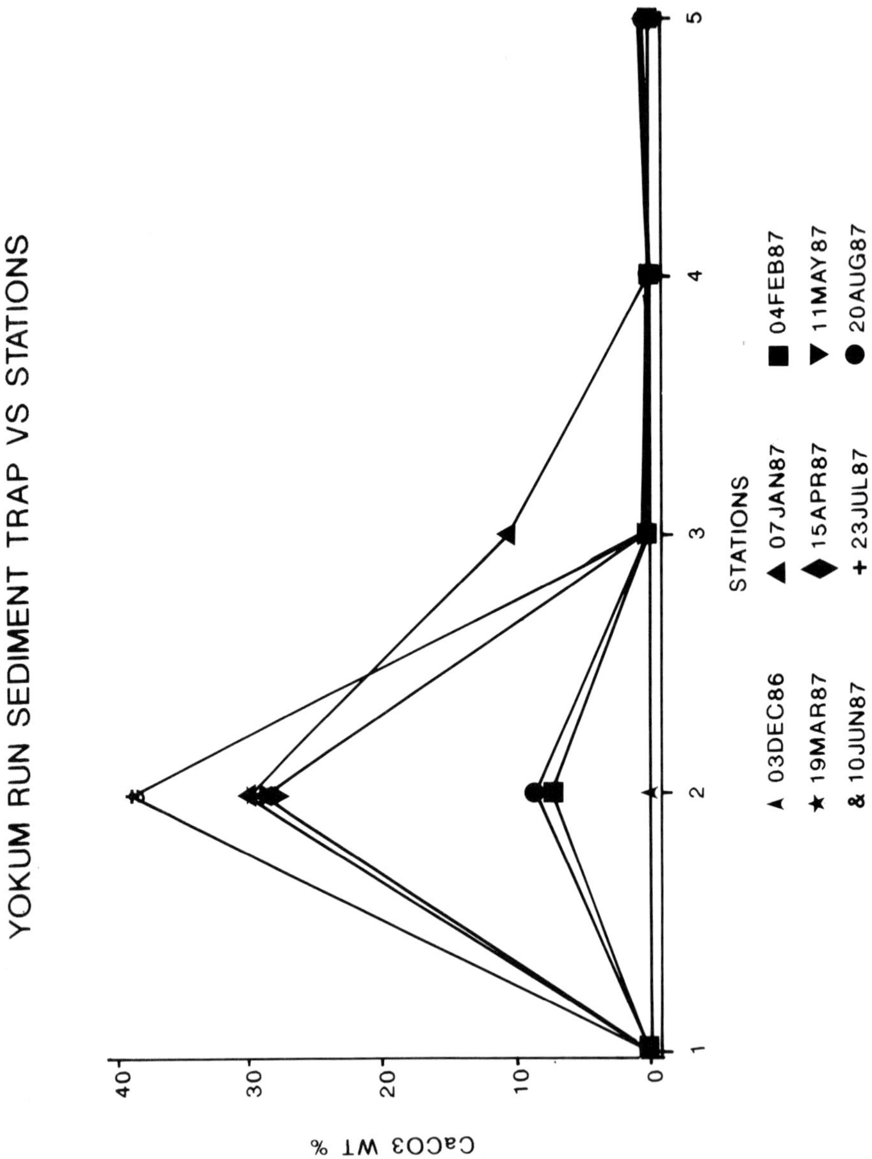
YOKUM RUN SEDIMENT TRAP VS STATIONS
CaCO3 WT %
0
10
20
30
40
1
2
3
4
5
STATIONS
03DEC86
19MAR87
10JUN87
07JAN87
15APR87
23JUL87
04FEB87
11MAY87
20AUG87

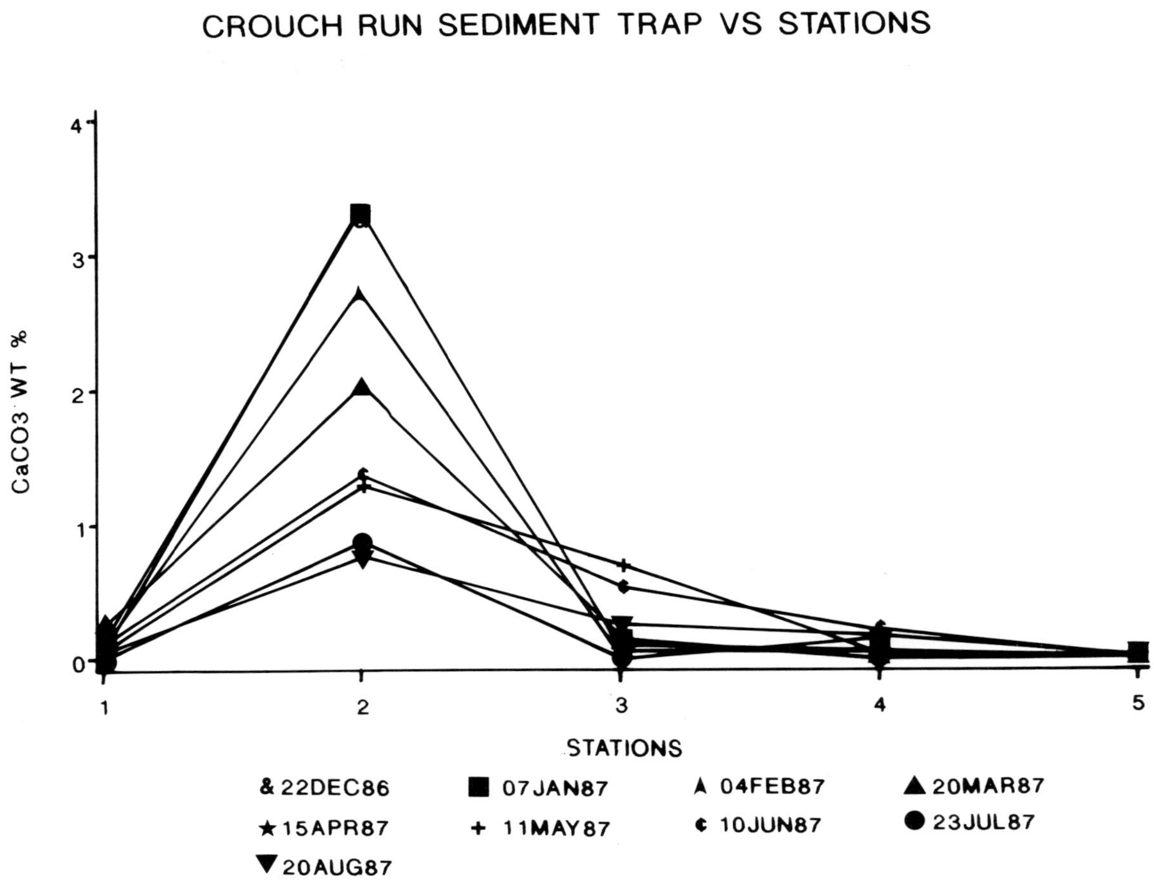

Figure 8. Natural sediment trap calcite values vs downstream stations, illustrating downstream transport of limestone.

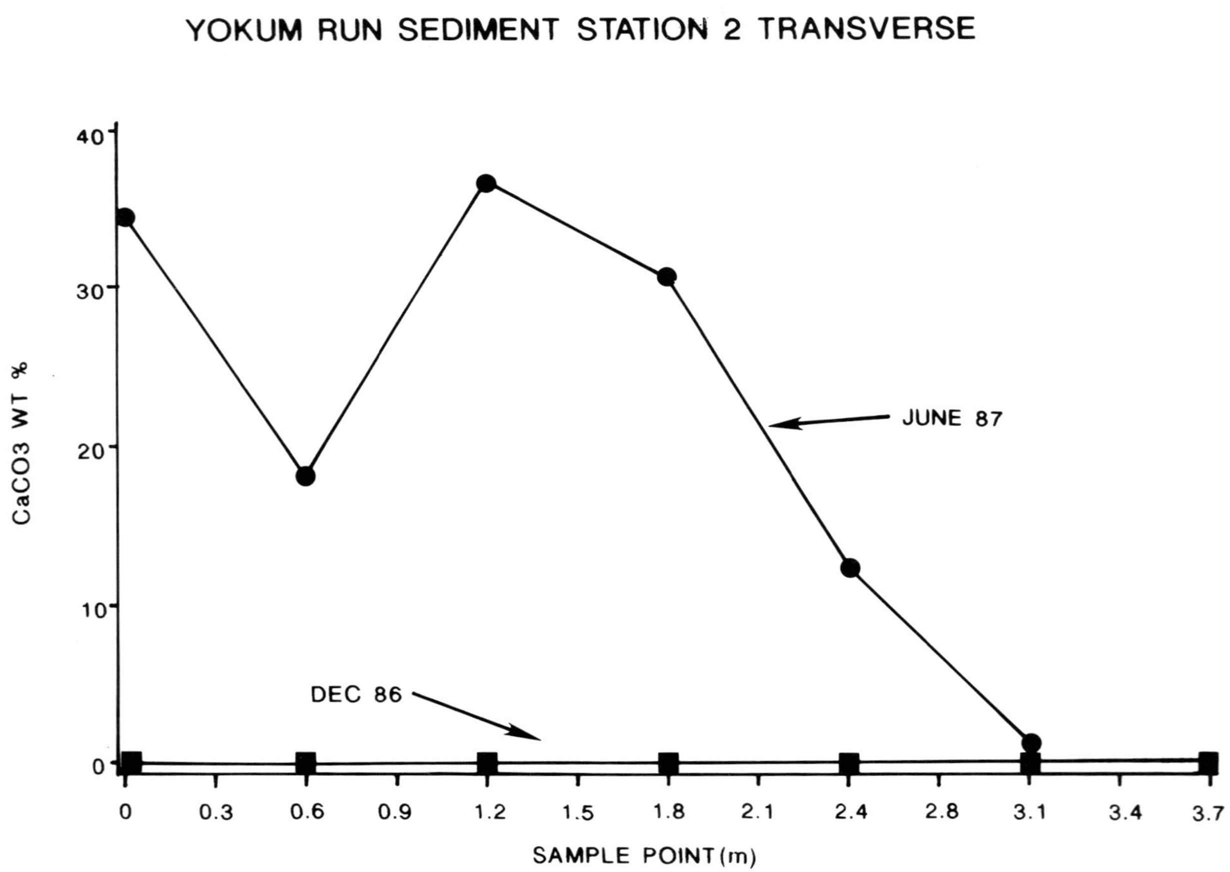

Figure 9. Cross-channel calcite values vs cross-channel sampling sites, illustrating limestone channel distribution.

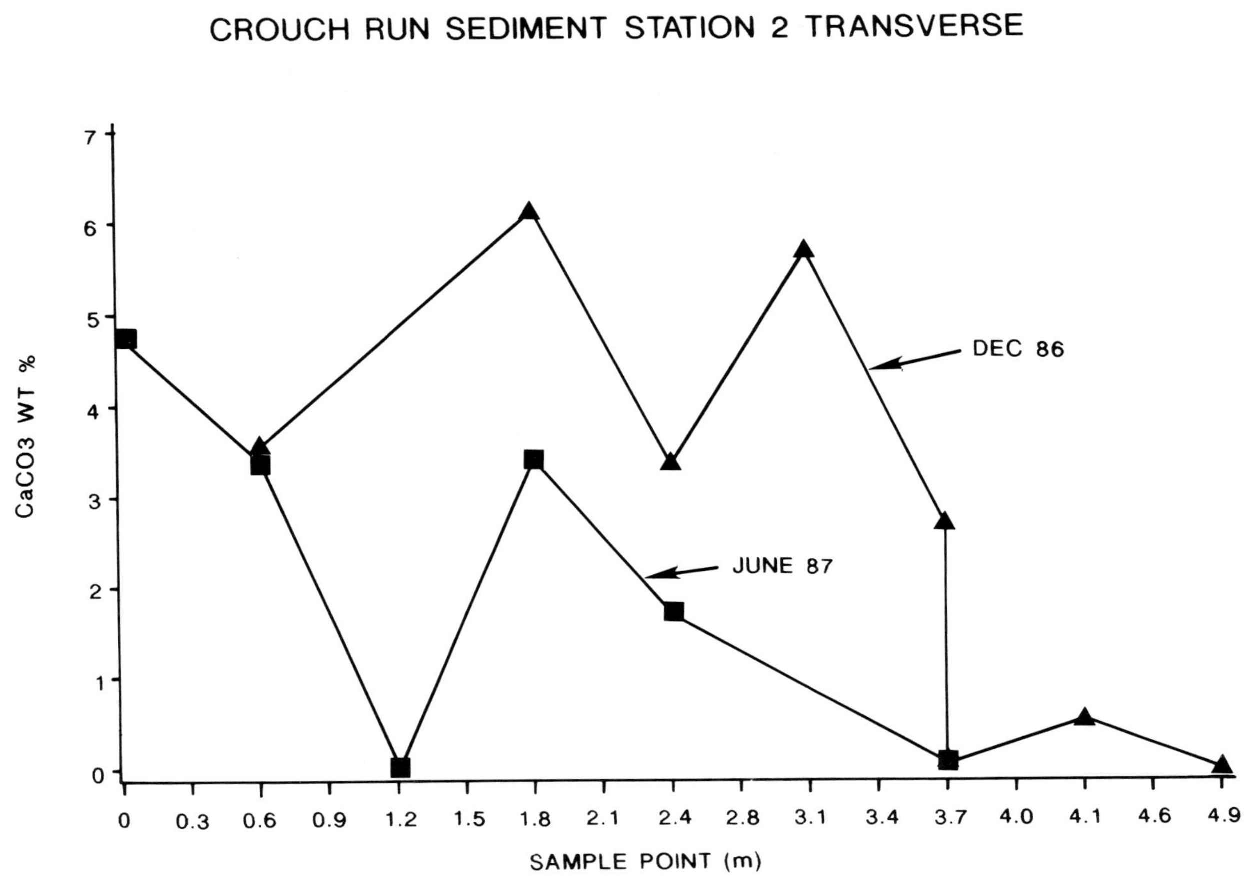

Figure 10. Cross-channel calcite values vs cross-channel sampling sites, illustrating limestone channel distribution.

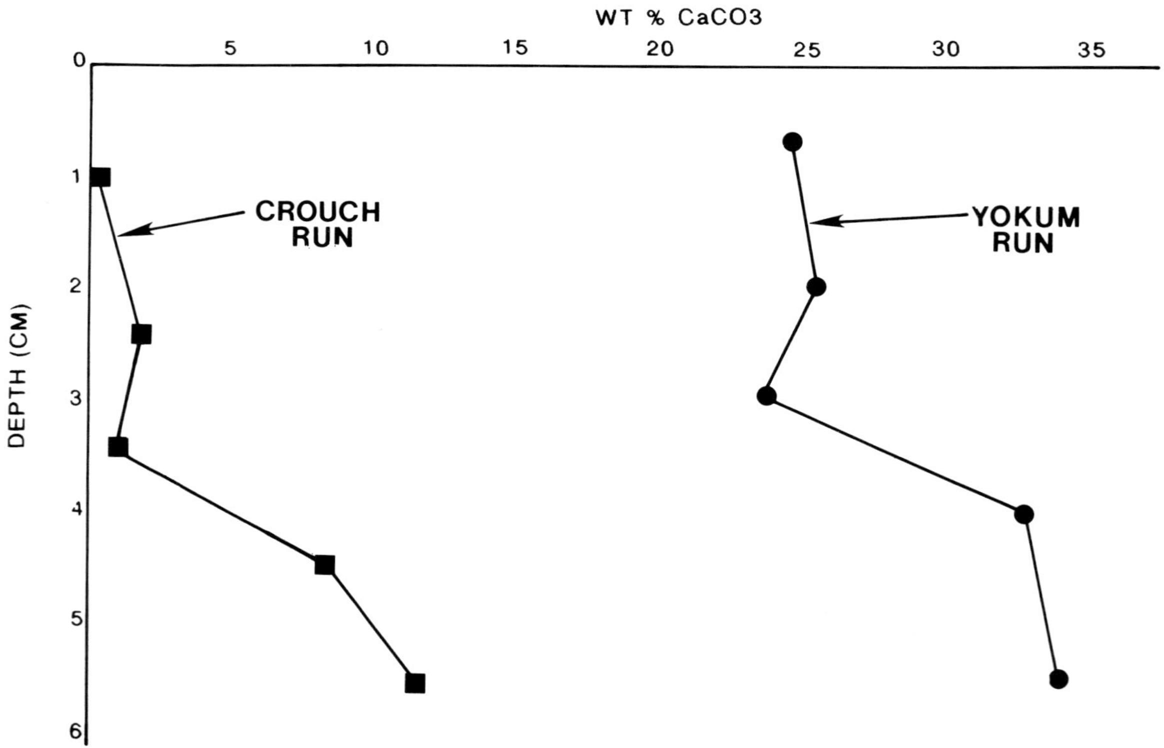

Figure 11. Calcite distribution with depth at station 2, illustrating results of the core analysis from Crouch Run and Yokum Run.

The limestone distribution within the channel is also variable. Rocks in the channel, especially in Crouch Run station 2, produce small sediment traps in which limestone fines may preferentially collect. Channel processes of erosion and deposition remove and deposit limestone fines and other sediments as described by Reading (1986). The trend of calcite distribution at Yokum Run station 2 reflects the erosion and deposition within a typical meander bend.

The results of the core sample analysis for Yokum and Crouch Runs are illustrated in Figure 11. The maximum core length was 6.5 cm with the majority of the calcite concentrated in the lower 3 cm of core. Yokum Run received nearly three times the amount of limestone than did Crouch Run which is reflected in the greater overall calcite content in the Yokum Run core. The calcite in the lower portion of the core is considered to be "in storage" until the next period of increased discharge when erosion will once again bring the limestone into contact with stream water. Movement of the sediments at an increased stream discharge will also abrade the particles, removing any precipitated deposit from the surface. The concept is very similar to that of Hasselrot and Hultberg (1984) who suggested liming near the lake shore, where wave action would "activate" the limestone particles and prevent loss of neutralization capacity through burial and coatings. Even when the limestone is buried, it is not totally lost to the system. Groundwater discharging upward through the sediments into the stream bottom during low discharge periods will dissolve the fines and contribute to the reduction of acidity. The buried limestone might be important by providing improved localized conditions for the survival of fish eggs.

3.3 Recommendations for Future Studies

The final objective of the study was to suggest a limestone treatment scheme and sampling methodology for future studies. Several possible variations of the point source treatment can be tried. Multiple inputs along the stream and more frequent treatments may achieve the same net result. Increasing the amount of limestone in a single treatment is also a suggestion. Changing the particle size to increase dissolution may also be an effective method for neutralizing streams. Ponding water downstream of an application site will increase the retention time at the stream water/particle interface and thereby allow for more efficient dissolution of the limestone particles. Although all the previous suggestions may be conducted as separate experiments, the best solution to the complex problem of neutralizing streams with a point source application may be a combination of the suggestions.

As for the sampling methodology, the frequency and number of samples collected often are a function of economics. Therefore, the net number of samples taken in a subsequent study will depend upon the allowance of time and funding.

4. SUMMARY

In this study, the point source limestone application showed promise as an ameliorant procedure for moderately acidic waters and seems to maintain water quality suitable for fisheries, provided methods can be found to increase effectiveness at discharges greater than 0.7 m^3 s^{-1}. Higher stream discharges resulted in pH values lower than pH 5.5. Calcium concentration, pH and ANC values increased downstream of the application point. A year after the treatment of Crouch Run, the values had not yet dropped to background levels. Yokum Run values of pH, Ca and ANC decreased within a month after application, but also had not dropped to the background levels. The data indicate that limestone reapplication would be required to maintain the stream as a fishery.

The limestone particles did not disperse throughout the entire length of the Yokum Run stream system, but were transported only a maximum distance of 562 m from the application site. Although the dispersal range of the material was short, the application was, in periods of low to normal discharges, effective in neutralization of the entire downstream waters. The limestone distribution in Crouch Run only reached a maximum distance of 122 m. The shorter transport distance could be because of the smaller initial input of limestone into Crouch Run and the sampling of the stream a year after treatment allowing time for limestone dissolution.

Although the calcite analysis of channel cross-section samples indicated variable distribution of limestone particles, calcite was found from one bank to the other. Calcite buried in the lower portions of the sediment traps is not considered lost to the system because of the possible "reactivation" of the materials during increased discharges and to the dissolution of the limestone particles by upward movement of groundwater during periods of low discharge.

Comparison of the results of XRF and HCl leach methods indicates that the leach technique, although less precise, is an economical and simple alternative to XRF in the determination of the calcite content of the stream sediment.

Suggestions for the improvement of the point source application include: multiple inputs, more frequent inputs, altering the particle sizes and damming the stream below the application site to increase the retention time. The best solution to point source application may be a combination of the previous suggestions. The sampling methodology in future studies will depend upon the amount of time and funding available to the researchers.

ACKNOWLEDGMENTS

The author would like to thank Living Lakes, Inc. and the West Virginia Department of Natural Resources for the funding of this project, and Jim Woodrum and Ray Menendez for their help in the field work.

THE EFFECTS OF STREAM LIMING ON WATER CHEMISTRY AND ANADROMOUS YELLOW PERCH SPAWNING SUCCESS IN TWO MARYLAND COASTAL PLAIN STREAMS

Anthony Janicki
Holly S. Greening

ABSTRACT

Automated stream dosers that deliver a wet slurry of calcite were installed in 1987 on two Maryland Coastal Plain streams subject to acidic pulses during rainstorms to evaluate the use of stream liming technology to maintain suitable water quality for early life stages of anadromous fish. Results of water quality sampling during baseflow conditions and hydrologic events indicated that significant changes in water chemistry occurred at the upstream (untreated) site during elevated flows on each stream. Observed responses in stream chemistry during the events included declines in pH, acid neutralizing capacity (ANC), and Ca, and increases in monomeric Al concentrations. With the addition of calcite, stream chemistry conditions at the downstream (treated) sites during each event were similar to those observed during baseflow. *In situ* bioassay experiments indicated that survival of yellow perch eggs and newly-hatched larvae may be enhanced by stream liming.

1. INTRODUCTION

1.1 Background

The effect of acidification on surface waters has become a major environmental concern in both North America and Europe. Acidification and the concomitant increases in the concentrations of toxic trace metals can adversely affect aquatic biota, especially fish (Baker, 1982; Haines, 1981; Mierle et al., 1986; Schindler, 1988). These adverse effects have led resource managers to consider the application of mitigation technologies to reduce the impact of acidification on aquatic resources (Fraser and Britt, 1982). Most notably, major mitigation programs have been established in Scandinavia and the United States (Rosseland and Skogheim, 1984; Lessmark and Thornelof, 1986; Schreiber and Britt, 1987; Brocksen et al., 1987).

Water, Air, and Soil Pollution **41** (1988) 359–383.

Recently, attention has been drawn to the potential effects of acidification on the anadromous fishes of the Atlantic Coastal Plain region (Hendrey, 1987). In Maryland, recent studies have shown that a high percentage of Coastal Plain streams is acidified or at risk from acidification (Janicki and Cummins, 1983; Janicki and Greening, 1988). In 1987, a statewide synoptic stream chemistry survey was conducted to provide estimates of the number and extent of stream resources at risk from acidification. Of 535 stream reaches sampled statewide, 32% had alkalinity values less than 200 *u*eq L^{-1}; however, in the South Coastal Plain region 74% had values less than 200 *u*eq L^{-1}. Values of pH less than 6.0 were recorded for 34.1% of the sampled research in the South Coastal Plain (Knapp et al., 1988).

Acidic pulses associated with hydrologic events have also been observed in several low order streams of the Maryland Coastal Plain (Correll et al., 1984; Janicki et al., 1986; Hall et al., 1985; Campbell et al., 1987; Greening et al., 1987). Laboratory bioassay experiments conducted with early life stages of anadromous fish have indicated that levels of pH observed during acidic pulses can adversely affect survival of striped bass (Mehrle et al., 1985; Buckler et al., 1987; Hall, 1987), blueback herring (Klauda and Palmer, 1986), and yellow perch (Correll et al., 1987). Anadromous fish stocks are at or near historically low levels in the Chesapeake Bay. Although direct causal mechanisms for the observed declines in fish stocks have not been established, acidic pulses in streams during the spawning season represent a potential obstacle to recovery of anadromous fish populations in those parts of the Coastal Plain where such episodes are common (Hendrey, 1987).

In Scandinavia, stream liming has been used as a management tool for the recovery of anadromous and non-anadromous salmonid populations in streams affected by acidic deposition (Abrahamsen and Matzow, 1984; Sverdrup, 1986). The stream liming technology currently available in Scandinavia is capable of maintaining stream water pH at specified target levels for extended periods of time despite rapid fluctuation in pH and flow, such as those observed in Maryland Coastal Plain streams. Based on the established success of stream liming technology for maintaining water quality, it was concluded that use of this technology may be effective in maintaining suitable water quality (i.e., moderate pH levels) in the spawning habitat of anadromous fishes in Maryland Coastal Plain streams.

Reported here are the results of the first year of a multi-year program designed to demonstrate the efficacy of stream liming to maintain suitable water quality for the early life stages of anadromous fishes in two streams in Maryland's Costal Plain. The two critical elements of this demonstration program include the selection of appropriate liming technologies and materials to meet the prescribed treatment objectives and an effective monitoring program to measure the success, if any, of the treatment program.

1.2 Description of Study Streams

The two streams chosen for treatment in 1987 were Bacon Ridge Branch and Mattawoman Creek (**Figure 1**). Both streams are located in the Coastal Plain physiographic province. The Coastal Plain is underlain by igneous and metamorphic rocks typical of the Piedmont physiographic province. The soils of the Coastal Plain are composed primarily of sands, silts and clays that have been eroded from the Piedmont and Valley and Ridge provinces, and have been deposited (often in layers greater than 30 m deep) within the Coastal Plain. The underlying bedrock has little effect on neutralizing acidity (Hendrey et al., 1987). Therefore, the base cation-poor soils of the Coastal Plain are the primary source of acid neutralizing capacity for the streams in this region.

Bacon Ridge Branch is a tributary of the South River, a sub-estuary location on the western shore of the Chesapeake Bay near Annapolis, MD (**Figure 1**). The drainage area of the Bacon Ridge Branch watershed is 17.9 km^2 with a 15-yr annual average discharge of 0.28 $m^3 s^{-1}$. Historically, Bacon Ridge Branch provided spawning habitat for yellow perch (*Perca flavescens*), white perch (*Morone americana*), and blueback herring (*Alosa aestivalis*) (O'Dell et al., 1975). However, the spawning population in this stream are currently nearly nonexistent (Greening et al., 1987). Given the manageable size of the stream flows, the loss of anadromous fish spawning, and poor water quality during the spawning season, Bacon Ridge Branch was chosen for treatment in the first year of this program (Greening et al., 1987).

Mattawoman Creek is a tributary to the Potomac River, located southeast of Washington D.C. (**Figure 1**). The drainage area of the Mattawoman Creek is 149.6 km^2 and the 22-yr annual average discharge is 1.5 $m^3 s^{-1}$. Currently, Mattawoman Creek supports spawning of

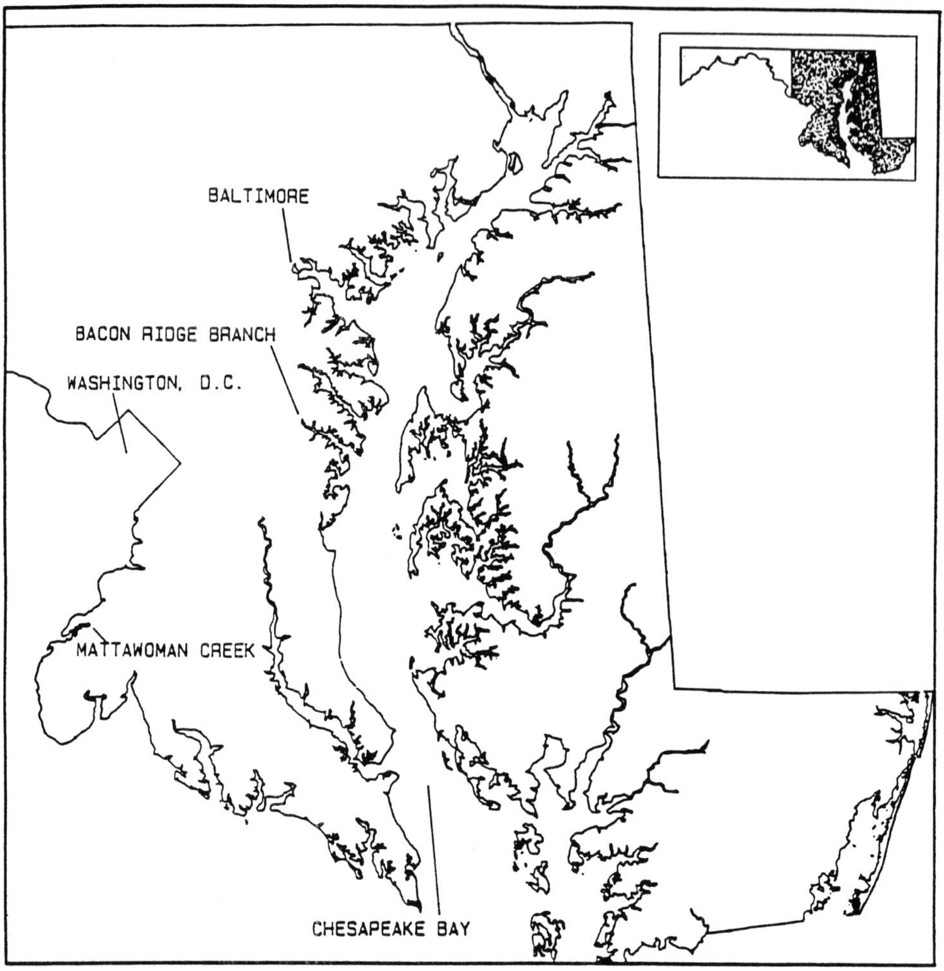

Figure 1. Location of the study streams, Maryland, USA.

anadromous fish in the tidal freshwater portions of the stream, however, much of the nontidal portion of the stream is not utilized by the spawning adult populations of anadromous fish (Janicki and Greening, 1988). In a survey of Coastal Plain streams conducted in the spring of 1983 (Ecological Analyst, 1983), pH levels less than 5.5 were commonly observed in Mattawoman Creek. These conditions and the desire to evaluate the potential benefits of stream liming on a much larger stream than Bacon Ridge Branch were the primary criteria for choosing Mattawoman Creek as a second study site (Janicki and Greening, 1988).

1.3 Description of Liming Technology

Selection of the appropriate technology for treatment of acidic pulses in flowing waters requires consideration of both the chemistry and hydrology of the streams chosen for treatment. The primary criterion used in choosing an appropriate liming technology in this program was the need to achieve water chemistry targets as rapidly and efficiently as possible, to provide treatment of the greatest extent of the spawning stream reach, and to preclude accumulation of neutralizing materials in the tidal portions of the treated streams. The second criterion was to use a technology capable of maintaining stream pH at target levels despite the wide variation in stream conditions that occurs during acidic episodes in Coastal Plain streams (Greening et al., 1987) Of the available technologies for stream liming, mechanical dosing devices provide the only method capable of adjusting the dosage of neutralizing material to maintain target water quality in streams with variable flow and pH values.

The DeAcid computer program (Saunders et al., 1985) was used to assist in the selection of an appropriate mechanical dosing technology. The critical feature of the DeAcid program is a model that simulates the dissolution of $CaCO_3$ (calcite) in natural waters. Model inputs include information on hydrology, water chemistry, the chemical and physical properties of the calcite, characteristics of the treatment site and the target pH and acid neutralizing capacity (ANC) to be achieved by calcite addition. After several runs, using a range of values for the input variables, the output from DeAcid suggested automated wet-slurry doser technology as the most appropriate stream liming method for use in the Maryland Coastal Plain study streams, primarily due to requirements for rapid, controlled acid neutralization.

Wet slurry doser technology requires an extremely fine calcite slurry with a particle size less than 5 *u*m for optimal performance. Coarser powders may clog the dispensing mechanisms and storage silos of these devices; they also tend to settle too quickly in the stream to provide sufficient ANC in downstream waters. The calcite selected for use in the dosers in this study was 98% $CaCO_3$ by weight, with a mean particle size of 1.9 *u*m.

The automated wet slurry doser units selected for use in Maryland are composed of a storage tank, a pump/control house, and an intake well. Located inside the pump/control house are a microprocessor that controls the calcite dose, a slurry pump, and a mixing tank. A control unit monitors stream stage using a pressure transducer located in the intake well and calculates stream discharge and the predicted stream water pH based on the stage datum. The relationship between stream pH and flow at both Bacon Ridge Branch and Mattawoman Creek is presented in Figure 2. The amount of calcite required to achieve the target pH, which was 6.5 in this study, is calculated and translated into a slurry pump duty cycle. The slurry pump transfers the concentrated calcite slurry to the mixing tank where the slurry is diluted with ambient stream water pumped from the intake well. This dilute slurry is pumped to a perforated discharge pipe located on the stream bottom immediately downstream of the intake.

1.4 Hypotheses and Study Elements

Three hypotheses are being tested in this study: 1) wet slurry doser technology will neutralize acidic pulses in Maryland Coastal Plain streams; 2) wet slurry doser technology will maintain suitable water quality for early life stages of anadromous fish species and 3) wet slurry doser technology will enhance the survival of early life stages of anadromous fish species. The first hypothesis provides for testing the short-term response capability of the stream liming technology with regard to episodic changes in water chemistry. The second hypothesis is related to reliability and maintainability of the stream liming technology over the long term, such as the duration of the spawning season. The third hypothesis provides for a

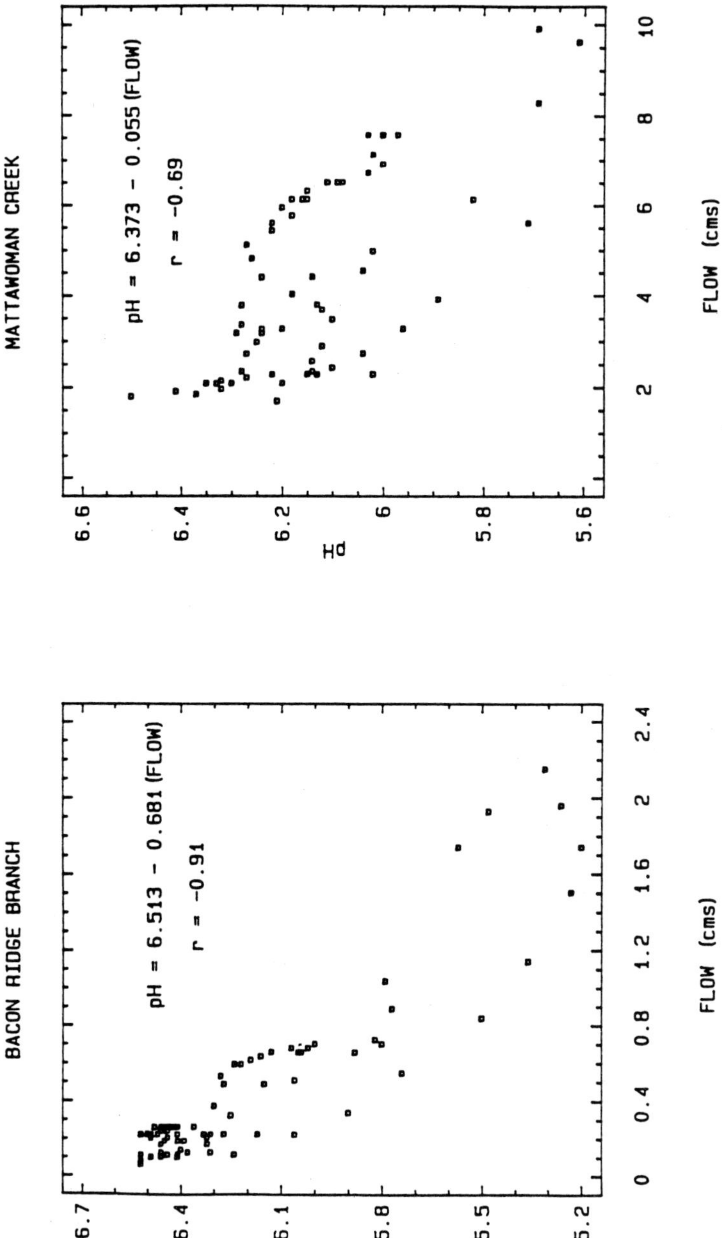

Fugure 2. Stream flow - pH relationships for Bacon Ridge Branch and Mattawoman Creek in 1987.

The study elements employed to test these hypotheses were: 1) water quality sampling during baseflow and elevated flow conditions; and 2) *in situ* bioassay experiments using early life stages of yellow perch (eggs and newly-hatched larvae).

2. STUDY METHODS

Water quality sampling was conducted in the two treated streams, Bacon Ridge Branch and Mattawoman Creek, during the spring of 1987. Sampling sites were located upstream of the point of calcite discharge (the upstream or control site) and at the lower end of the treated stream reach (the downstream or target site) on each of the treated streams. At Bacon Ridge Branch, the length of treated stream was ca. 1.5 km; at Mattawoman Creek the length of treated stream was ca. 9.6 km.

The field sampling and laboratory analysis protocols followed those prescribed for the Living Lakes Inc. demonstration program (Living Lakes, 1987). In general, these methods are comparable to those recommended for the U.S. Fish and Wildlife Service Acid Precipitation Mitigation Program (APMP) (Saunders et al., 1985) and the U.S. EPA National Surface Water Survey Stream Survey (Hillman et al., 1986)

Hydrologic event sampling was conducted at the two treated streams to characterize the short-term responses in stream chemistry to elevated stream flows. Sampling began at or just prior to the onset of precipitation. On the rising limb of the hydrograph, the temperature was measured hourly and grab samples were collected every 4 hr. Over the course of the event, samples were collected for analysis of the complete suite of analytes over every 24 hr, in addition to a limited suite sample collected every 8 hr for water chemistry analysis of pH, ANC, turbidity, conductivity, major anions and cations, dissolved organic carbon (DOC), dissolved inorganic carbon (DIC) and the monomeric Al fractions. Monthly sampling for complete water chemistry was also conducted during baseflow conditions at both streams.

In situ bioassay experiments with yellow perch eggs were conducted on Bacon Ridge Branch and Mattawoman Creek in 1987. In each of three experiments, four flow-through chambers, each containing 200 to 300 fertilized eggs, were placed at the upstream and downstream sites selected for stream chemistry sampling. Eggs used in the bioassay experiments were stripped from ripe females, dry-fertilized with milt from several males, and water- hardened in water from the test sites. Four females were used for each experiment. Mortality counts were made every second day. The three experiments were begun at 5 to 7 day intervals, and continued until 4 days after hatching (i.e., free-swimming larvae were present) was observed. At the termination of each experiment, all eggs and larvae were collected, preserved, and final counts were made in the laboratory. Comparison of the total mortality estimates observed upstream and downstream of the point of calcite addition was accomplished using a paired-sample t-test.

3. RESULTS

3.1 Bacon Ridge Branch - Water Quality

The monthly baseflow stream chemistry data for Bacon Ridge Branch are presented in Table I. Throughout the March to June sampling period, of the pH observed at the upstream site was consistently between 6.45 and 6.5, while at the downstream site pH ranged from 6.65 to 6.81. Both ANC and Ca concentrations were also consistently lower at the upstream site. There was no apparent difference between the concentration of either total monomeric Al or the inorganic monomeric Al fraction observed upstream and downstream at Bacon Ridge Branch under baseflow conditions.

During the first event sampled on Bacon Ridge Branch (30 March to 2 April), 13.6 mm of precipitation fell resulting in an increase in stream flow from 0.08 to 0.25 $m^3 s^{-1}$ which 12 hr of the onset of the event (**Figure 3**). Concomitantly, the upstream pH declined from 6.52 to 6.35; the downstream pH remained above 6.6 throughout the event. ANC values remained above 200 μeq L^{-1} downstream, but declined slightly from 190 to 150 μeq L^{-1} at the upstream sampling site. Consistently higher downstream Ca concentrations also reflect the addition of calcite. Both total monomeric Al and inorganic monomeric Al exhibited slightly higher values (0.01 to 0.02 mg L^{-1}) at the upstream site throughout the event (Figure 3).

Table I

Monthly Baseflow Chemistry Data for Upstream and Downstream Sites on Bacon Ridge Branch, 1987

SITE	10 March Upstream	10 March Downstream	9 April Upstream	9 April Downstream	13 May Upstream	13 May Downstream	4 June Upstream	4 June Downstream
Flow (cms)	0.126		0.103		0.126		0.065	
pH	6.46	6.65	6.50	6.72	6.50	6.48	6.84	6.81
ANC (ueq/L)	173.2	208.4	171.4	219.1	208.9	338.4	291.3	330.1
Calcium (ueq/L)	406.0	414.6	400.4	417.5	362.6	454.6	440.4	481.1
Total Dissolved Al (mg/L)	0.016	0.009	0.011	0.005	0.127	0.028	0.009	0.008
Total Dissolved Al (mg/L)	N/A	N/A	0.012	0.014	0.014	0.013	0.009	0.013
Inorganic Monomeric Al (mg/L)	N/A	N/A	0.007	0.007	0.002	0.000	0.003	0.002

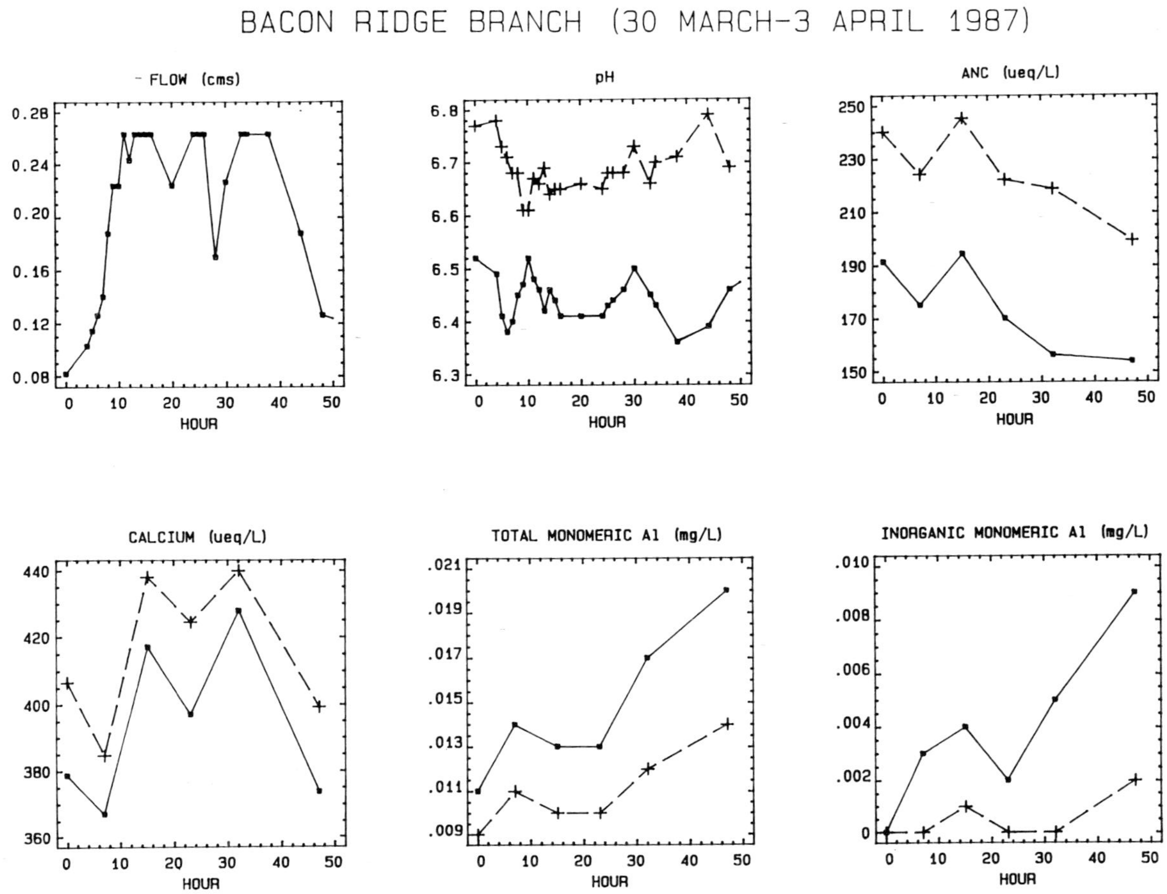

Figure 3. Upstream (o) and downstream (+) flow, pH, ANC, Ca, total monomeric Al, and inorganic monomeric Al values on Bacon Ridge Branch: 30 March - 3 April, 1987.

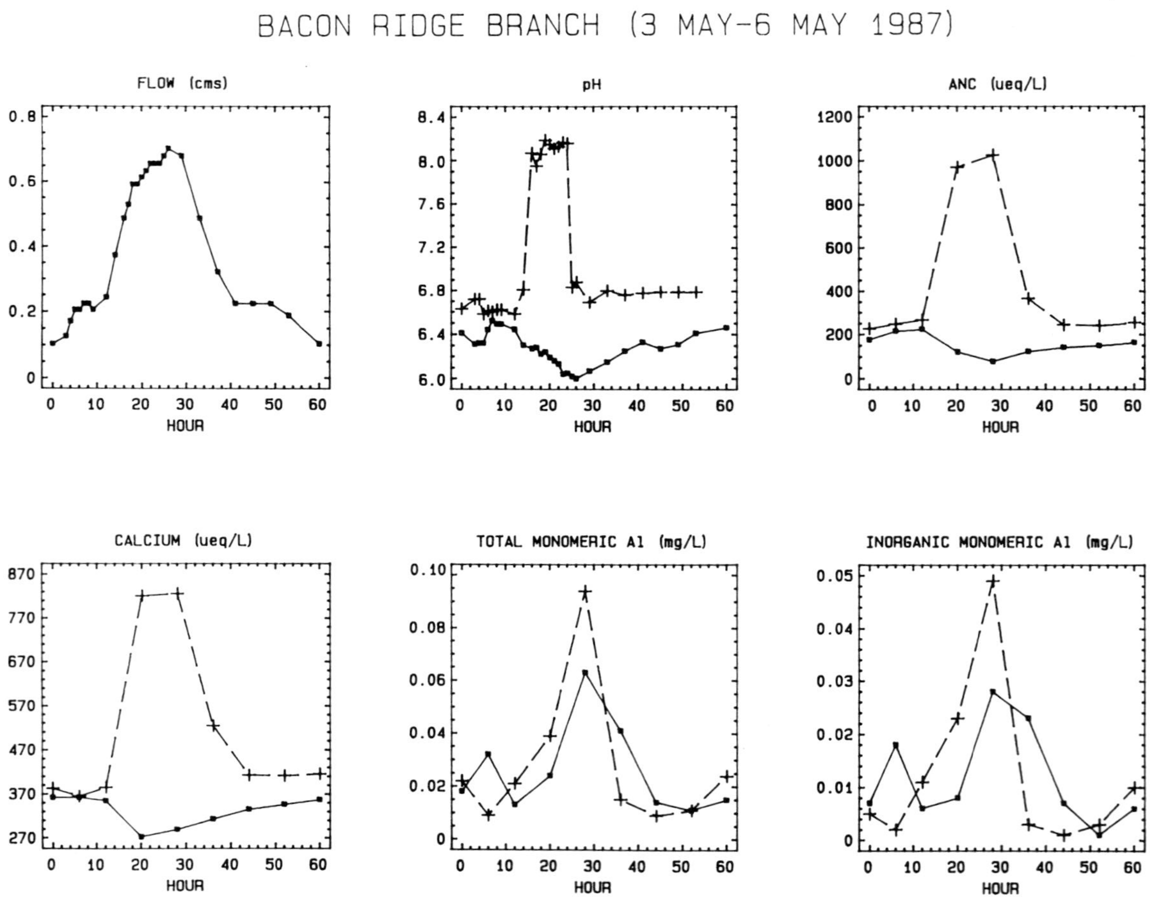

Figure 4. Upstream (o) and downstream (+) flow, pH, ANC, Ca, total monomeric Al, and inorganic monomerica Al values on Bacon Ridge Branch: 3 May - 6 May, 1987.

The second event sampled at Bacon Ridge Branch (3 to 6 May) was slightly larger with 28.6 mm of rain falling and an increase in flow from 0.1 to 0.7 $m^3\ s^{-1}$ within 24 hr (**Figure 4**). Despite this large increase in flow, upstream pH declined only slightly from 6.4 to 6.0 coincident with the increase in flow. As a result of the increase in flow with a smaller than expected decease in pH, the amount of calcite added was considerably higher than required and the downstream pH increased from 6.6 to 8.1 in a 12 hr period. During this period, ANC and Ca concentrations also increased significantly at the downstream site. Maximum total and inorganic monomeric Al concentrations were observed at both sites at the peak in the hydrograph for this event.

A total of 41 mm of rain fell during the third event sampled at Bacon Ridge branch, 4 to 7 June, resulting in more than an order of magnitude increase in stream flow (**Figure 5**). The upstream pH declined from 6.5 to 5.2 on the rising limb of the hydrograph. At the downstream site, pH declined from 6.8 to 6.1 before recovering to values between 6.2 and 6.4 for most of the event. ANC remained above 180 *u*eq L^{-1} at the downstream site, while a minimum of 19.6 *u*eq L^{-1} was observed at the upstream site. Calcium concentrations were also consistently higher downstream during this event. Major between-site differences in the concentrations of both monomeric Al fractions were observed during the course of this event. Maxima of 0.124 mg L^{-1} total monomeric Al and 0.097 mg L^{-1} inorganic monomeric Al were recorded at the upstream site. In contrast, the total and inorganic monomeric Al concentrations at the downstream site increased only slightly to values of 0.032 and 0.007 mg L^{-1}, respectively **(Figure 5)**.

3.2 Mattawoman Creek - Water Quality

At Mattawoman Creek, the March monthly baseflow water quality samples were taken prior to installation of the wet slurry doser. The upstream pH was 6.12, while the downstream pH was 6.25 **(Table II)**. Both ANC and Ca concentrations were similar between the upstream and downstream sites. After the doser was installed and operating, the baseflow pH and concentrations of ANC and Ca were higher at the downstream site (**Table II**). The baseflow

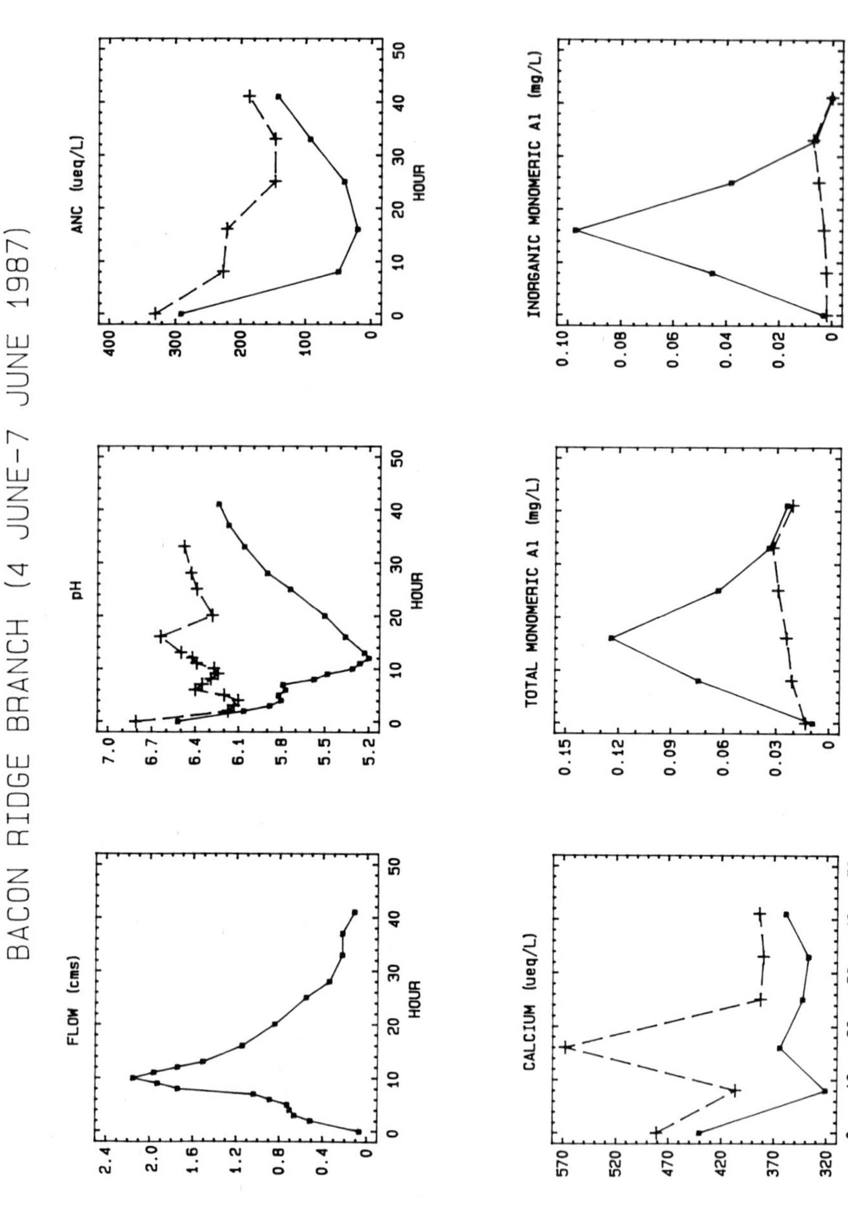

Figure 5. Upstream (o) and downstream (+) flow, pH, ANC, Ca, total monomerica Al, and inorganic monomeric Al values on Bacon Ridge Branch: 4 June - 7 June, 1987.

Table II

Monthly Baseflow Chemistry Data for Upstream and Downstream Sites on Mattawoman Creek, 1987

	10 March		**9 April**		**13 May**		**4 June**	
SITE	Upstream	Downstream	Upstream	Downstream	Upstream	Downstream	Upstream	Downstream
Flow (cms)	2.90		2.95		1.55		1.70	
pH	6.12	6.25	6.47	6.90	6.70	N/A	6.19	6.33
ANC (ueq/L)	48.9	55.7	72.8	160.1	160.2	271.2	129.8	174.2
Calcium (ueq/L)	274	272.8	241.0	349.3	231.4	346.7	225.6	277.0
Total Dissolved Al (mg/L)	0.154	0.121	0.087	0.087	0.146	0.118	0.098	0.247
Total Dissolved Al (mg/L)	N/A	N/A	0.024	0.021	0.019	0.023	0.046	0.044
Inorganic Monomeric Al (mg/L)	N/A	N/A	0.006	0.008	0.002	0.007	0.000	0.000

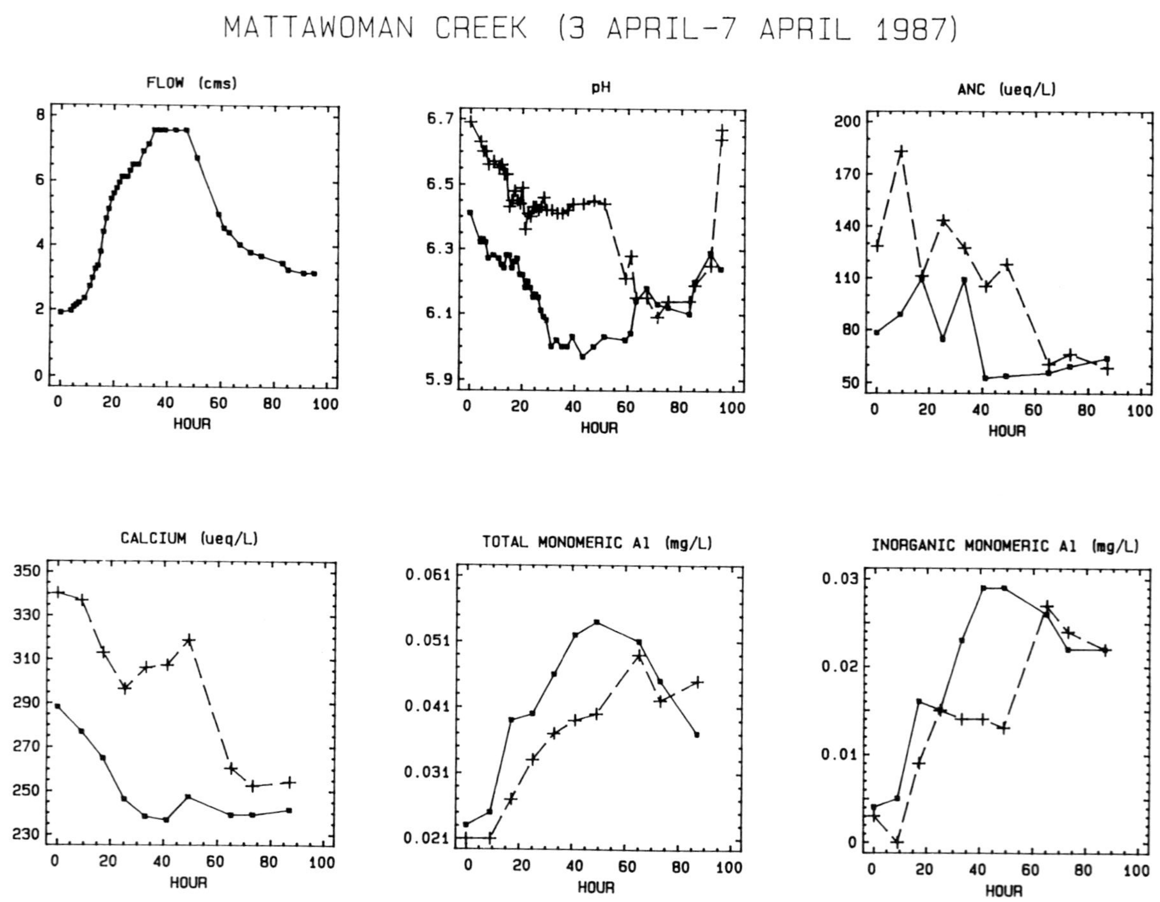

Figure 6: Upstream (o) and downstream (+) flow, pH, ANC, Ca, total monomeric Al values values on Mattawoman Creek: 3 April - 7 April 1987.

monomeric Al concentrations were similar at both sites on Mattawoman Creek from April through June.

The first event sampled on Mattawoman Creek occurred from 3 April to 7 April; the total precipitation during this event was 34.1 mm. There was a ca. four-fold increase in flow during this event, with the peak flow occurring 36 to 42 hr after the onset of the event (**Figure 6**). Upstream pH declined from 6.4 to 6.0 and closely tracked the rising limb of the hydrograph. As expected, upstream ANC and Ca concentrations also declined during this period. Downstream pH was maintained above 6.4 until the calcite discharge pipe became clogged at hr-55 of the event. The pH then declined to levels similar to those observed upstream until the pipe was unclogged at hr-90. Downstream ANC and Ca concentrations closely followed the changes in pH, declining to values obtained at the upstream site during this period (**Figure 6**). During the first two days of this event, the monomeric Al concentrations were consistently, but only slightly higher upstream than downstream. As with ANC and Ca, upstream and downstream monomeric Al concentrations were similar during the period when calcite was not discharged.

Two consecutive events were sampled on Mattawoman Creek from 21 June through 25 June; both of these events resulted from short, intense convective storms. A total of 26.1 mm of rain fell within 2 hr on 21 June, and an increase in stream flow was recorded within 1 hr of the onset of the event (**Figure 7**). After a maximum of 2.3 cm was observed at hr-6, flow began to decline. A second, very intense storm totaling 48 mm of additional precipitation occurred 24 hr after the initial rainfall, resulting in a sharp and immediate response in flow, increasing five-fold within a 4-hr period. The upstream pH declined from 6.4 to 5.6 during these events, while the downstream pH remained about 6.4 (**Figure 7**). ANC values were above 180 *u*eq L-1 at the downstream site, while upstream a minimum of 60 *u*eq L-1 was recorded. Calcium concentrations were consistently higher at the downstream site throughout these events. Total monomeric Al concentrations were also higher at the upstream site, especially during the peak in the hydrograph when a value of 0.086 mg L^{-1} was observed upstream. Inorganic monomeric Al was not detected at either site during these events on Mattawoman Creek (**Figure 7**).

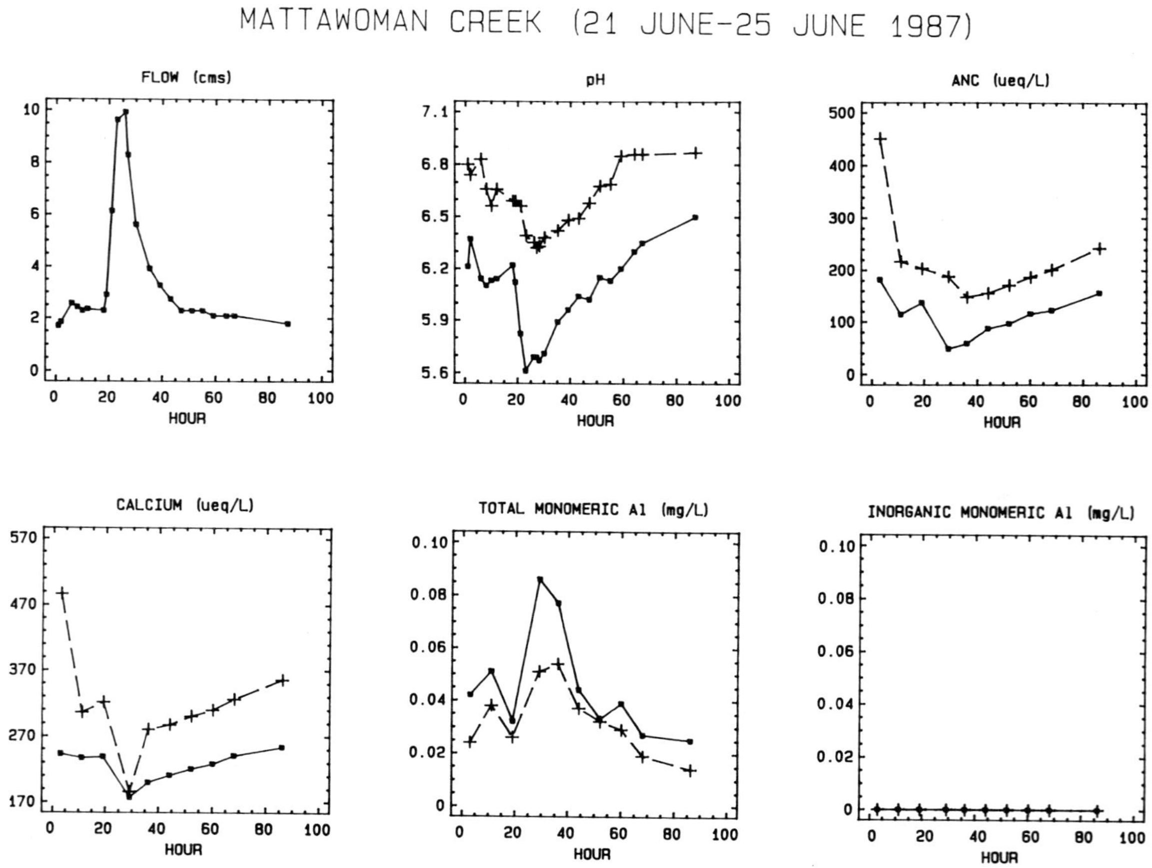

Figure 7: Upstream (o) and downstream (+) flow, pH, ANC, Ca, total monomeric Al, and inorganic monomeric Al vlaues on Mattawoman Creek: 21 June - 25 June, 1987.

3.3 Yellow Perch In Situ Bioassays

The results of the three yellow perch *in situ* bioassays conducted at Bacon Ridge Branch are presented in Table III. Observations of mortality were made for three developmental stages: undeveloped eggs, eggs with clearly visible developed embryos (termed "developed eggs"), and yolk-sac larvae.

In the first experiment, the total mortality at the upstream site was 72.2%, the majority being observed in the undeveloped egg stage. At the downstream site, the total mortality was significantly lower (18.4%) with most of the mortality observed in the developed egg stage.

During the second experiment on Bacon Ridge Branch, the total mortality observed at the upstream site was also significantly greater than at the downstream site (**Table III**). Unlike the first experiment, most of the total mortality was observed in the developed egg stage. Similar results were obtained in the third experiment, however, the difference between the two sites was not statistically significant due to high variability in the mortality estimates at the upstream site (**Table III**).

The organisms examined during these experiments in Bacon Ridge Branch were exposed to a single hydrologic event that occurred on 30 March and continued through 3 April (**Figure 3**). The minimum upstream pH observed during this event was 6.38 as compared to a minimum of 6.6 at the downstream site. Total and inorganic monomeric Al concentrations were consistently higher at the upstream site during the event.

The results of the yellow perch bioassays conducted in Mattawoman Creek are presented in Table IV. There were relatively greater numbers of larvae hatched in the Mattawoman Creek as compared to the number observed in Bacon Ridge Branch. As a result, the majority of the mortality observed in Mattawoman Creek during all three experiments occurred in the yolk-sac larval stage. Unlike Bacon Ridge Branch, the total mortality observed at the upstream and downstream sites did not differ during the first two experiments at Mattawoman Creek. During the third experiment, the total downstream mortality was significantly greater than at the upstream site.

The results obtained at Mattawoman Creek were likely due to the response in the stream chemistry during the hydrologic event that occurred from 3 through 7 April (**Figure 6**).

Table III

Mattawoman Creek Yellow Perch In-Situ Bioassay Experiments. Mean percent mortality by experiment. Totals for each experiment tested with a paired-sample t-test.

Experiment	Date	Life Stage	% Mortality Upstream	% Mortality Downstream	p>t
1	3/13-4/9	Eggs	0	0	
(n=4 pairs)		Developed eggs	8.4	1.1	
		Yolk-sac Larvae	8.5	23.4	
		Total	16.9±9.5	24.5±14.0	0.593
2	3/20-4/10	Eggs	0	0	
(n=4 pairs)		Developed eggs	7.0	6.3	
		Yolk-sac Larvae	1.4	3.3	
		Total	8.4±5.6	9.6±10.4	0.830
3	3/25-4/15	Eggs	0	0	
(n=4 pairs)		Developed eggs	8.8	1.6	
		Yolk-sac Larvae	0.2	22.3	
		Total	10.4±5.0	22.5±3.6	0.042

Table IV

Bacon Ridge Branch Yellow Perch In-Situ Bioassay Experiments. Mean percent mortality by experiment. Totals for each experiment tested with a paired-sample t-test.

Experiment	Date	Life Stage	% Mortality Upstream	% Mortality Downstream	p>t
1	3/18-4/13	Eggs	50.0	0	
(n=4 pairs)		Developed eggs	20.9	12.4	
		Yolk-sac Larvae	1.3	6.0	
		Total	72.2±32.3	18.4±18.6	0.007
2	3/24-4/13	Eggs	17.8	1.4	
(n=4 pairs)		Developed eggs	23.1	11.9	
		Yolk-sac Larvae	0.3	0.7	
		Total	41.2±27.3	13.3±20.0	0.083
3	3/26-/15	Eggs	0.3	1.9	
(n=4 pairs)		Developed eggs	19.3	1.9	
		Yolk-sac Larvae	2.0	3.3	
		Total	21.6±23.7	7.1±1.3	0.330

During this event there was a disruption in the addition of calcite mid-way through the event. This resulted in a very rapid decline in pH at the downstream site, more rapid than that observed at the upstream site on the rising limb of the hydrograph. During this period, there was also a rapid decline in the downstream Ca concentration and an increase in the concentrations of both total and inorganic monomeric Al.

4. DISCUSSION

The results obtained in 1987 represent the initial evaluation of the chemical and biological responses of two Coastal Plain streams to the addition of calcite during hydrologic events. The stream chemistry data indicate that both streams become more acidic during these events. pH and Ca declined and monomeric Al concentrations increased dramatically during the initial hours of elevated flow at the upstream sites of both streams. With few exceptions (such as the clogged discharge pipe experienced at Mattawoman Creek), downstream water quality during elevated flows was maintained at levels similar to those observed during baseflow conditions. It appears, therefore, that the wet slurry doser technology is capable of mitigating the episodic increases in stream acidity which commonly occur during hydrologic events.

Results of the *in situ* yellow perch bioassay experiments indicate that neutralization of the episodic increases in stream acidity by calcite addition may enhance the survival of the early life stages. At Bacon Ridge Branch, significantly greater mortality was recorded at the upstream site in two of three experiments. Much of the mortality observed in the Bacon Ridge Branch experiments occurred during the stage when well-developed embryos were readily apparent within the egg membrane. These observations compare well with those of Schofield and Driscoll (1987) in which yellow perch embryos exposed to acidic water apparently developed normally yet were unable to hatch and ultimately died.

The between-site differences in the mortality of the early life stages of yellow perch may be due to the observed between-site differences in stream chemistry. During the bioassay experiments at Bacon Ridge Branch, baseflow pH was maintained at consistently higher levels at the downstream site. A hydrologic event occurred during the experiments;

during this event upstream pH and Ca concentration declined and the monomeric Al concentrations increased, while downstream pH and Ca were consistently higher and monomeric Al concentrations were lower. Figure 8 presents the relationship between pH and the concentration of inorganic monomeric Al observed at Bacon Ridge Branch. These data indicate that maintaining pH above 6.0 via addition of calcite resulted in much reduced concentrations of this potentially toxic form of Al (Baker and Schofield, 1980). Additionally, earlier research has shown that neutralization of acidic water with $CaCO_3$ provides added protection to fish from toxic Al and other metals due to the competition between the Ca^{+2} ion and other divalent and trivalent metal ions for binding sites on the gill surface, thereby reducing the uptake of the toxic metal ions (Brown, 1983; Hunn et al., 1985).

The yellow perch bioassay results at Mattawoman Creek did not show the same pattern as those at Bacon Ridge Branch. In two of the experiments there were no significant between-site differences in mortality of the early life stages and in one experiment there was significantly higher mortality at the downstream site. These results were likely due to the rapid decrease in pH at the downstream site after calcite addition was disrupted during the hydrologic event that occurred while the bioassay experiments were being conducted. The downstream changes in pH were more rapid than the changes in pH observed upstream on the rising limb of the hydrograph for this event. Also, at the lower pH levels experienced during the bioassay experiments at Mattawoman Creek, total and inorganic monomeric Al concentrations both were elevated (**Figure 8**). It is yet unknown, however, whether these Al concentrations are toxic to the yellow perch early life stages. The results of these experiments indicate the need for maintaining relative constant pH above some critical threshold to allow successful completion of the early life stages by yellow perch.

Some problems remain that must be considered before wide scale application of wet slurry doser technology to mitigate acidic conditions in Coastal Plain streams is recommended. First, the disruption of calcite addition observed at Mattawoman Creek indicates the need for very thorough routine maintenance of the dosers to preclude clogging or other malfunctions. The anomalous response in the pH of Bacon Ridge Branch during one event resulted in a higher than required calcite dose that increased the pH to levels greater than 8.0 and led to

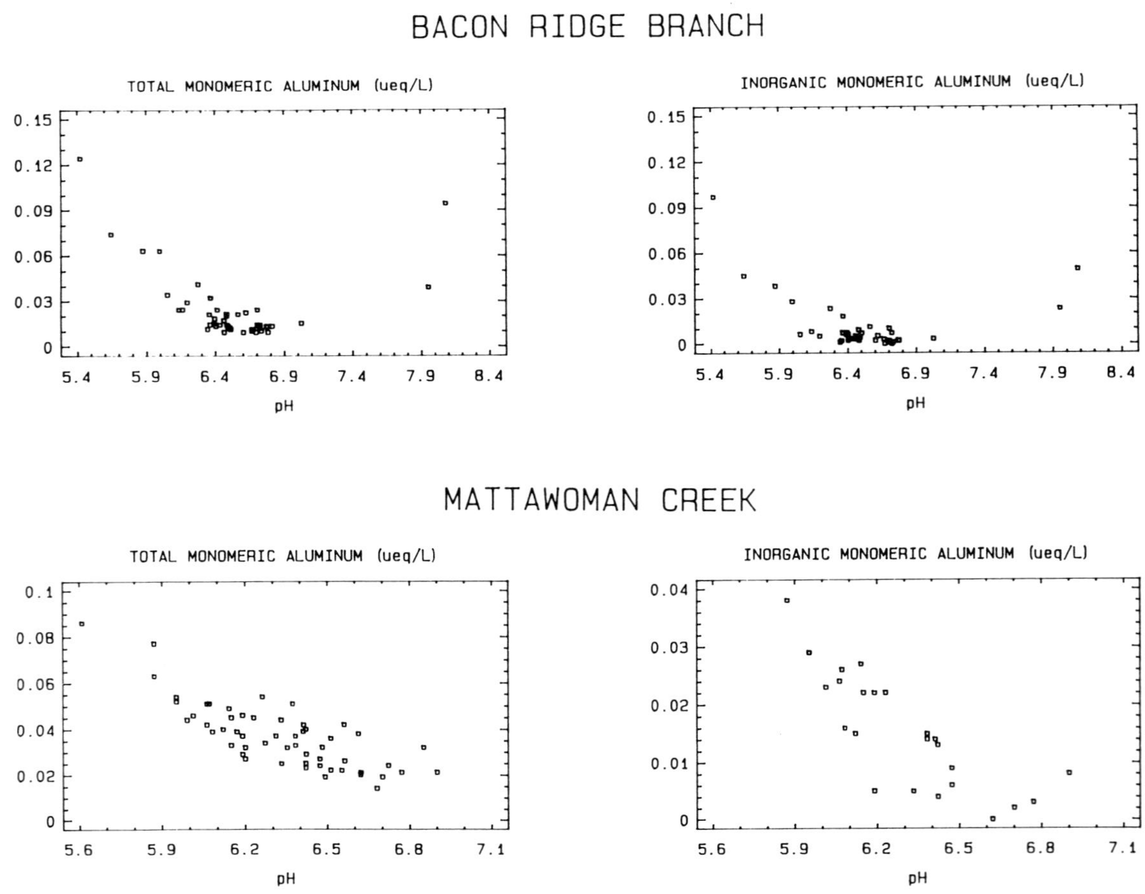

Figure 8. Relationships between pH and total and inorganic monomeric Al concentrations in Bacon Ridge Branch and Mattawoman Creek.

elevated concentrations of monomeric Al. Finally, evaluation of the effects of calcite addition on other anadromous fish species is also required.

ACKNOWLEDGMENTS

The authors would like to thank Robert Brocksen and Michael Bowman for the thoughtful consideration of this project and their helpful comments and critiques. We would also like to especially thank B. Killen, M. Pinder, K. Bugley, G. Filbin and P. Saunders for their efforts on this project. This project was jointly funded by Living Lakes, Inc. and the Maryland Power Plant Research Program.

EVALUATION OF THREE EXPERIMENTAL LOW-TECHNOLOGY APPROACHES TO ACID MITIGATION IN HEADWATER STREAMS

Dean E. Arnold
William D. Skinner
David E. Spotts

ABSTRACT

Although powered dosing devices may be effective in regulating pH and acid neutralizing capacity in acid-sensitive streams, their high costs and maintenance requirements may make them unsuitable for most sites. Three alternative approaches were tried on a mountain trout stream with pH 4.5 to 5.3 and essentially no acid neutralizing capacity: (1) A barrier of wire gabions across the stream, filled with limestone chunks, raised pH about 0.5 unit for about 2 weeks. Its effect then became undetectable. Effectiveness could have been improved by using high-Ca (>95%) stone of more uniform size (5 cm in diameter), and improving the control of leaf accumulation. No biological effects of the device were detected. (2) Metal mesh baskets fed by gravity with soda ash (sodium carbonate) briquettes worked well in raising pH by as much as 1.5 units and acid neutralizing capacity by as much as 40 $\mu eq.L^{-1}$ for several months. However, it proved difficult to provide adequate replenishment to keep these flow-regulated devices operating during high-flow periods. (3) To reduce consumption of soda ash, we tested devices that dissolved the briquettes in an enclosed chamber and fed the concentrated sodium carbonate solution into the stream. These provided adequate neutralization when operating properly but frequently failed due to pipe clogging and to the hardening of the briquettes into a solid mass. Studies of benthic invertebrate drift and abundance associated with the soda ash treatment showed no significant effect in a few hours or in 60 days.

1. INTRODUCTION

Interest has been high in developing ways to mitigate acidity in streams and lakes since the beginning of the current awareness of acid precipitation and its adverse effects on aquatic ecosystems (Oden, 1968, 1976; Likens et al. 1972; Cowling, 1982). Although few observers consider mitigation to be an ultimate solution or a panacea for the acid precipitation problem, many view it as necessary to protect or rehabilitate waters that otherwise would become fishless, or to restore fish populations to waters now too acid to support them. Since

Water, Air, and Soil Pollution **41** (1988) 385–406.

even immediate action to prevent acid precipitation at its source would involve much delay (a common estimate is 10 yr from the decision date to the time when acidity could be significantly reduced), mitigative action seems desirable.

Fraser and Britt (1982) produced the first comprehensive review of mitigative methods either in use or proposed for combating the effects of acid precipitation. They noted that a number of treatment programs were already operational in Scandinavia but that such treatment was in its infancy in North America.

Many of the methods reviewed by Fraser and Britt (1982), although effective, were either very expensive or required frequent maintenance, or both. Explored are the development of low-cost, low-maintenance approaches to the problem by performing a series of low-cost, low-technology mitigation experiments over a period of several years on a stream system on public land with limited access but high recreational interest. Because much of the materials and labor were donated, we were unable to make useful cost comparisons among methods. However, chemical and biological data were collected and several independent biological studies were connected with the experiments, enabling us to make some general conclusions about the relative effectiveness and desirability of three devices: (1) limestone flow-through barrier, (2) soda ash briquette baskets, and (3) soda ash dissolution tanks.

2. STUDY SITES

The Mosquito Creek drainage basin (165 km^2) is enclosed almost entirely within publicly owned forest land in Clearfield and Elk counties of north-central Pennsylvania (**Figure** 1). It is underlain by insoluble sandstone bedrock and very thin soils that contribute little acid neutralizing capacity to associated waters. Water chemistry was monitored in Mosquito Creek and several of its tributaries, including Gifford Run (a primary study site), frequently in 1975-1988. Summaries of pH, acid neutralizing capacity, and conductivity of Gifford Run in 1975-1988 are shown in Figure 2. These values are typical of streams in the area, which tend to be quite acidic and to have little or no acid neutralizing capacity. The watershed of Gifford Run above the study site is 16.8 km^2 and its channel slope is 0.025. Hale (1978) gave a detailed description of the Gifford Run study site. Related experiments were conducted in

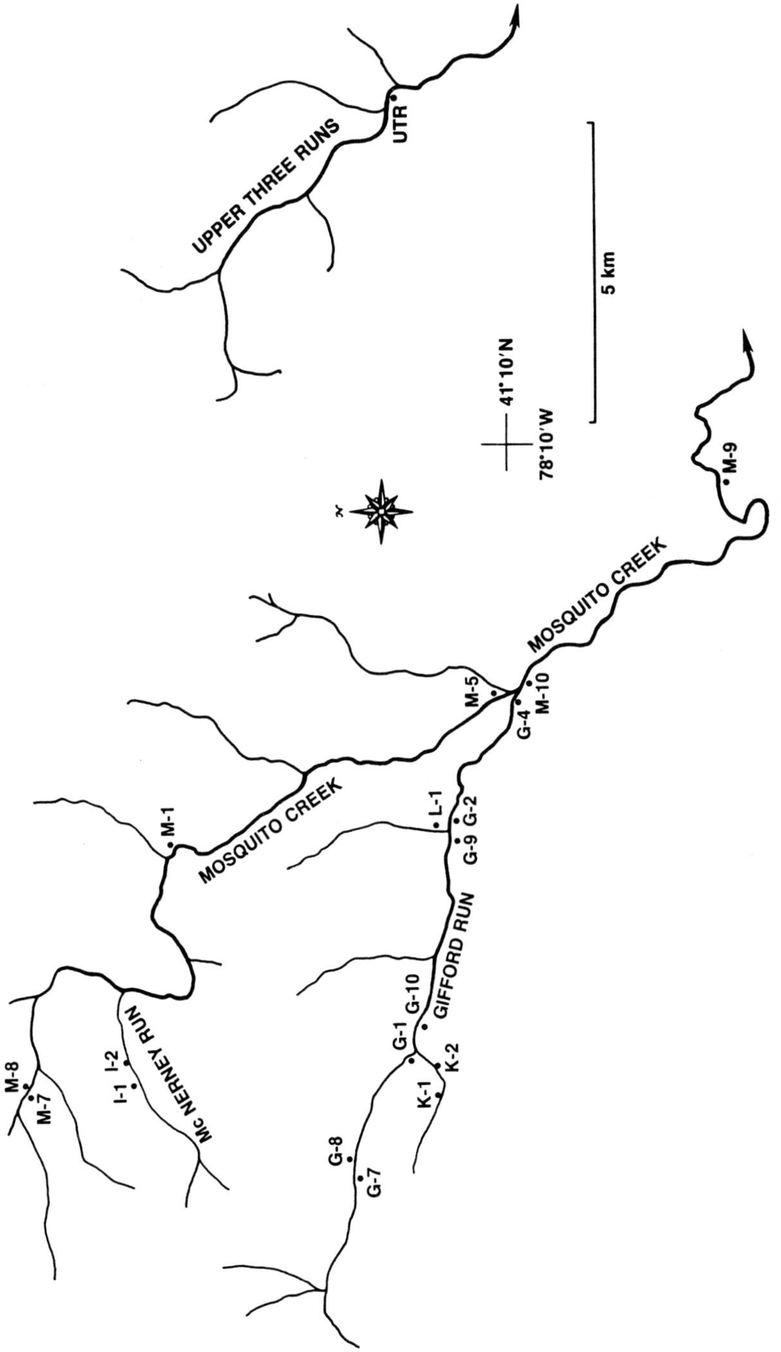

Figure 1. Mosquito Creek and Upper Three Runs watersheds showing sampling stations mentioned in text.

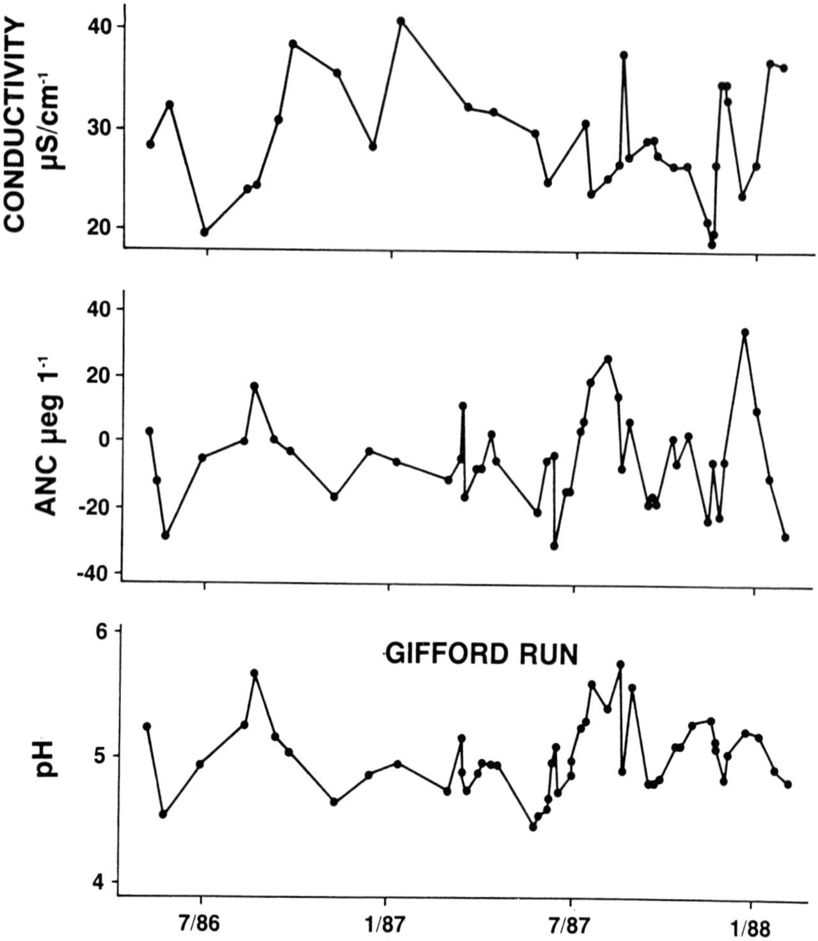

Figure 2. Conductivity, acid neutralizing capacity and pH in Gifford run at station G-1, 1986-1988.

McNerney Run, another tributary of Mosquito Creek, and nearby Upper Three Runs. These streams are smaller than Gifford Run but have similar chemical and other characteristics. Fish in all these streams -- mostly brook trout (*Salvelinus fontinalis*) -- are few in number and exhibit poor growth and reproduction (D. E. Arnold and R. W. Light, unpublished data).

Rain chemistry analyses, under both state and national network programs, indicated that strongly acidic rainfall was characteristic of the study area (Lynch et al., 1985; National Atmospheric Deposition Program, 1987.) These reports and our own occasional measurements indicated that rain pH in the study area was near 4.0 during most of the year, and occasionally was near 3.8. Although the Mosquito Creek watershed is not near a major population center, it is heavily used for recreation, and would undoubtedly be used more if water quality were more conducive to fish production or could be made so by a practical mitigation program.

3. METHODS

The pH in the field was measured with portable meters and also collected samples that were returned to the laboratory in polyethylene bottles (which were filled to overflowing and capped underwater to exclude air bubbles) and refrigerated until analyzed. All analyses were accomplished within 48 hr. Because readings from field instruments showed considerable disagreement, only data from the laboratory pH analyses are reported here. All laboratory pH readings were taken with a Corning* model 125 m and Orion 81-04 Ross electrode. Acid neutralizing capacity was determined from samples collected and transported as described above and analyzed by the double-endpoint titration method (American Public Health Association et al., 1981). Arnold et al., (1985) showed that this method yielded comparable results to those given by the Gran-plot method for low ionic strength waters in the study area. Conductivity was measured in the field with a Cole-Parmer hand-held digital

* Reference to trade names or manufacturers does not imply Government endorsement of commercial products.

conductivity meter, and checked in the laboratory with the same instrument. Readings were generally in agreement within 10%.

For collecting samples of invertebrate drift, three nets (round mouth opening 15 cm in diameter, net length 60 cm, 255 *u*m mesh) were placed at equal intervals across the stream. Velocity measurements were taken in the mouth of each net with a Marsh-McBirney current meter before and after each sampling to check for backwash and to estimate the abundance of larvae (no.$\cdot m^{-3}$) in the drift. Stream bottom invertebrates were collected in a modified Surber sampler (275 *u*m mesh). The contents of sampling nets were preserved in 10% formalin and returned to the laboratory for analysis. Invertebrates were identified, measured, and counted under a dissecting microscope. The taxonomic keys of Merritt and Cummins (1984) was used to identify the invertebrates. All sorted specimens were archived at our laboratory.

4. THREE EXPERIMENTAL NEUTRALIZATION METHODS

4.1 Limestone Flow-Through Barrier

Wire gabions are a common device for constructing structures in and along waterways. A commonly-used commercial gabion unit is about 1 m x 1 m x 2 m, comes with the correct wire for joining and strengthening units, and cost less than $40 in April 1988 (Maccafferri Gabions, Inc., Williamsport, MD, USA). A well-built gabion structure should last for several decades. Such units are usually filled with native rock from the site, but can be filled with limestone chunks of appropriate size to help neutralize acidic water.

Small-scale experiments were conducted, using water from Gifford Run and samples of "cupola fluxing stone," a high-calcium (>95% $CaCO_3$) limestone product, in which the pieces average about 5 cm in diameter. The field experiment consisted of packing 2 pipes, 10 cm in diameter and 5 m long, with stone -- one with cupola fluxing stone and one with similar-size native stone from the creek bed -- allowing stream water to flow through them, and measuring pH at the end. pH of the limestone pipe water rose 0.6 unit almost immediately and 0.3 unit after 34 days (the initial pH was 5.0 to 5.3).

A barrier construction of 6 gabions (**Figure 3**) across Gifford Run (below G-1, **Figure 1**). Unfortunately, the rock obtained by the cooperators was not the cupola fluxing stone specified, but was what is commonly known as "dolomite riprap," which contains more Mg and less Ca and stones of nonuniform size. Under the circumstances it was necessary to use this material, which showed a much lower reaction rate than the high-calcium stone. More important, the uneven-sized pieces (from ca. 1 to 2 to 20 cm in largest dimension) caused two problems: (a) porosity was reduced to the point where the barrier began to act as a dam, when small pieces filled in between large pieces; and (b) reactivity was reduced further because of the lower surface to volume ratio of the larger pieces.

Since the material as placed in the barrier contained a considerable amount of fine particles, initial reaction rates were relatively good, and pH of the stream water increased significantly. However, the fine particles soon disappeared and net treatment effect decreased rapidly after about 2 weeks. For the rest of the life of the structure, differences in pH and alkalinity above and below it were insignificant (**Table I**).

Bender (1978) and Hale (1978) studied plant and insect life, respectively, above and below the barrier, but found no significant effects (Arnold et al., 1981). Coating of the rock was observed inside the structure by a film which appeared to be bacterial. After 1 yr the barrier was rebuilt by removing the stone from one gabion chamber at a time, scrubbing off the bacterial coating, and replacing only rocks within a size range of about 8 to 12 cm in largest dimension. This change improved porosity, but acid neutralization did not improve enough to significantly increase pH.

Probably the most serious problem with the barrier device -- one that would have occurred even if the stone had been of uniform, optimum size and high chemical quality-- was the clogging of the barrier by leaves. A trash rack consisting of sloping bars parallel to the streamflow and extending at a 45-degree angle from the stream bottom to the top of the barrier was intended to direct leaves over the top (**Figure 3**), but did not do so effectively. The leaves further reduced the already poor porosity and impaired the performance of the barrier. Mechanical leaf protection devices are available which can be driven by water power, but such devices would be costly. It is believed that this type of neutralizing device-- properly constructed -- has potential for treating mildly acidic streams in remote areas, particularly if installed in multiple.

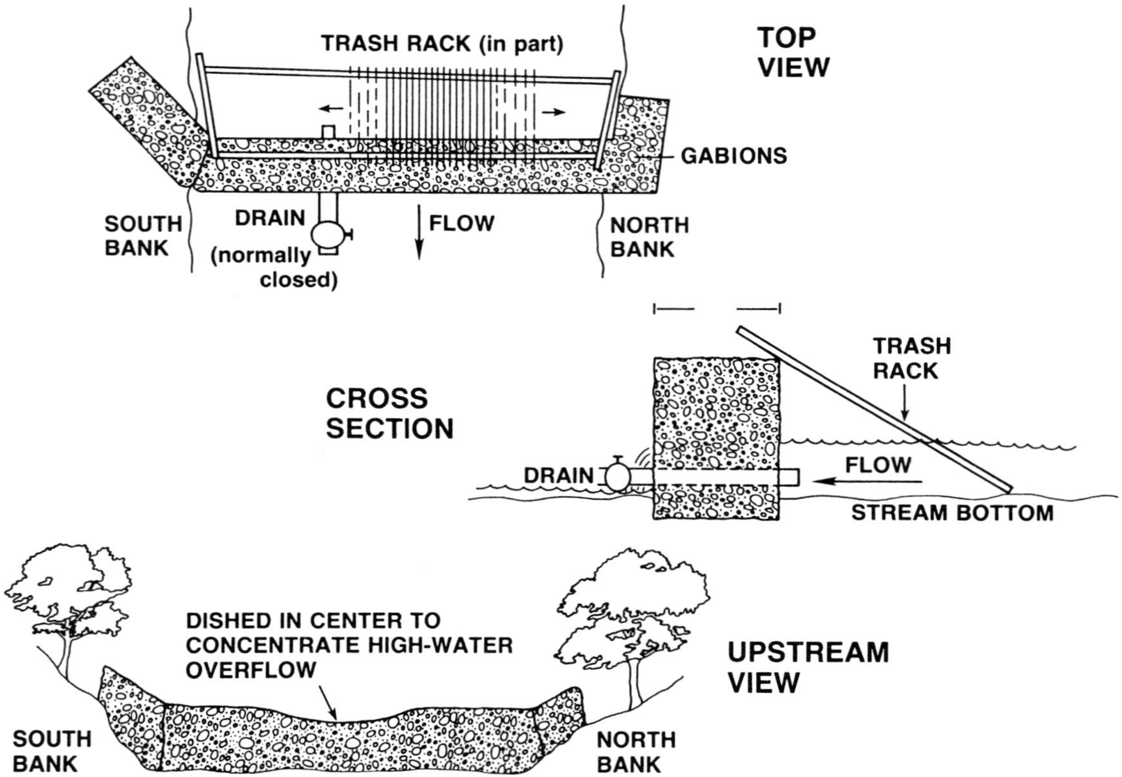

Figure 3. Gabion-limestone barrier installed at Gifford Run below station G-1, October 1976.

Table I

Effect of limestone barrier on pH of Gifford Run, Fall 1986. (Means calculated using H^+ concentration and converted back to pH.)

	Above barrier	**Below barrier**
Pre-operational period 8/19 - 10/16 (n=9)		
Mean pH	5.3	same
Range	4.7 - 5.9	same
Post-operational period 10/17 - 12/31		
Mean pH	5.3	5.5
Range	4.8 - 5.7	5.1 - 5.9
Mean difference (pH below - pH above) 0.18		
Range	0 - 0.4	

4.2 Soda Ash Briquette Baskets

Mine drainage control workers have recognized that there is an inverse relation between the cost of basic materials and their effectiveness per unit of weight or volume -- sodium hydroxide (NaOH) being most effective, but most expensive, and limestone ($CaCO_3$) being least effective, but also least expensive (Lovell, 1971). Sodium carbonate or soda ash (Na_2CO_3) is intermediate in cost and effectiveness, and is available in briquette form (longest dimension ca. 5 cm) which facilitates storage and handling.

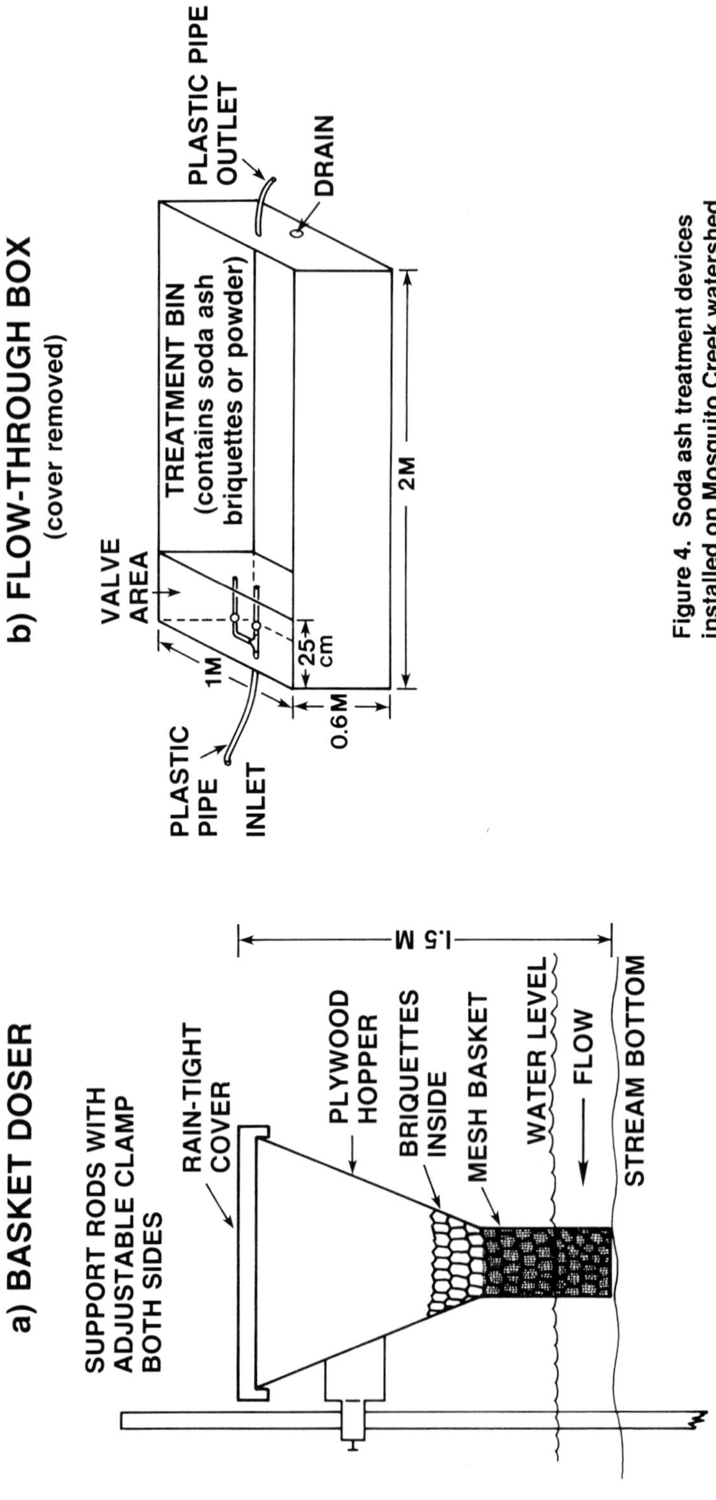

Figure 4. Soda ash treatment devices installed on Mosquito Creek watershed in 1985.

Three tapered hoppers with expanded metal mesh baskets at the lower end (**Figure 4a**) were installed at three locations (G-7, M-7, I-1, **Figure 1**) on the Mosquito Creek watershed during the week of March 20, 1985. A fourth device was constructed in the form of a flow-through box (**Figure 4b**) and installed on another location (K-2, **Figure 1**) in the same watershed. Stream water was sampled above and below each basket site and at downstream locations, one week before and several weeks after the beginning of the treatment (**Table II**). Although the use of the baskets effectively raised the pH and acid neutralizing capacity over a range of flows (as water rises, more of the briquettes in the basket section are submerged), it proved difficult for the cooperators to keep them supplied with briquettes. The soda ash dissolved more quickly than expected, and the capacity of the hoppers (ca. 90 kg) was not adequate; semiweekly refilling was required even at normal flows. There was little difficulty with clogging or failure of the soda ash briquettes to feed properly into the water, but flow through the basket was occasionally reduced by debris lodged against the mesh. However, the blockage did not seem to interfere with dissolution of the briquettes. A simple V-shaped screen upstream of the basket might prevent such accumulation and improve mixing. These basket devices have potential for relatively low cost treatment of acidic streams. A larger supply bin should be incorporated into the design to reduce replenishment visits.

4.3 Soda Ash Dissolution Tanks

In an effort to reduce replenishment needs, a second type of device (**Figure 5**) was placed in the Gifford Run watershed (locations G-1, K-2, G-2, L-1, **Figure 1**) in spring 1986. These followed the general principle of the flow-through box used at one location in 1985 (**Figure 4b**), but were constructed from cylindrical fiberglass fish hatchery tanks ca. 1 m high and 70 cm in diameter. A perforated pipe crossed the upper end of the tank, and a drain with hose leading into the stream was installed near the bottom. Unfortunately, although this device did not have to be filled as often as the baskets, it lost the advantage of good material flow. The briquettes, when wet by water from the perforated pipe, tended to adhere to each other and eventually coalesced into a mass that was extremely hard and did not sink into the tank

Table II

Effect of soda ash briquette baskets on pH of study streams, Spring 1985. (See Figure 1 for map locations.) Pretreatment monitoring period was 3/2 - 3/14; post-treatment period was 3/31 - 5/22. (Means calculated using H^+ concentration and converted back to pH.)

	Mean pH above device	Mean pH below device
Location G-7		
Pretreatment (n=2)	4.85	same
Posttreatment (n=10)	5.01	5.61
Location I-1		
Pretreatment (n=2)	4.95	same
Posttreatment (n=10)	5.20	6.83
Location M-7		
Pretreatment (n=2)	4.60	same
Posttreatment (n=9)	4.92	6.35
Location K-1 (flow box, not basket)		
Pretreatment (n=2)	5.50	same
Posttreatment (n=8)	5.73	6.40

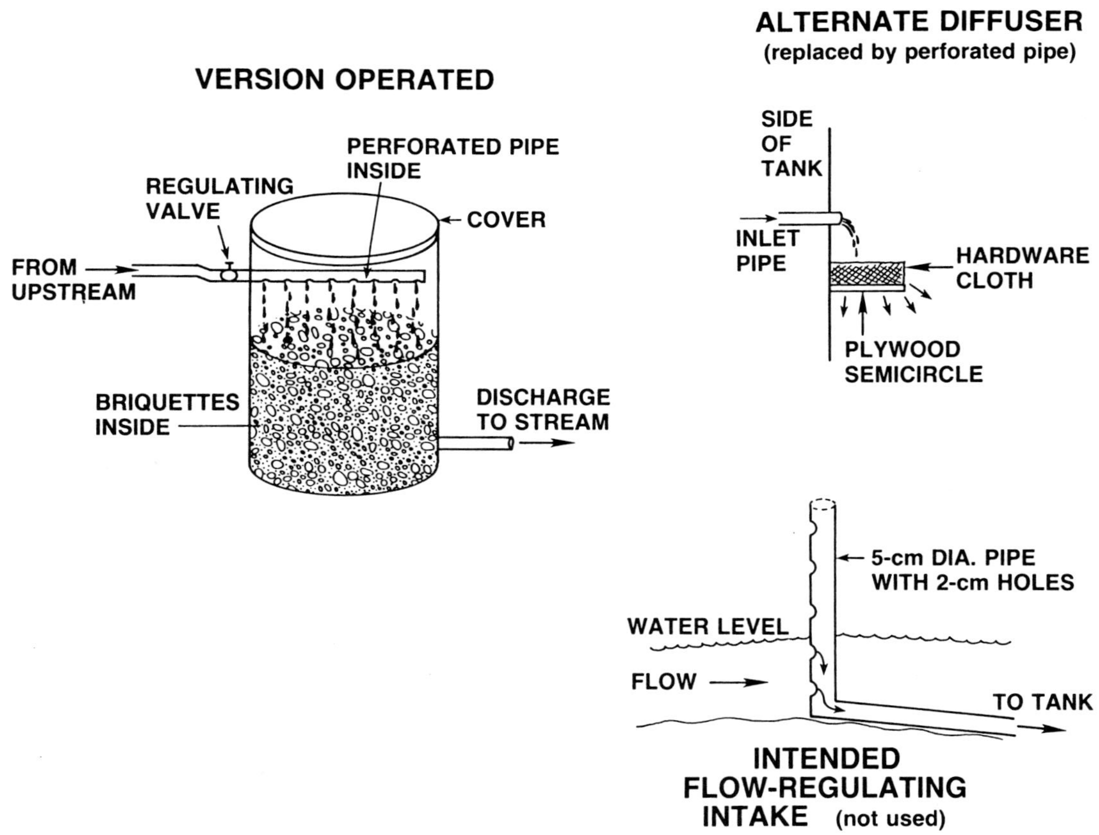

Figure 5. Soda ash dissolution tanks as installed on Mosquito Creek watershed in 1986 and 1987.

as the material slowly dissolved. Consequently, we consider the use of these devices to be an unacceptable treatment method. The soda ash produced adequate increases in pH and acid neutralizing capacity when the dosers were working properly (**Table III**), but they either clogged, or were allowed to empty too often. During the 1986 experiments, 4.5 t of soda ash briquettes were used at a total cost of $975 or $216 per t -- a somewhat lower cost than in

Table III

Effect of soda ash dissolution tanks on pH and acid neutralizing capacity (pH / ANC) of Gifford Run, Spring 1986 and Spring 1987. (pH means calculated using H^+ concentration and converted back to pH.) ANC in $u$eq·L^{-1}.

	above treatment	below 2 tanks	below 4 tanks
1986 experiments: treatment period 4/3 - 5/27			
Preoperational	5.0/-29	5.1/ -	5.2/7
During treatment (n=9)	5.2/-7	5.8/4	5.8/21
After treatment	5.4/5	5.7/23	5.8/24
1987 experiments: treatment period 4/1 - 5/17			
Preoperational	5.0/-9	5.3/5	5.7/4
During treatment (n=8)	5.1/-6	5.7/10	6.0/23
After treatment	4.5/-21	- / -	5.3/6

the previous year. About 8 t were used during the 1987 experiments, but cost was not determined. This tank design is much less promising than the barrier and the briquette basket described above. Table III also illustrates the slight natural increase in pH and alkalinity that we typically note in progressive downstream samples from these streams -- a trend that must be kept in mind when interpreting treatment data.

5. EFFECT OF SODA ASH TREATMENT ON BENTHIC INVERTEBRATES

Despite the problems experienced with the various types of soda ash dosers, we believed that this material had promise as a mitigation agent for streams. As a result, two experiments to evaluate the biological effect of such treatments. Inasmuch as soda ash is a potent acid neutralizing chemical, stream reaches immediately below application points could have a high pH, which would decrease to a lower pH as stream discharge increased downstream. This high pH zone immediately downstream from dosing devices concerns biologists, but if mortality is not observed immediately downstream, it would not be expected in target areas far downstream. If the addition of soda ash to streams kills many invertebrates, the number of individuals in the stream drift should increase after soda ash was applied. Alternatively, high pH may influence benthic invertebrates more subtly, by reducing population abundance slowly over a longer period of time.

In our first experiment, we examined the influence of soda ash induced pH increase on the initiation of invertebrate drift; in the second, we followed benthic invertebrate populations through a 2-mo treatment of a stream with soda ash.

5.1 Effect of Increased pH on Invertebrate Drift

Sites on two streams were used to examine the influence of soda ash on the initiation of invertebrate drift: Gifford Run on September 6, 1986 and Upper Three Runs on May 24, 1988; both are near the previous treatment sites (**Figure 1**). Our approach involved sampling water and invertebrate drift before, during and after the application of soda ash. At each site, about 20 kg of soda ash, in briquette form, was applied to the stream. The briquettes were

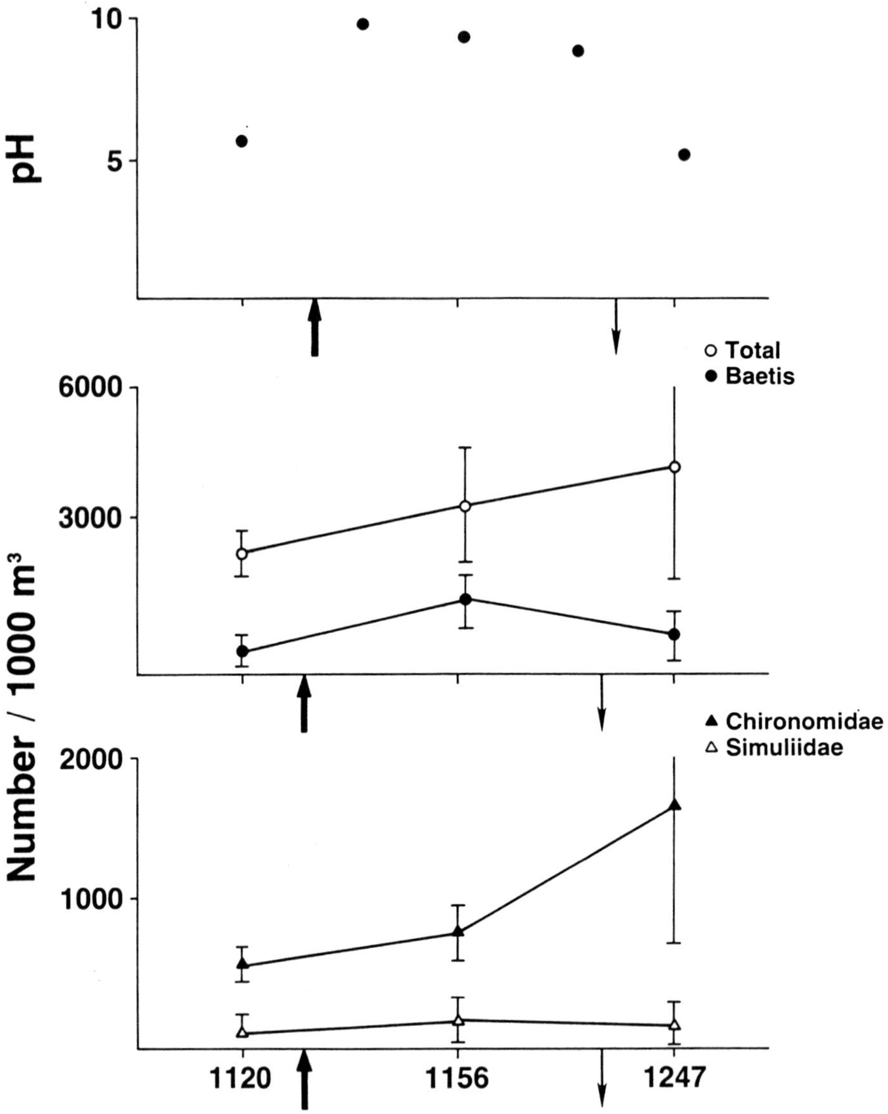

Figure 6. Invertebrate drift and pH before, during and after soda-ash application at Gifford Run station G-9. The up arrow indicates time of soda ash introduction, the down arror indicates its cessation. Vertical bars are 95% confidence intervals.

placed in a basket (2.5 X 6 cm mesh) anchored in the center of the stream. Water flowed directly through the basket, dissolving the soda ash. Three drift nets were placed about 15 m downstream from the soda ash basket. At Gifford Run, we collected 3 drift samples about 30 min before the soda ash was applied, 3 samples during the 1 hr application, and 3 after the application period. At Upper Three Runs, we collected 3 drift samples each at about 1 hr and 0.5 hr prior to application (6 samples), 3 samples each at about 0.5 hr and 1 hr after the application of soda ash was started (6 samples), and a total of three samples about 0.5 hr after application had ceased. Nets were left 30 min for each sampling in Gifford Run, and 15 min in Upper Three Runs. Water samples were collected at about 20-min intervals at Gifford Run and 25-min intervals at Upper Three Runs.

Soda ash increased the pH of Gifford Run by about 4 pH units during the 1-hr application; thereafter the pH returned to pre-application levels within 20 min (**Figure** 6). Soda ash increased the pH of Upper Three Runs by about 2 pH units, and it had not yet returned to pre-application levels 25 min after soda ash application had ceased (**Figure** 7).

Invertebrate drift densities generally did not change after the application of soda ash (**Figures 6 and** 7). The variables we examined at both sites included densities of all invertebrates combined, Chironomidae, and Simuliidae. *Baetis* spp. (Ephemerop-tera: Baetidae) were collected at Gifford Run only. No other taxa occurred frequently enough to warrant analysis. All variables but one clearly did not differ with respect to time of sampling (**Figure** 7; $p > 0.1$, one-way ANOVA). *Baetis* spp. had a slightly higher density during soda ash application ($p = 0.1$), but pre- and post- application densities were about equal.

5.2 Benthic Population Changes During Soda Ash Treatment

The second, longer experiment was conducted in Rock Spring Run, a first order tributary of Gifford Run. One of the fiberglass tanks fabricated for the 1986 treatments described above was used. The treatment period extended from March 29 to May 22, 1986. Stream bottom invertebrates were sampled in Rock Spring Run about 20 m downstream from the tank and at two points on Gifford Run -- one about 400 m above its confluence with Rock Spring Run (G-1) and the second about 50 m downstream from the confluence (G-10). The flow through tank

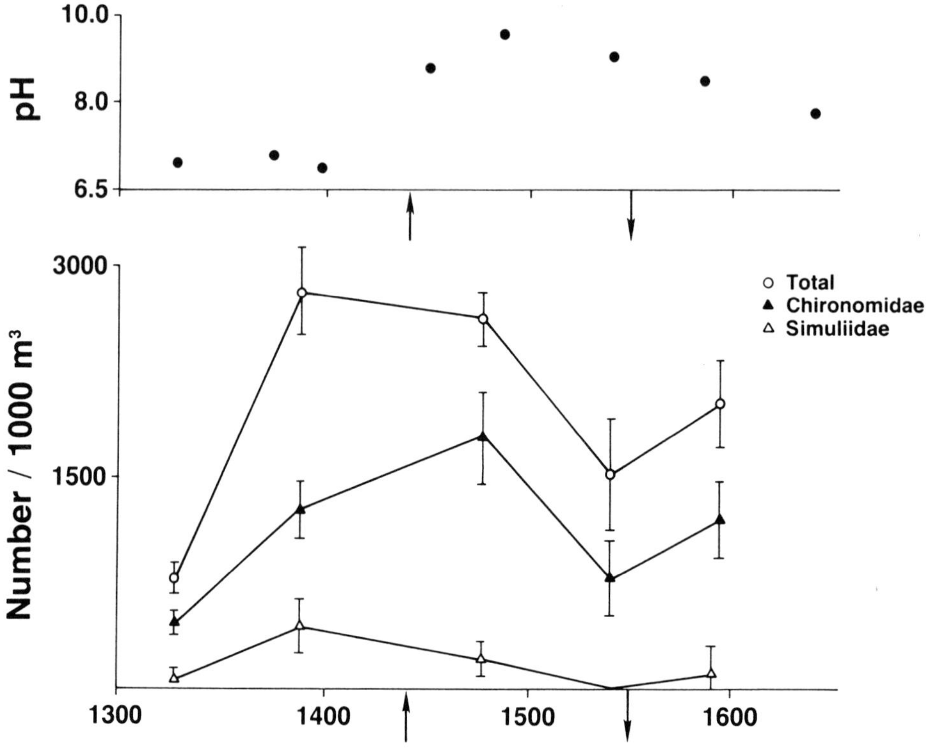

Figure 7. Invertebrate drift and pH before, during and after soda ash application at Upper Three Runs. The up arrow indicates time of soda ash introduction, the down arrow indicates its cessation. Vertical bars are one standard error of the mean.

device increased the pH of Rock Spring Run by as much as 2 pH units during the experiment; we noted no detectable increase at G-10.

Patterns of invertebrate change during the soda ash application period with respect to location, having one station unaffected (G-1) and two stations treated (K-2 and G-10), were examined. Specifically, the evaluated changes in total number, richness of taxa, *Leuctra* (Plecoptera: Leuctridae), *Amphinemoura* (Plecoptera: Nemouridae), *Eurylophella* Ephemeroptera: Ephemerellidae), Chironomidae and Oligochaeta were observed. Changes discovered at K-2 and G-10 that were not also evident at G-1 (**Table IV**). Nearly all taxa showed no significant difference in density among the four sampling times (**Table IV**); those that did (e.g., *Amphinemoura* and Oligochaeta) closely reflected the patterns at the reference station.

5.3 Discussion of Effects on Invertebrates

Limited research has been done on the influence of pH change on benthic invertebrates in streams; pH depression has been the main approach (e.g., Hall et al., 1980; Raddum and Fjellheim, 1987; Kratz et al., 1988). Hall et al. (1980) observed overall drift to increase by nearly eight times within 3 days after artificially acidifying a New Hampshire stream. Kratz et al. (1988) found 14 times more drifting *Baetis* after exposing artificial channels in the Sierras to H_2SO_4 or HNO_3 for 8 hr, in an effort to simulate a naturally occurring acid rain episode. Raddum and Fjellheim (1987) found the drift rate (no. drifting / no. in benthos) of *B. rhodani* in experimental channels to increase after pH decreased from 5.8 to 5.4.

No significant change was observed in invertebrate drift after increases of 2 to 4 pH units. An increase in pH, particularly from 5 to 9, does not have the same effect on the toxicity of certain metals (such as Al) as a decrease in pH from 6 to 4.5 (for instance). Hall et al. (1980) proposed three mechanisms by which pH depression may be harmful to aquatic organisms: (1) physiological, (2) through mobilization of metals, and (3) through ecological pathways. Each was reviewed by Wiederholm (1984). The invertebrate drift response to a sudden decrease in pH must be a result of the first two mechanisms stated by Hall et al. (1980), not the third. No immediate drift response when pH was increased occurred. Drops in pH of even less than 0.5 pH unit (Raddum and Fjellheim, 1987) seemed to have more potential

Table IV

Chemical Results from soda ash dissolution tank experiments, Spring 1986.
*** = more than two tanks (dosers) malfuntioning. Station G-1 is above all dosers, K-2 below one doser on tributary, G-9 below two dosers, G-4 below all four dosers at end of treatment section.**

Date	pH				ANC (ueq . l^{-1})			
	G-1	K-2	G-9	G-4	G-1	K-2	G-9	G-4
March 25	5.0	5.9	5.1	5.2	-29	16	-	7
START OF TREATMENT								
April 3	5.3	10.7	6.3	6.5	0	1235	-5	43
April 10	4.9	8.8	5.4	5.5	-7	198	7	26
April 15	5.6	9.5	6.3	6.4	-5	299	4	20
April 24*	4.8	8.4	5.3	5.4	-15	285	-22	-3
May 1	5.1	7.8	6.3	6.2	-9	163	13	15
May 8*	5.1	5.8	5.5	5.3	-9	102	2	19
May 16	6.0	7.1	6.7	6.8	-18	99	9	36
May 21*	4.6	5.2	5.0	4.6	-5	123	9	8
May 27*	5.0	6.3	5.3	5.6	3	44	18	21
END OF TREATMENT								
June 2	5.4	6.1	5.7	5.8	5	158	23	24

for causing effects on invertebrates than increases in pH. This is probably because the pH of typical, low order streams is low and very near the point where metals, such as Al, become very toxic. However, this low pH also prevents pH increases -- e.g., those resulting from liming -- from attaining levels where alkalinity and alkalinity-dependent compounds become toxic.

If immediate drift response does not exist, perhaps a more subtle, slower change occurs in benthic communities. Much research has been done on the communities of naturally acidified streams in England (e.g., Minshall and Kuehne 1969; Sutcliffe and Carricke, 1979; Minshall and Minshall, 1978) and in Sweden (P. Lingdell, unpublished data described in Wiederholm, 1984). These studies suggested that communities adjust to changing acid conditions through time. Although pH depressions may cause "spike" drift increases, probably more than 8 hr to 8 days are required for acidification to alter an invertebrate community.

Our second experiment indicated that a pH increase of up to 2 units did not significantly affect benthic invertebrates over about 60 days. Either a longer time period -- probably more than one life cycle -- is required for pH increase to alter benthic invertebrate population densities, or pH increase does not influence benthic invertebrates in such situations.

In summary, our working supposition was that application of soda ash to a stream would influence benthic invertebrates, most probably decreasing the populations. The results of our two experiments, one addressing immediate effects, the other more long term effects, suggest that pH increase, within the range that we studied, does not reduce invertebrate numbers. Continued research on long term population trends resulting from soda ash treatment seems warranted. This might include experiments lasting a year or more, although few mitigation treatments would likely operate continuously for such a long period.

ACKNOWLEDGMENTS

For assistance in the field and laboratory we thank Patricia Bender, Lynne Conner, Edward Dunstan, Martin Friday, Alan Hale, Bruce Hollender, Deborah Humphrey, Rick Lorsen, Mark Silvestri, Robert Wilberding, and others. The Mosquito Creek Sportsmen's Association, especially President Raymond Savel, and the Allegheny Mountain Chapter of Trout Unlimited contributed considerable manpower, materials, and support. Financial and material support was provided by the Pennsylvania Fish Commission, the U. S. Fish and Wildlife Service, and the Pennsylvania State University. Support for publication was provided by Living Lakes, Inc.

SECTION IV: WATERSHEDS

The extensive discussions and presentations in the previous papers have described the application of liming materials directly to the surface water of lakes and streams. A common theme to all of the papers is that liming must, under most circumstances, be repeated. the time interval between applications varies with the types of water body, source of acidity and many other physical, chemical and biological factors. The papers presented in this section are based on the simple premise that lakes and streams receive their water, to a great extent, from the surrounding watersheds. Therefore, by placing calcitic materials on the soils of the watershed, it is hypothesized that the resulting neutralization of lakes and streams can be sustained over a longer time period. The benefits that could be derived are described in the following papers and truly represent a systems approach to restoration.

THE LOCH FLEET AND OTHER CATCHMENT LIMING PROGRAMS

D.J.A. Brown

ABSTRACT

The circumstances under which catchment liming is an advisable economic alternative to direct water liming are described, as are research and operational experiences of catchment liming, drawing mainly on the Loch Fleet Project results and other European data. The arguments for and against catchment liming as a means of restoring acidified aquatic habitat are presented. In general, it is concluded that there are conditions, such as lakes with short retention times and/or where stream spawning fish species are involved, where catchment liming is a sensible option.

1. INTRODUCTION

The technology for the direct addition of neutralizing agents (e.g., limestone) to acid surface waters to alter water quality, and thus enable fisheries to be improved/restored is now fairly well-established as confirmed by numerous examples elsewhere within this volume. Ways of determining the inter-treatment intervals are described, and it can be seen that the effective duration depends upon several factors, least of which is the hydraulic retention time of the lake (a measure of the time required for the lake volume to be replaced by water inputs). Experience and calculations suggest that treated lakes maintain improved water quality for between one and four multiples of their hydraulic retention time before retreatment becomes necessary (Sverdrup and Warfvinge, 1985). Thus, lakes (e.g., many in Sweden) with retention times of 2 yr or longer are suited to treatment by this direct application technique. However, there are lakes in other parts of the world for example, Scotland, where a survey of 64 lochs (Maitland et al., 1987) revealed that more than half had retention times of less than 1 mo. Direct application to these sites is clearly not suitable. For the treatment of such lakes, it is sensible to consider the application of lime to the catchment in order to get

Water, Air, and Soil Pollution **41** (1988) 409–415.

a more long-lasting result. It can also be postulated that such a method may have the added advantages of reducing both the severity of episodic acidic conditions, and the leaching of toxic Al from the edaphic to the aquatic environment which can sometimes give rise to biological problems after direct liming (Nyberg, 1984). The published experiences of catchment liming that exist are reviewed below.

2. CATCHMENT LIMING THROUGHOUT THE WORLD

Soil liming to overcome surface water acidity has been conducted on an experimental basis in Sweden since an exploratory liming program was initiated in 1976. Indeed, of the 207,000 t of limestone spread between 1976 and 1982, 26% was spread on land. However, according to Warfvinge and Sverdrup (1988a) different combinations of limestone application have often been used, and although a few stream data are available, there is no Swedish set of data available to compare, for example, model predictions on the effect of terrestrial liming on whole lake water quality. In general, Swedish experiences with terrestrial liming to mitigate acidity in surface waters has not been very positive. For example, Lessmark (1987) lists 52 lake and stream catchments treated by terrestrial liming; although, in 59% there was a positive effect on pH of the surface water. This effect was of short duration, and in only 6% of the cases was the pH held above 6 and the alkalinity above 50 uE L^{-1} for more than 1 yr. However, it should be noted that many application rates, when calculated on a whole catchment basis, were low (75% were less than 1 t ha^{-1}), and the chances of success were much higher if either the application rate was higher and/or a considerable percentage of the total catchment area was treated.

Elsewhere in the world, there are only a few examples of catchment liming. Experience in Norway is restricted to a catchment at Tjonnstrond (Johannessen and Hindar, 1987) where 75 t of limestone were evenly spread at a dose of 3 t ha^{-1}. In spite of a theoretical retention time of 0.2 yr for the two lakes within the catchment, the liming effects after 5 yr are "still highly acceptable and are expected to be so for some years to come" (A. Hindar- pers. comm.).

As far as can be ascertained, there is little experience of catchment liming at present within North America (K. Schreiber - pers. comm.) and within the UK. Experiments are in progress at four locations. In the Esk Valley in S.W. Cumbria, the North West Authority spread 3,000 t of limestone at a local rate of 5 t ha^{-1} onto agricultural land (equivalent to a total catchment rate of 0.45 t ha^{-1}) which caused an immediate and significant increase in pH in the river which has been sustained (Crawshaw et al., 1987). In the Llyn Brianne catchment in Wales, the Welsh Water Authority spread 60 t of Mg limestone at a local rate of 10 t ha^{-1} (equivalent to a whole catchment rate of 0.6 t ha^{-1}) which resulted in only a transient increase in pH, which fell to pre-treatment values during the first storm events (Stoner, 1987). In Scotland, the Solway River Purification Board added 80 t of limestone powder to an area of 0.3 ha (equivalent to a whole catchment rate of 0.15 t ha^{-1}) and found it had a negligible effect on water quality (Welsh -pers. comm.). The final example from the UK is the Loch Fleet Project (Brown et al., 1988). In this experimental catchment liming project, a variety of application rates (up to 30 t ha^{-1}) to individual subcatchments has been made, and the response of the lake to a total watershed application rate of c 3 t ha^{-1} has been very positive and has allowed a fishery to be reinstated.

The results of these stream, subcatchment and whole lake terrestrial liming exercises are summarized in Figure 1. Included in these data are results from applications of a variety of different materials (e.g., Mg as opposed to pure Ca limestone) of a variety of different particle sizes (from fine powder to coarse granules to pelletized material) onto catchments with widely different hydrologies; thus, it is not surprising that similar application rates onto similar portions of catchments can have very different effectiveness. Nevertheless, it can be seen that the chances of success are greater if the application rate (calculated to the whole subcatchment/catchment) is in excess of 1 t ha^{-1}. Also, success is more likely if either the local application rate is higher and/or a high percentage of the catchment area is treated. This conclusion should be compared with the theoretical suggestions of Warfvinge and Sverdrup (1988b) that to be successful the dose should exceed the corresponding exchangeable acidity in the soil surface layer and, at least, 50% of the discharge area active at high flow rate be treated.

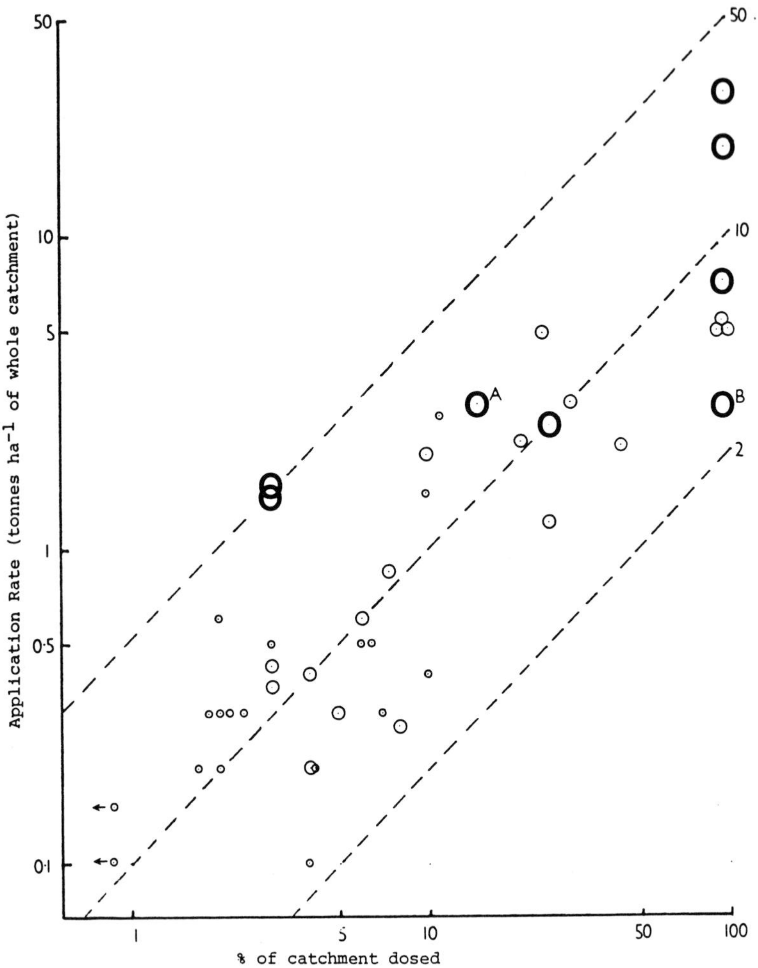

Fig 1. The results of terrestrial liming experiments throughout the world. (For sources of data see text). Diagonal lines indicate average local application rates of 50, 10 and 2 tonnes ha^{-1} respectively. o indicates no effect of treatment on surface water quality, O some short term positive effect on pH, **O** indicates pH being held above 6.0 for longer than 1 year. A and B are the whole lake experiments at Loch Fleet in Scotland and Tjonnstrond in Norway respectively. Other long term neutralisations are subcatchments within Loch Fleet and 2 streams on the Swedish West Coast.

It is interesting to note that successful whole lake catchment liming -- using the somewhat arbitrary, but perhaps not unreasonable definition of Lessmark (1987) of success (i.e., pH held above 6 for more than 1 yr) -- appears to be restricted to the example of Tjonnstrond in Norway (Johannessen and Hindar, 1987) and Loch Fleet in Scotland (Brown et al., 1988).

3. DISCUSSION

The major disadvantage of terrestrial liming that can be readily identified is the potential deleterious effect of the relatively high doses of limestone needed on what is, by definition, acidophilus vegetation. Details are given in Brown et al. (1988) of the immediate effects on vegetation of the lime applied to the Loch Fleet catchment. The most dramatic of these is the local killing of the dominant *Sphagum* species, but there are still large areas of even this species, together with other more tolerant species which, at the moment at least, are still alive. In the Swedish study at Bravattenbacken, 6 yr after the lime was applied, the *Sphagum* in the limed area had completely disappeared (Hasselrot - pers. comm.).

Another possible related disadvantage of catchment liming which has been raised is the possibility that soil liming in forests may cause reduced growth rate of trees (Derome et al., 1986) and that this may have commercial implications. Whether this happens with the high application rates used at Loch Fleet is being studied. At this early stage, no adverse effects are apparent.

With regard to the relative costs of catchment liming compared to direct lake liming, Warfvinge and Sverdrup (1988a) have estimated that the annual cost for watershed liming to maintain a pH above 6.0 is five times more expensive than direct lake application. These estimates assume a lake with a retention time of 1 yr and a reacidification time of 3 yr and 6 yr for direct and catchmenf liming, respectively. Should the ratio between these latter figures prove to be in error, then the relative costs would need to be altered.

Despite the economic comparison just described, there are circumstances under which direct lake liming is unlikely to work as well or be as economic as catchment application. These include lakes with short retention times (e.g., less than 6 mo) where to repeat the treatment at frequent intervals would change the economics and, if delayed due to any one of a variety of reasons, could have disastrous biological consequences. However, perhaps a more powerful argument in favor of catchment liming is its use in ensuring adequate chemical conditions in feeder streams for lake dwelling and stream spawning fish species such as brown trout. There are available devices such as flow related dosing silos which can be used to protect flowing water, but their installation and maintenance in remote locations can present problems, and the costs of such an exercise (i.e., being able to fill a silo at the appropriate intervals) together with the cost of the device itself, would need to be considered in the economic argument. It is most unlikely that they could be made totally fail-safe, and problems can still be encountered with aluminum toxicity. Also, dosing from silos affects the water running over the stream bed, whereas the early life history stages of the fish are buried in the gravel, and it is here that the improved water quality is required. Catchment liming apparently is satisfactory at achieving this improvement, including a reduction in the toxic Al concentration and in the severity of acid episodes, (Brown et al., 1988; Rosseland and Hindar, 1988).

4. CONCLUSIONS

Although there are only two examples in the published literature of catchment liming having been successfully employed to mitigate lake water acidity, there are a few other instances where acid streams have been treated in this way. Provided the rate of application is sufficient, and the treated area is large enough, it would appear that application of limestone to the land can be a satisfactory way of making acid waters chemically acceptable to fish. Although the economics remain to be worked out in detail, there are instances (where lakes have very short retention times and/or where stream spawning fish are involved) when catchment liming may represent the only viable option. It may not be possible to apply limestone in sufficient quantities to catchments without some adverse effects on the acidophilus terrestrial vegetation, but limited experience suggests that this need not be extensive.

ACKNOWLEDGMENTS

The contributions to this paper by many people in the form of replies to a questionnaire are gratefully acknowledged. Especially, I would like to thank Mr. E. Degerman (Swedish Data), Dr. M. Johannessen (Dr. A. Hindar - Norwegian data), Dr. K. Schreiber (North American data), Dr. D. Crawshaw (Cumbrian data), Dr. J. Stoner (Dr. A. Donald - Welsh data) and Mr. W.T. Welsh (Dr. F. Lees - Loch Dee data). I am also indebted to Dr. T.R.K. Dalziel, Dr. K. Sadler (CERL) and Dr. G.D. Howells and other members of the Loch Fleet Management Committee for helpful discussions at various stages throughout the preparation of this paper. The paper is published with the permission of the Loch Fleet Project sponsors, the Central Electricity Generating Board, the South of Scotland Electricity Board, the North of Scotland Hydro Electric Board and British Coal.

HYDROCHEMICAL BUDGET CALCULATIONS FOR PARTS OF THE LOCH FLEET CATCHMENT BEFORE AND AFTER WATERSHED LIMING

T.R.K. Dalziel
M.V. Proctor
A. Dickson

ABSTRACT

Input/output hydrochemical budgets have been calculated for areas within the Loch Fleet catchment, Galloway, S.W. Scotland and the UK which were limed in April 1986. Bulk precipitation inputs and streamwater outputs are dominated by sea salt components. Calcium outputs over the first year post-liming increased dramatically and H^+ ion outputs dropped. After the second year post-liming, Ca outputs have declined, although H^+ outputs have remained very low and streamwater pH is still circumneutral. The continuation of budget calculations over the next few years, coupled with streamwater and soil chemistry analysis data, should allow estimates to be made of the duration over which of these land-based liming treatments may remain effective in acid mitigation.

1. INTRODUCTION

The Loch Fleet Project is a Research Watershed Liming Project being carried out at Loch Fleet in the Galloway district of southwest Scotland (Brown et al., 1988). The aim of the Project has been to improve the water quality of the loch so that it is once again able to support a brown trout (*Salmo trutta*) population. To meet these objectives, limestone (and other) treatments have been applied to parts of the catchment, not to the loch water directly. It is hoped that, as a result of the work, practical land, lake and stream management procedures will be developed which may be applicable elsewhere.

Hydrochemical budget (or flux) calculations show whether a catchment is a net source or a net sink for various chemical constituents (Likens et al., 1977). These calculations can also serve to help the understanding of acidification processes and how these may vary between areas with different land use (e.g., forest and moorland). Most importantly, so far as the Loch Fleet Project is concerned, Ca budgets, in conjunction with monitoring of drainage

Water, Air, and Soil Pollution **41** (1988) 417–434.

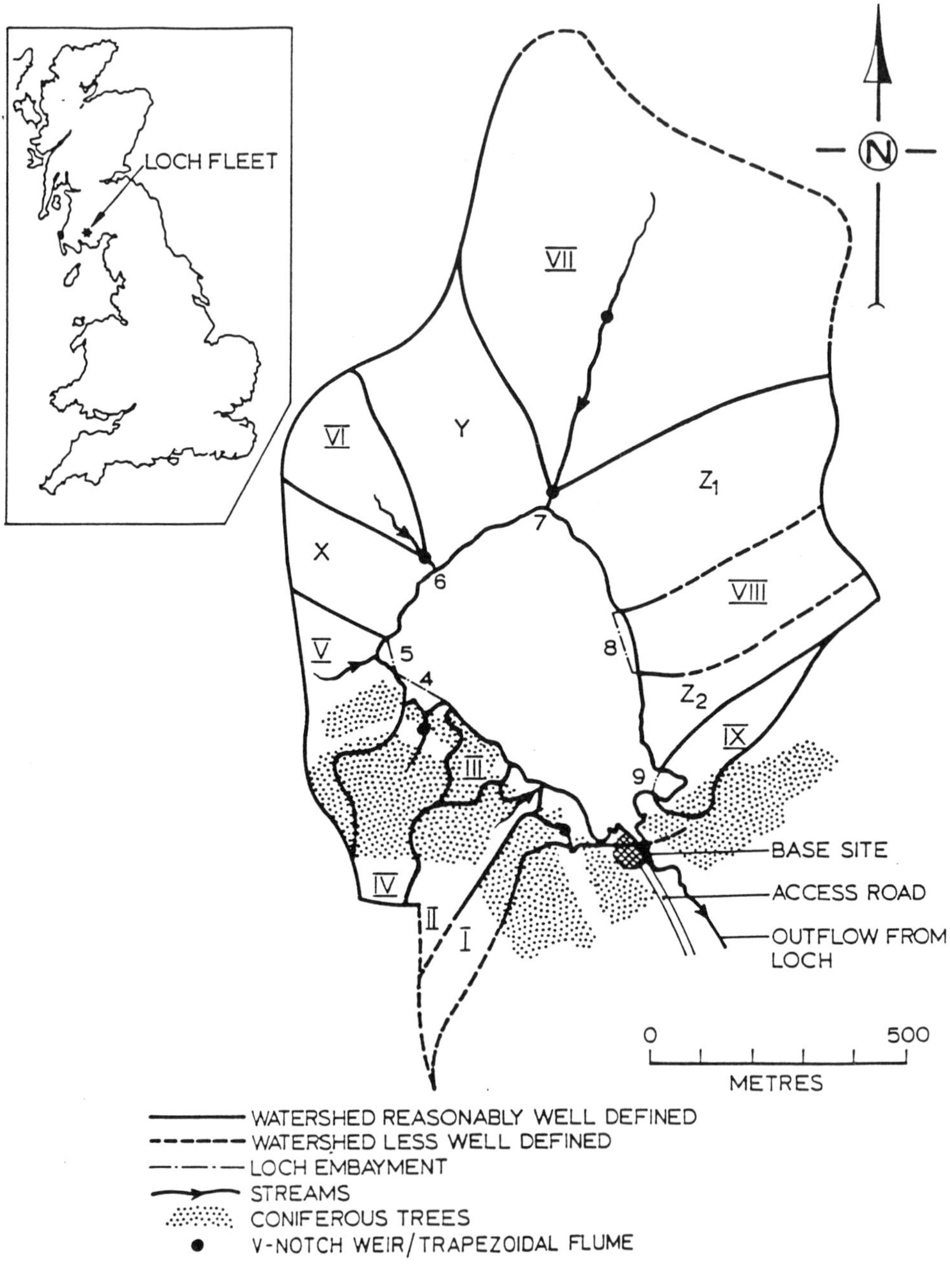

Figure 1. Loch Fleet Catchment showing subcatchments (sectors).

water quality and soil chemistry over time, will allow estimates to be made of the duration and effectiveness of the applied treatments in acid mitigation.

2. METHODS AND MATERIALS

Site description and project plans, etc., are given in some detail by Brown et al. (1988).

In order to test a variety of limestone applications, the catchment was separated into several fairly distinct sub-catchments or "sectors" on the basis of on-site assessment and the use of aerial photographs and detailed contour maps of the site (**Figure 1**). Streams drain into the loch from three of the sectors (IV, VI and VII); flow measuring apparatus was installed on each of these streams. Streamflow is measured automatically every 15 min using a still well and pressure transducer linked to either a V-notch weir (IV and VI) or a trapezoidal flume (VII). A trapezoidal flume was also installed at the loch outlet to measure total flow from the loch (whole catchment). Automatic measurement of pH, conductivity and temperature of the streamwater is made simultaneously with the flow at each site.

The 15 min automatic streamflow measurement system on sectors IV, VI and VII and at the loch outlet has a resolution of $\pm$4 mm; any error in streamflow measurement due to resolution is around 2 to 5% relative to flow conditions. Errors introduced into the calculations due to resolution limitations should be less than this since they can be both positive and negative. In addition to automatic flow measurement, manual water height measurements ("spot flows") are made daily at each of the sites. These heights, converted into flow, have confirmed the suitability of the system resolution. Calibration of the pressure transducers has shown progressive drift over time; this was monitored at each site and equations correcting it were applied in the conversion of head of water (pressure) to streamflow.

Streamwater samples are taken approximately daily at each site for chemical analysis.

Four rain gauges and two bulk collectors were installed in the catchment to measure both rainfall quantity and quality; these have shown good consistency in both amount of rainfall collected and its chemical composition.

Table I

Limestone Treatment Applied in the Loch Fleet Catchment, April 1986

SECTOR	DESCRIPTION	TREATMENT	APPLICATION
IV	86% Mature Coniferous Forest	Limestone Slurry	20 t ha^{-1}
VI	Open Moorland	Limestone Dust	30 t ha^{-1}
VII	Open Moorland and Bog	Limestone Dust	10 t ha^{-1} selectively to bog areas. Equivalent to 2.6 t ha^{-1} over whole sector

Full details of data collection can be found in Howells and Dalziel (1988) and Paterson (1987).

2.1 Treatments Applied - 1986

Following extensive chemical data collection, which started in autumn 1984 to provide "baseline" information, limestone applications were made to each of the three monitored sectors in April 1986. These are given in Table I.

On sector IV, loch water was pumped to the top of the sector and mixed with powdered limestone (2.75 t limestone to 4.55 m^3 water) to produce a slurry. This was then piped downhill and sprayed as uniformly as possible on the ground beneath the tree canopy. The water used in this process was equivalent to 3.3 mm of rainfall over the period of application. Inspection after application showed a reasonably uniform distribution beneath the tree canopy, even in the most dense areas of forest.

Manual distribution of limestone powder on Sector VI was made difficult by rain and sleet at the time of application. Some of the bags of limestone suffered an ingress of water which made the powder too wet to spread, although uniform distribution of bags of limestone (approx. 50 kg in each) over the whole sector was achieved.

Bog areas at the top of Sector VII were identified as the main stream recharge areas for the stream draining the largest sector in the catchment. This stream was identified as a likely spawning place for any trout introduced into the loch. In order to provide conditions amenable to newly hatched trout fry, the limestone powder was applied selectively to the bog areas according to a plan recommended by Nyberg and Hultberg (pers. comm.). The intention was to maintain high pH, high Ca and low Al levels during increased streamflow condition; before applying the limestone there were high acid and Al episodes occurring here.

The pre-treatment streamwater chemical conditions and the effects of the limestone applications made in 1986 ón these are given in Brown et al. (1988).

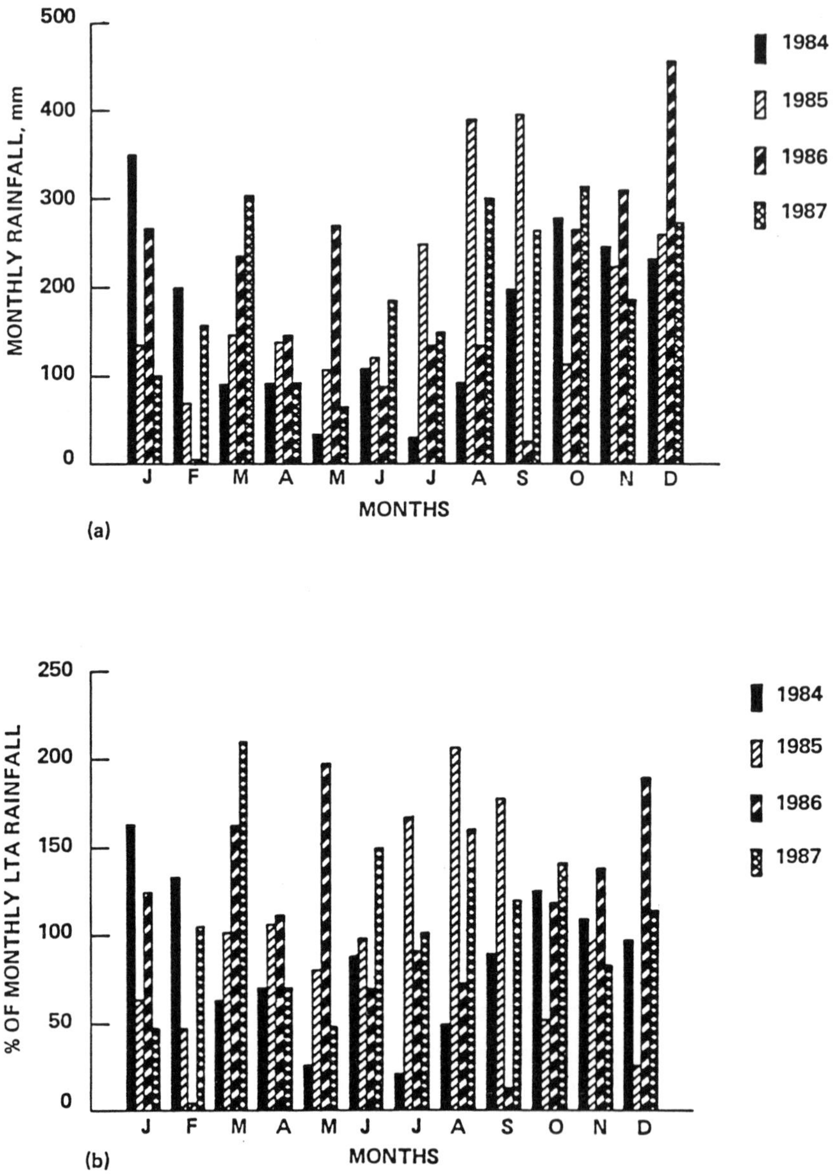

Figure 2. Rainfall (a) monthly, (b) as % of monthly long-trem average.

2.2 Methods

Over two year have elapsed since these treatments were applied. Measurement of rainfall and streamflow quantity and quality have continued. Annual hydrochemical budgets for each of the three treated sectors, as well as and for the whole catchment, have been calculated for three year.

It was recognized that any water budget calculated has to satisfy the equation:

Input = Output $\pm$ Δ Storage

or Precipitation = Runoff (comprising overland flow, lateral throughflow and groundwater flow)
- evapotranspiration

$\pm$ Δ soil moisture (or loch) storage

Streamflow contains components of both overland flow, i.e., that over and through the soil surface, and deeper throughflow pathways, although it is possible that some throughflow may not be included in streamflow measurements. Studies have shown that the flow from the Loch Fleet catchment is predominantly (>90%) by overland rather than by other pathways (Nisbet in Howells, 1986). It seems reasonable to assume, therefore, that any throughflow and groundwater flow not accounted for in streamflow measurements are small enough to be ignored. A further assumption made is that sectoral soil moisture storage at the start of the budget period is essentially equal to that at the end of the period and thus, that Δ storage in the terrestrial component is negligible. Differences in loch water height, however, were used to account for Δ storage in the aquatic component over each budget year, since this can fluctuate more rapidly than soil moisture.

Annual rainfall input is calculated by totalling over the year the daily rainfall quantities measured in a standard Meterological Office 5 inch rain gauge. Annual ion inputs from wet deposition are calculated by multiplying rain ion concentrations in samples from the bulk collectors by the corresponding rain volumes over the time of sample collection and totalling these over the year. Rainfall quantity over the last four year is shown in Figure 2.

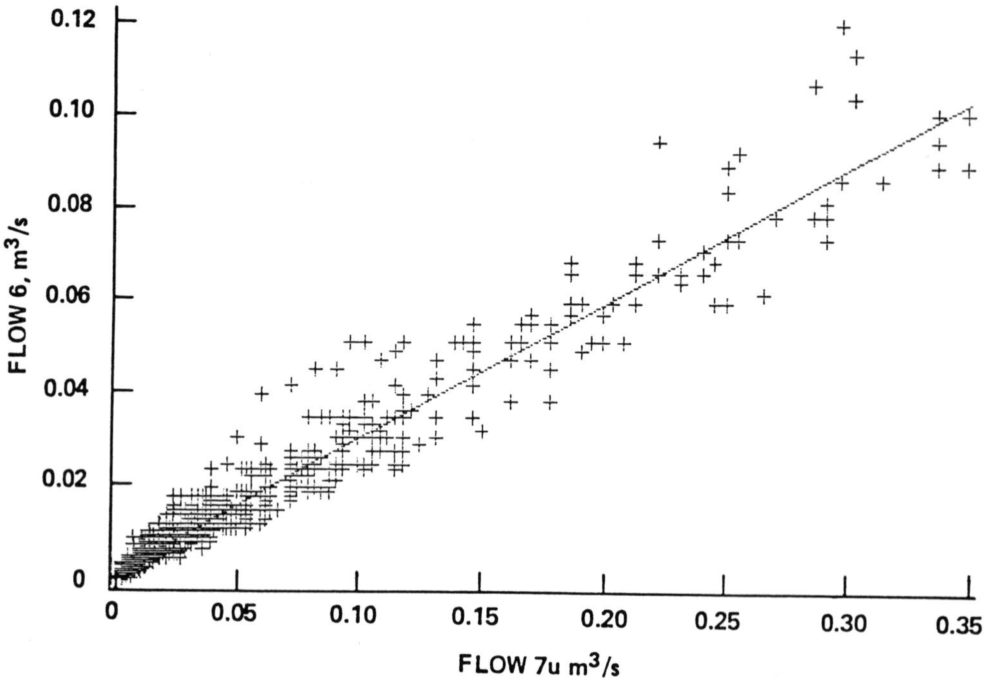

Figure 3. Correlation between streamflow on Sector VII and Sector VI. Equation used to infill missing flow data on Sector VI between 1 to 13 August 1985. EQUATION: 3.3085 x flow 7u + 0.0024 (^{2}=0.95)

Annual streamwater (or sectoral drainage) output is calculated from the automatic continuous streamwater flow measurements. Annual ion outputs are calculated using the volume of water flow between times t_1 and t_2 (V_{t1-t2}) at which daily streamwater ion concentrations between t_1 and t_2 are measured. These are then totalled over the year.

The topographical boundaries between sectors within the catchment are not clear and consequently it is difficult to quantify the exact sector areas from which water drains through the v-notch weirs and trapezoidal flumes. Actual sector areas were assessed using measured rainfall quantity, streamflow volume and estimated evapotranspiration values for the vegetation types in each sector.

As a prerequisite of any calculations, data quality was initially assured by screening out anomalous data. Rainfall quantity and quality were found to be reliable without further recourse to substitution with synthesized data. The 15 min streamflow data, however, were found to be erroneous or missing over a number of periods mainly through malfunction of the recording apparatus, for example, during frost periods. Infilling of missing periods was achieved by fitting a linear statistical relationship between the flows of different sectors, not greater than 300 m apart (**Figure 3**). The relationship is fitted around the time of the missing period ensuring that transient effects of soil moisture storage are minimized. In general, a very good correlation between reliable flows from different sectors is found (r^2 usually > 0.90) and so an estimate can be made to infill the missing 15 min flow records (**Figure 4, Table II**).

3. RESULTS

On the basis of the foregoing and using sector areas re-evaluated using measured rainfall quantity, streamflow volume, and estimated evapotranspiration values for the vegetation types in each sector, hydrochemical budgets for sectors IV, VI, and VII and the whole catchment were calculated and are shown in Table III.

Limestone is comprised of 40% Ca by weight. Based on the amount of limestone, and hence Ca applied (**Table I**), and the amount of Ca calculated as having left the treated sectors in the two year since April 1986, the amount of Ca remaining on each sector can be

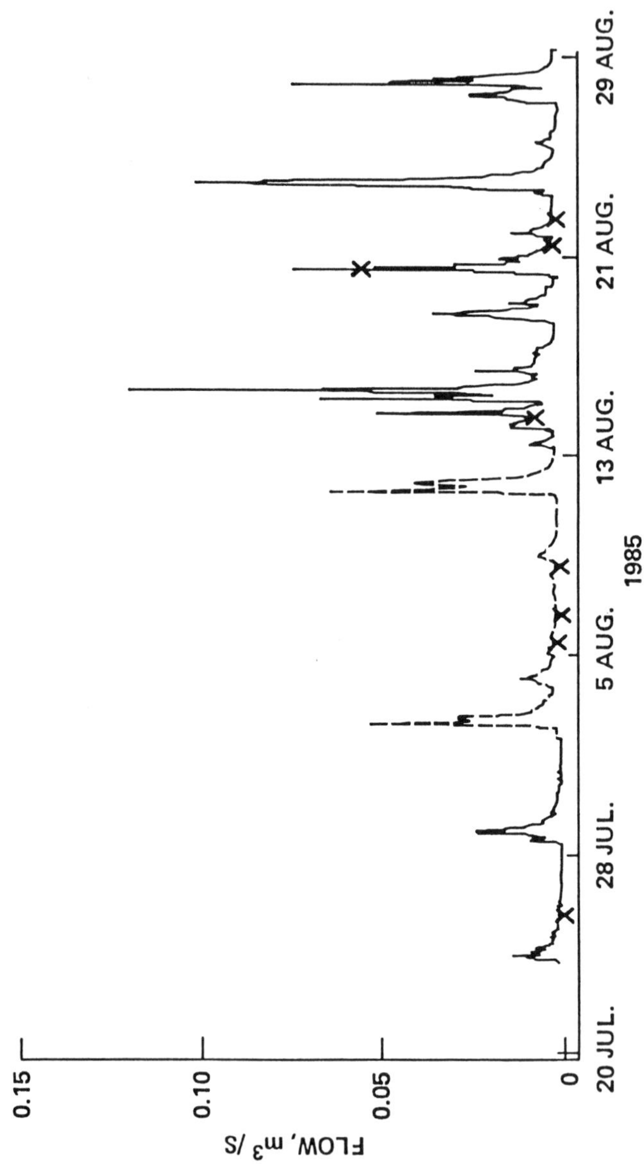

Figure 4. Hydrograph of flow on Sector VI showing infilled flow by cross-correlation (---) with flow on Sector VII. Also shown are manual "spot" flow measurements (X) taken over this time.

Table II

Number of Days in Annual Records Over Which Flow Was Infilled Using Cross-Correlation on Sectors IV, VI and VII.

SECTOR	Total Days Infilled	Days Infilled Using	
IV		**VI**	**VII**
1985-86	116	59	57
1986-87	87	63	24
1987-88	140	130	10
VI		**IV**	**VII**
1985-86	91	10	81
1986-87	45	4	41
1987-88	18	0	18
VII		**IV**	**VI**
1985-86	49	28	21
1986-87	40	23	17
1987-88	23	0	23

Table III

Hydrochemical Budgets for 1985–86 (Pre-treatment) and 1986–87 and 1987–88 (Post-treatment)

(a) 1985 to 1986. Rainfall = 2336 mm

	IN	OUT SECTOR IV (4.6 ha)	OUT SECTOR VI (6.3 ha)	OUT SECTOR VII (32.1 ha)	OUT SECTOR Whole Catchment (128 ha)
Water	2.34×10^7	1.84×10^7	2.13×10^7	1.98×10^7	1.89×10^7
Cl^-	2.78	3.92	3.62	3.45	3.42
SO_4^{2-}	1.15	1.81	1.82	1.86	1.84
NO_3^-	0.28	0.20	0.32	0.30	0.30
Na^+	2.23	3.42	3.28	3.17	3.18
Mg^{2+}	0.50	1.05	0.92	0.91	0.88
Ca^{2+}	0.18	1.42	0.76	0.99	0.86
H^+	0.41	0.91	0.99	0.67	0.55

Units: water, liters
ions, keq ha^{-1}

(b) 1986 to 1987. Rainfall = 2406 mm

	IN	OUT SECTOR IV (4.6 ha)	OUT SECTOR VI (6.3 ha)	OUT SECTOR VII (32.1 ha)	OUT SECTOR Whole Catchment (128 ha)
Water	2.41×10^7	1.67×10^7	1.91×10^7	2.08×10^7	1.88×10^7
Cl^-	2.83	4.31	3.88	4.16	3.93
SO_4^{2-}	1.08	2.40	2.25	2.00	1.75
NO_3^-	0.42	0.54	0.69	0.56	0.44
Na^+	2.47	3.86	3.55	3.83	3.43
Mg^{2+}	0.48	1.37	0.98	0.97	0.83
Ca^{2+}	0.27	20.20	16.02	8.84	2.67
H^+	0.50	0.003	0.003	0.007	0.04

Units: water, liters
ions, keq ha^{-1}

Table III (con't)

(c) 1987 to 1988. Rainfall = 2354 mm

	IN	OUT SECTOR			
		IV (4.6 ha)	VI (6.3 ha)	VII (32.1 ha)	Whole Catchment (128 ha)
Water	2.35×10^7	1.38×10^7	1.85×10^7	1.77×10^7	1.93×10^7
Cl^-	2.54	3.00	2.97	2.73	2.77
SO_4^{2-}	0.90	1.38	1.45	1.31	1.59
NO_3^-	0.37	0.28	0.40	0.22	0.46
Na^+	2.10	2.55	2.63	2.56	2.68
Mg^{2+}	0.40	0.96	0.73	0.67	0.69
Ca^{2+}	0.24	14.32	11.24	5.74	3.34
H^+	0.46	0.001	0.001	0.002	0.005

Units: water, liters
ions, keq ha^{-1}

Table IV

Calcium Applied to and Lost from Sectors Receiving Limestone Applications in April 1986.

Sector	Ca Applied, keq ha^{-1}	Ca Lost, Yr 1 keq ha^{-1}	Leaving keq ha^{-1}	Ca Lost, Yr 2 keq ha^{-1}	Leaving keq ha^{-1}
IV	399.2	18.5 (4.6%)*	380.7	12.7 (3.3%)*	368.0
VI	598.8	15.0 (2.5%)*	583.8	10.2 (1.7%)*	574.0
VII	51.9	7.6 (14.6%)*	44.3	4.5 (10.2%)*	39.8

*The % lost values are calculated on the basis of amount of Ca available at the beginning of each year.

calculated. In addition, the emerging pattern of Ca loss from these sectors can be established (**Table IV**).

4. DISCUSSION

It is recognized that precipitation chemistry measured in samples from the bulk collectors most likely contains only a part of the total dry or occult deposition. Two methods are given by Wright and Johannesson (1980) to correct for errors due to this in total ion inputs. One relies on measurements of SO_2 gas, SO_4 aerosol concentrations and deposition velocities; the other on chloride inputs equalling outputs. Both require a number of assumptions to be made, the validities of which are difficult to test. Consequently, only the directly measured (bulk precipitation) ion inputs have been considered in this study.

Loch Fleet rainfall inputs and runoff chemistry are dominated by sea salt components, in particular Na and chloride, and in addition the outputs of these components are greater than the inputs. Without a significant catchment source, this indicates a contribution by other forms of deposition not fully accounted for in rainfall chemical analysis. The results from each sector are similar despite the 86% forest cover on sector IV; it might have been expected that some ion outputs would have been higher than from the other sectors due to an atmospheric scavenging effect of the trees.

Outputs of Na and chloride relative to inputs varied over the three year; these were greatest in the year immediately following liming (1986-87), but then dropped in the second year post-liming to lower levels. The difference between the years may be due to differences in the magnitude of dry and occult deposition, although variations in exchangeable levels of Na caused by liming may cause some short term changes in streamwater chemistry. This effect, however, would most likely be undetectable over a year. Moreover, the fact that the Na/chloride ratio in inputs and outputs remains similar suggests climatic or atmospheric influences.

The next most plentiful ion input was that of sulphate, another component for which there is likely to be a dry or occult deposition contribution. The inputs of sulphate declined over the three year, a finding also observed by Farley and Werritty (1988) at Loch Dee over the period 1981 to 1985. This decline may reflect changes in industrial activity and the reduced S emissions in the UK over the same period (RGAR, 1987), but climatic changes may also be important.

The calculation of Ca budgets for the years following liming is important in assessing the duration of effectiveness of the lime treatments applied to the three sectors. For the whole catchment, the results show that Ca output increased between the first and second years post-liming due to the application of further limestone treatments to other sectors in the catchment in April 1987 (Brown et al., 1988); outputs from the three limed sectors decreased over this period.

Calcium outputs from the limed sectors appear to be declining in a non-linear way. This is illustrated in Figure 5 using the results obtained from Sector VII (**Table IV**) as an example. Considerable amounts of limestone remain on the catchment (**Table IV**). Despite this, the availability of the calcite to maintain circumneutral conditions is being curtailed in some way, possibly through inorganic or organic coating of limestone particles. The time will come, ultimately, when the limestone particles become either exhausted or inaccessible, and this would be indicated in Figure 5 by an approach to a zero slope. At present, it is expected that the Ca export will continue to decline over time, but further data collection and interpretation will provide improved clarity.

The decline in Ca outputs observed at Loch Fleet has been similar to those experienced in terrestrial liming projects in Scandanavia. For example, at Tjonnstrond (with overall application of 3 t ha^{-1}) 3.5% and 11% of the original dose had dissolved after one and four years, respectively (Johannessen, 1987). At Bravattenbacken and Nordbacken (each with a total of 30 t Ca locally applied) the losses of original Ca dose in subsequent years was 11.3, 6.1, 5.1, 4.9 and 3.7% and 12.5, 5.3 and 4.0%, respectively (Lessmark, 1987). Hence, the occurrence of a non-linear decline in total Ca loss following liming has also been observed elsewhere.

In order to make estimates of the likely durations and effectiveness of the treatments in maintaining water quality acceptable for fish survival, the results of budget calculations need to be compared with streamwater chemistry changes and with limed sector soil chemistry analysis. Water chemistry plots from the three treated sectors (and the loch outlet) are given

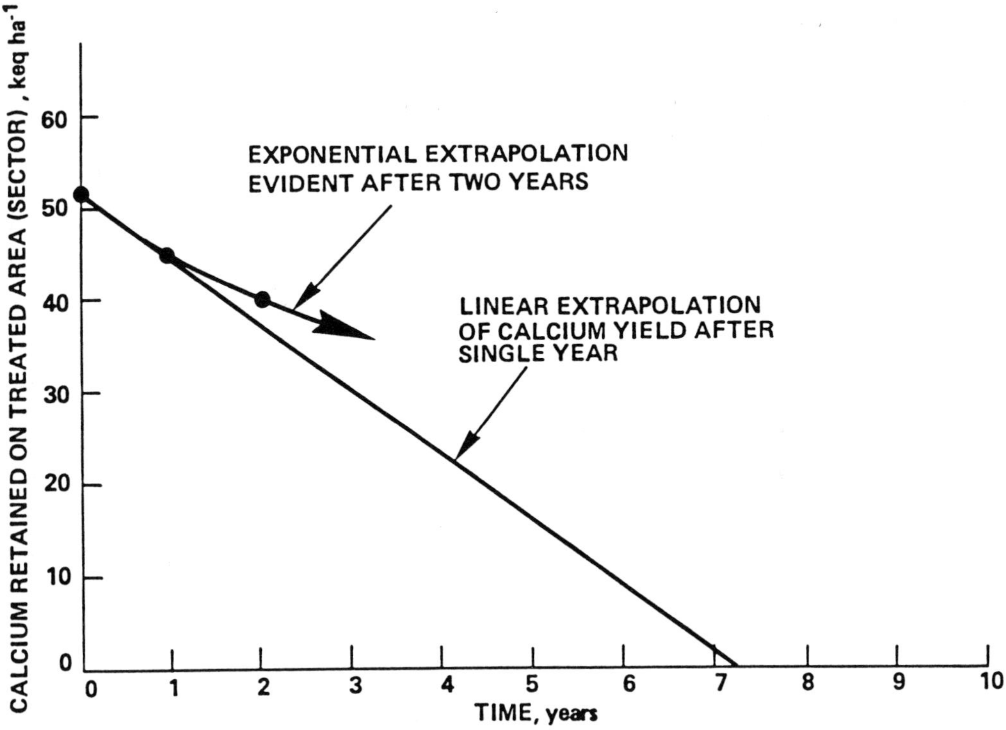

Figure 5. Calcium retained on Sector VII.

in Brown et al. (1988). These show that the pH of runoff water increased from around pH 4.2 to 4.5 to circumneutral in a few days following liming, coinciding with increased Ca concentrations. Labile monomeric Al (the fraction of total Al most toxic to fish) declined significantly over the same time. Since then, runoff pH has remained circumneutral and the toxic Al fraction has remained at low levels although the Ca concentration in each case is showing gradual decline.

Soil sampling for soil chemistry analysis was carried out six months after liming (Wilson et al., 1988). It was found that the variability between the samples in each sector was so great that conclusions regarding the amount of free lime remaining after this time were difficult to make with any confidence. It did seem though, from this work that most of the applied limestone had remained free and had not penetrated the soil to a great depth. An improved soil sampling strategy was implemented and further soil samples were removed for analysis a year later (18 mo following liming); analyses of these samples are not yet available.

ACKNOWLEDGMENTS

This paper is published with the permission of the Central Electricity Generating Board, the South of Scotland Electricity Board, the North of Scotland Hydro Electric Board and British Coal, for which we are grateful. We are also grateful to Ms. K. Paterson and members of Technical Services (SSEB), Dr. G.D. Howells (University of Cambridge), Dr. D.J.A. Brown and Ms. E.J. Wilson (CERL) for supplying data and for offering advice and assistance with the preparation of this paper.

THE USE OF SIMULATION MODELS AS RESOURCE MANAGEMENT TOOLS FOR RESTORING ACIDIFIED WATERS

James E. Davis

ABSTRACT

Several computer simulation models are available to resource managers for predicting the response of watersheds and adjacent surface waters to acidification and subsequent neutralization. Some of these models are useful in designing and implementing liming restoration programs. The simpler models emphasize in-lake biochemical processes involved in the dissolution of $CaCO_3$ and dilution hydrology; these are best suited for simulating straightforward direct applications to surface waters that have minimal hydrologic interaction with the terrestrial watershed. The more sophisticated models integrate various complex biogeochemical processes in land catchments in addition to in-lake and in-stream chemistry. These are best suited for rigorous, process-oriented research applications, for systems which have significant hydrologic interaction with the terrestrial catchment, and for simulating catchment liming applications. The approach and results from a recently published liming simulation study employing the modified Integrated Lake-Watershed Acidification (ILWAS) model are used to illustrate the methodology for using simulation models to complement field research and operational liming programs. Also, modifications to existing models are identified for increasing their suitability for more robust management applications under a wider variety of conditions in different regions.

1. INTRODUCTION

Significant attention has focused recently on using computer models for simulating surface water responses to neutralization mitigation strategies involving the application of lime to acidic lakes, streams and adjacent land catchments. Such simulations can be integral to the planning and project design of effective fisheries restoration projects. Moreover, models can assist in the interpretation of field data and the interaction of natural processes governing lake and watershed acidification and neutralization. However, our current understanding of the mechanisms and biogeochemical processes governing such responses is limited, especially with regard to neutralizing North American lakes. As a result, our ability to accurately

Water, Air, and Soil Pollution **41** (1988) 435–454.

predict whole ecosystem (i.e., including forest) responses with models is also limited, but nevertheless evolving.

The major expense and difficulty of conducting large scale, long-term field experiments and management programs make the use of computer models an especially attractive complement to field and laboratory efforts. Models can test hypotheses in a fraction of the time required by field experimentation alone and provide a valuable theoretical tool for analyzing data. As such, modeling is an integral component of several experimental and operational surface water liming programs in the United States (Porcella, 1988; Adams and Brocksen, 1988; Brocksen, 1988; Schreiber, 1988).

In addition, simulation models have been employed to evaluate the efficacy of liming terrestrial catchments for neutralizing adjacent surface waters at an Adirondack mountain lake in the U.S. (Davis and Goldstein, 1988), at Loch Fleet in Great Britain (CEGB, 1988; Brown et at., 1988), and in Sweden (Warfvinge and Sverdrup, 1984; 1988). These studies suggest that terrestrial liming might be advantageous in some acidic systems compared with direct treatment to surface waters.

Although these and other modeling studies have been encouraging, further model development is necessary before results can be widely applied to acidic systems in different regions and under various conditions. Currently, none of the available models is sufficiently robust for comparing the effectiveness of different liming strategies (i.e., lake, sediment, or catchment dosing) for a variety of watersheds (i.e., drainage vs seepage, lakes vs streams).

This paper provides a brief comparative overview of several neutralization models which employ different approaches to simulate the effects of liming under various conditions. The approach and results from a recent liming simulation study employing a slightly modified version of the ILWAS model are used to illustrate the methodology for using simulation models to complement field research and operational liming programs. Finally, recommendations are offered for combining the strengths of existing models to develop a robust and widely applicable management tool for resource managers.

2. NEUTRALIZATION MODELS

The goal of neutralization models is to simulate the response of a given system over time to a given treatment. As a result, these models generally share the characteristics of a mechanistic approach and a time series response. A mechanistic representation of the neutralization/reacidification processes is necessary to account for dramatic changes in system parameter values that occur during a sudden chemical manipulation such as liming. Likewise, a time series response is desirable for simulating not only the long term (months to years) duration of the treatment, but also the short term (hours to weeks) behavior of the response which may be equally important in the selection of a treatment technique.

There are three basic types of models that have been developed and applied to liming: 1) lake models, 2) soil models, and 3) integrated watershed models.

2.1 Lake Models

Lake neutralization models focus on *in situ* lake processes, such as the kinetics of calcite dissolution, dilution hydrology, and water-sediment interactions. Perhaps the simplest of these, DeAcid (Warfvinge et al., 1984; Saunders et al., 1985), was developed as a dose calculation model for treatments directly to the water column. It is a titration-based model which uses lake area, mean depth, pre-treatment water quality data and the particle size distribution of the calcite material to back-calculate the dose required to reach a target pH, ANC, and Ca concentration. Although this model relies primarily on calcite dissolution to control the response, a rudimentary sediment dose calculation is also included. DeAcid has been used quite successfully for straightforward lake treatments (Adams and Brocksen, 1988), but its simplicity limits its usefulness for more complex applications involving two or more consecutive treatments and/or a mix of treatment material types. Moreover, Al chemistry is not accounted for, which may further limit its utility for some applications.

A complementary lake reacidification model developed by Sverdrup and Warfvinge (1985), utilizes many of the calcite dissolution principles included in DeAcid, but places more emphasis on hydrologic residence time and calcite-sediment interactions to control the dynamic reacidification response following liming. Alkalinity inputs from the watershed are provided by a simple regression function for tributary acidity. One version of this model also accounts for seasonal stratification and mixing in lakes. This model has been used to accurately predict liming responses in Scandinavian lakes (Sverdrup and Warfvinge, 1985), but is limited by many of the simplifications found in DeAcid.

The most comprehensive of the lake models, Acid Lake Reacidification Model (ALaRM) (Depinto et al., 1987), is based on dynamic water column and sediment mass balances of ANC and dissolved inorganic carbon (DIC). It incorporates a detailed multi-layer sediment submodel which includes many more biochemical processes than the previous two models. ALaRM also includes thermal stratification and mixing in the lake and biochemical lake processes not included in the simpler models; notably, Al and organic acid chemistry are represented. Inputs from the terrestrial catchment are provided by a separate model, Watershed Acidification Mitigation (WAM) (Booty, 1983), which represents a few terrestrial processes mechanistically, but since it is not used interactively with ALaRM it functions primarily as a regression-based alkalinity production model that does not respond dynamically to changes in catchment parameters. ALaRM has been successfully applied to a series of lime treatments on two Adirondack lakes (Depinto et al., 1987), and owing to its comprehensive representation of many in-lake processes appears to offer a relatively robust model for evaluating a variety of direct lake applications. However, the simplistic representation of watershed processes does not lend itself to catchment manipulations.

These lake models generally yield good results in systems where only direct whole-lake liming is being evaluated, but they do not account for many important and dynamic terrestrial processes which influence the acid-base status of a water body. Therefore, applications of these models are limited to systems having relatively constant chemical inputs from groundwater and other portions of the terrestrial watershed, and are not suitable for evaluating catchment liming treatment techniques relative to lake liming. Nevertheless, these models provide sophisticated representations of important in-lake processes which may be incorporated into more comprehensive watershed models as discussed below.

2.2 *Soil Models*

Warfvinge and Sverdrup (1984 and 1988) have developed a dynamic soil model, SLIM, that focuses on calcite dissolution and a simplified representation of cation exchange chemistry in catchment soils to predict the dynamic response of groundwater to soil liming. No surface water component is included in the model, so that aquatic processes are absent and lake and stream quality must be inferred from groundwater conditions; alternatively, soil model outputs may be used as input for the lake model described above (Sverdrup and Warfvinge, 1985), but this is not an easily integrated approach. Although soil Al chemistry is included, the model does not include cation exchange reactions involving major cations other than Ca, nor does it include processes such as N mineralization, deammonification reactions, sulfate adsorption reactions and weathering of silicate minerals. These omitted processes were not found to be important for liming in the Scandinavian systems where the model has been applied (Warfvinge and Sverdrup, 1988), but may significantly influence the acid-base chemistry of soils in other regions (Chen et al., 1983).

This model yields accurate simulations for systems dominated by physical-chemical terrestrial processes, such as streams. However, SLIM does not account for many terrestrial and aquatic biochemical processes which may be significant in some systems. Like lake only models, it does not have the versatility necessary for application to a wide variety of systems and treatment approaches involving lake and catchment neutralization, but it does contain important soil liming chemistry that is lacking in more comprehensive watershed models as discussed below.

2.3 *Integrated Watershed Models*

Several general watershed acidification models have been developed to simulate responses of catchments and surface waters to acid deposition. However, only one, the ILWAS model (Gherini et al., 1985), has been adapted to simulate the response to lake and catchment neutralization (Davis and Goldstein, 1988). Therefore, this discussion of watershed models for neutralization focuses on the process representation in ILWAS.

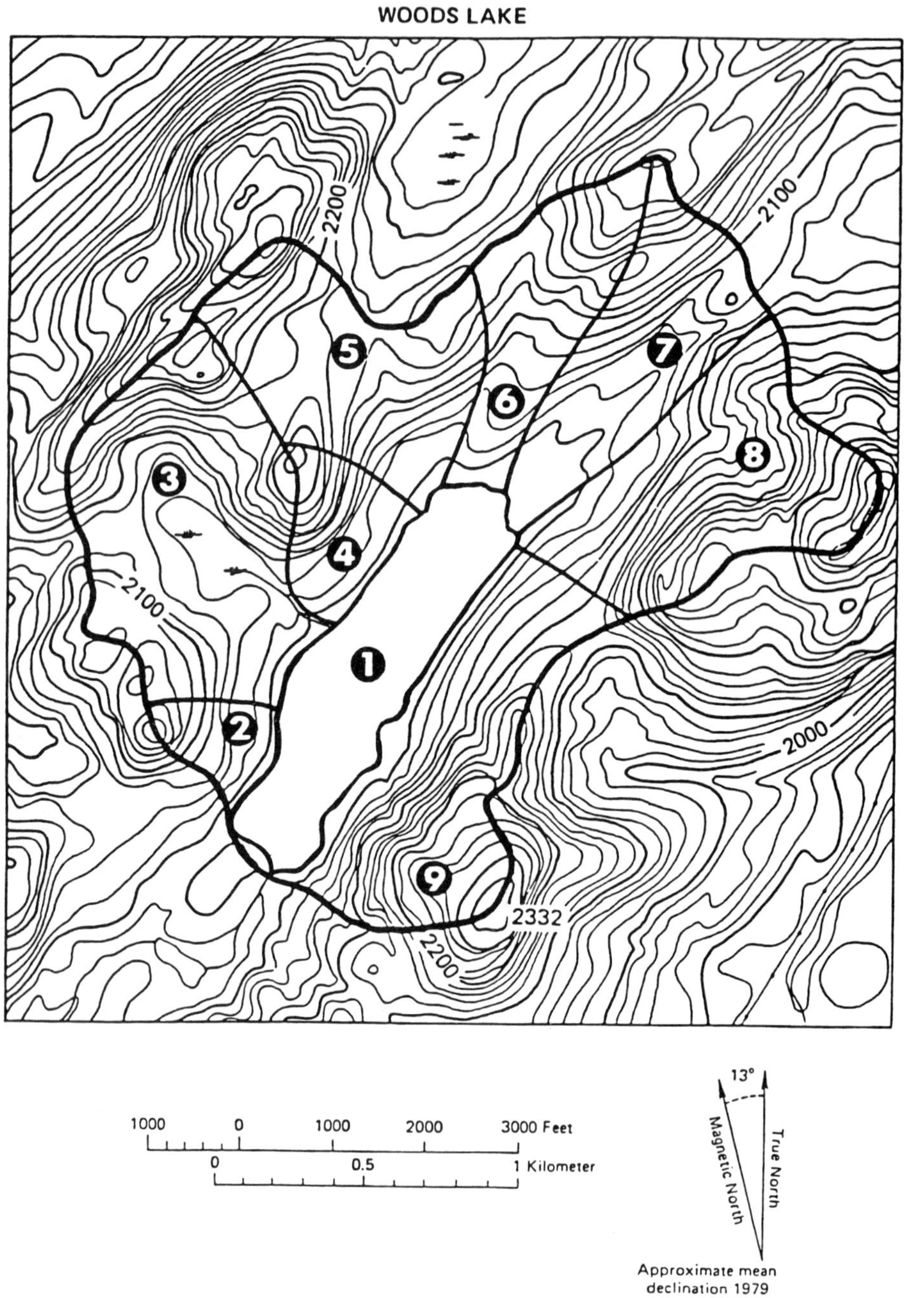

Figure 1: Horitzontal segmentation of Woods Lake basin illustrating spatial resolution of ILWAS (Chen _et al_., 1983; Figure 5-6).

The ILWAS computer model (Gherini, et al., 1985) was originally developed to quantify a general theory of acidification by integrating biogeochemical processes occurring in hydrologic basins. The basic model inputs are meteorological data (precipitation quantity, ambient air temperature, relative humidity), precipitation quality, ambient air quality (for calculating dry deposition loads), and various physical parameters that characterize the basin. The model outputs include hydrologic data (throughfall volumes, snowpack water storage, percolation and lateral flows in the soil, stream flows, and lake outflows) and the volumetric flow rates and concentrations of major cations (H^+, Ca^{2+}, Mg^{2+}, K^+, Na^+, and NH_4^+), anions (SO_4^{2-}, NO_3^-, Cl^-, and F^-), monomeric Al, total inorganic C, and an organic acid analog throughout the entire lake-watershed system (Chen et al., 1983).

All ILWAS formulations consist of deterministic equations based on mechanistic principles, so that model parameters represent meaningful physical parameters, many of which may be measured and verified in the field. Consequently, the input and output data may be measured in the field, estimated from survey maps, or extrapolated from existing data at similar or nearby sites.

The ILWAS model segments heterogeneous lake-watershed systems into cascades of homogeneous compartments representing sub-watershed units, or subcatchments (**Figure** 1). Within each subcatchment, precipitation is routed along the hydrologic pathway through the forest canopy, snow pack, soil horizons, streams, bogs, and lakes using hydraulic formulations and the principle of continuity (**Figure** 2). As the water is routed along this pathway, the concentrations of dissolved and solid phase associated chemical constituents are calculated by simulating the complex array of integrated biogeochemical reactions which occur in each compartment. As with other models, alkalinity (ANC) is used as a master variable, so that the major alkalinity producing and consuming reactions are simulated (Chen et al., 1983).

Although ILWAS was not specifically designed to simulate lake-watershed neutralization and does not include calcite dissolution, it does include many related acidification processes ignored by other models and also provides a comprehensive framework for integrating the important neutralization processes found in the other models discussed above.

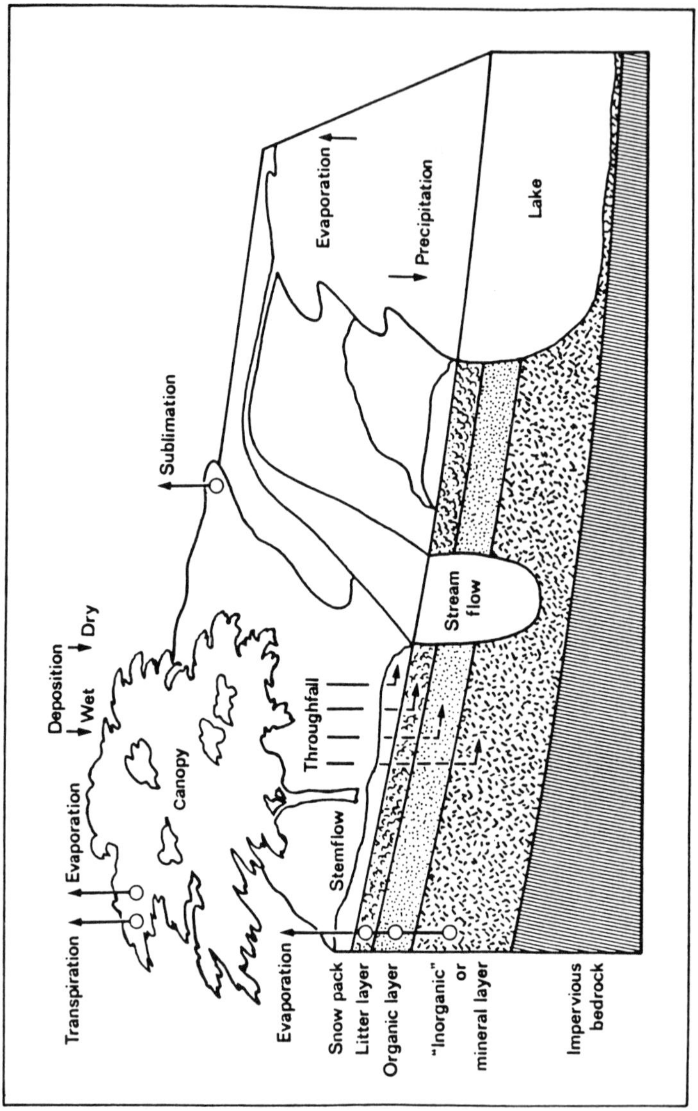

Figure 2: Pathways of water tributary to a lake (Chen <u>et al</u>., 1983; Figure 2-3).

3. IMPLEMENTING SIMULATION MODELS AS MANAGEMENT TOOLS

Once the necessary data have been collected for a given site, model applications for evaluating management decisions involve three basic steps: 1) model calibration, 2) dosing and treatment design, and 3) model validation and interpretation of results.

3.1 Calibration

Before reliable predictions can be made, a model must be calibrated to the observed pre-treatment conditions in the catchment. Even for models having parameters derived from measured data, some degree of calibration is usually necessary. Calibration is best achieved by fitting observed and simulated time series values by adjustment of model coefficient values. While there are autocalibration programs available, the manual procedure of adjusting coefficients for best fit can provide important insights about system behavior (Chen et al., 1983).

3.2 Dosing and Field Application Design

Once a reliable, calibrated model representation has been developed, the modeling effort can focus on assisting the field team in designing the liming application treatment plan. Several planning questions need to be addressed to ensure effective treatment and measurable field responses.

i) Quantity: How much lime should be applied to achieve both optimal buffering of seasonal/episodic acidification events and maximum neutralization against long-term reacidification?

ii) Timing: Does the seasonal timing of lime application significantly affect the magnitude and/or duration of the system response?

iii) Application Strategy: Should the treatment cover the entire lake or catchment evenly, or be concentrated in certain subcatchments of the watershed or sections of the lake?

iv) Lime Materials: Which "lime" material and/or particle sizes would be most effective for liming?

In order to answer these management questions, trial model runs can be made to predict the response to varying appropriate model inputs such as (i) dose, (ii) application timing, (iii) catchment coverage, and (iv) material specific dissolution parameters. The relevant parameters can be varied both independently and in different combinations. These simulation results can then be entered on a decision matrix which will assist the project team in evaluating the treatment effectiveness in terms of:

1) immediacy of initial response in soils and surface waters;
2) magnitude of initial response in soils and surface waters;
3) long-term average reacidification rate;
4) short-term buffering capacity in response to episodic events; and
5) estimated treatment cost (a function of material, particle size, and application strategy).

This Simulation Result Matrix can then be used to evaluate the efficacy of each treatment strategy for achieving the stated project goals.

3.4 Model Validation and Interpretation of Results

In the post-treatment phase of the project, the focus shifts to validation and field verification of the assumptions, formulations and resulting response predictions from the pre-liming model predictions.

The generic approach is outlined below:

i) Calibration: Additional model calibration trial runs may be needed to adjust model parameters to account for unforeseen field conditions during or after liming.

ii) Model Validation and Verification: The pre-treatment liming predictions (and subsequent refinements based on additional calibration) can be compared with observed field data to evaluate the accuracy and precision of the simulations.

iii) Data Evaluation and Process Determination: With a mechanistic model such as ILWAS, detailed simulations and analysis of field data can then be undertaken to assist field team members in interpreting the interaction of watershed processes governing the liming response.

iv) Predictions: Once the model has been validated and verified for the system, additional hypothetical treatments and applications to other systems may be simulated to compare a variety of management strategies for different watershed types. In this way, models can be useful in transferring liming technologies to potential users for other research and/or management programs.

4. EXAMPLE: WOODS LAKE LIMING SIMULATIONS

Davis and Goldstein (1988) used a calibrated ILWAS model representation of Woods Lake watershed to analyze ecosystem response to several hypothetical applications of lime. The details of the application and results will not be repeated here, but selected examples are used to illustrate the types of parameters of interest to the resource manager which may be simulated.

The computer code was modified to accept pulse inputs of calcium carbonate. Using repetitions of a calibrated 1979-1981 dataset for Woods lake, separate lime applications to both the lake and land were performed at different dose levels. The approach taken involved several assumptions which effectively simulated a controlled aerial application of fine-powdered lime or lime slurry (e.g., from a helicopter) (Davis and Goldstein, 1988).

Although the lime dose applied was comparable, Davis and Goldstein (1988) did not attempt to compare the results from this hypothetical liming simulation with response data from a more recent actual liming of Woods Lake (Porcella, 1988). The intention was to qualitatively examine the mechanisms of catchment neutralization compared to whole lake neutralization, not to quantitatively verify model predictions.

The model allowed Davis and Goldstein (1988) to compare pH responses to lake and land treatments (**Figure 3a**), which indicated that lake liming generated a more immediate but shorter duration neutralization effect relative to catchment liming. They also found that varying the dose level to the land catchment (**Figure 3b**) had a significant effect on the magnitude of the pH response, but not necessarily the duration of the response.

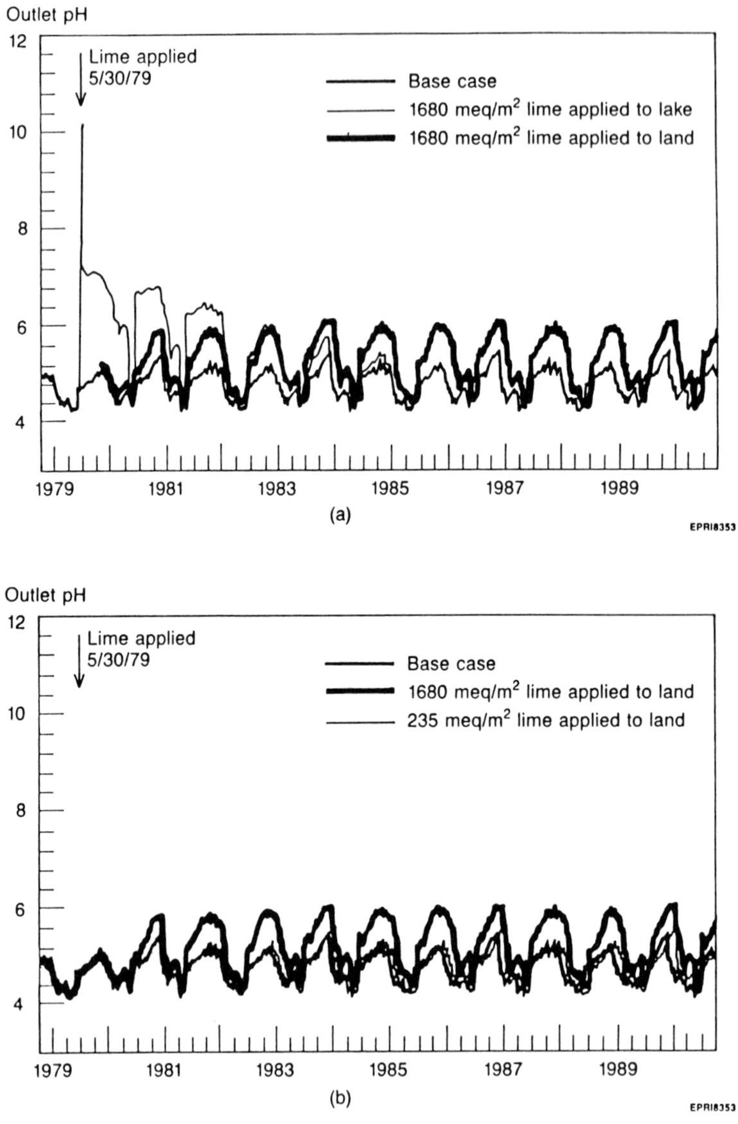

Figure 3: Simulated pH responses to spring liming: (a) comparison of equal area rate application to lake and land; (b) effect of reducing the total mass of lime applied to the land (David and Goldstein, 1988; Figure 2).

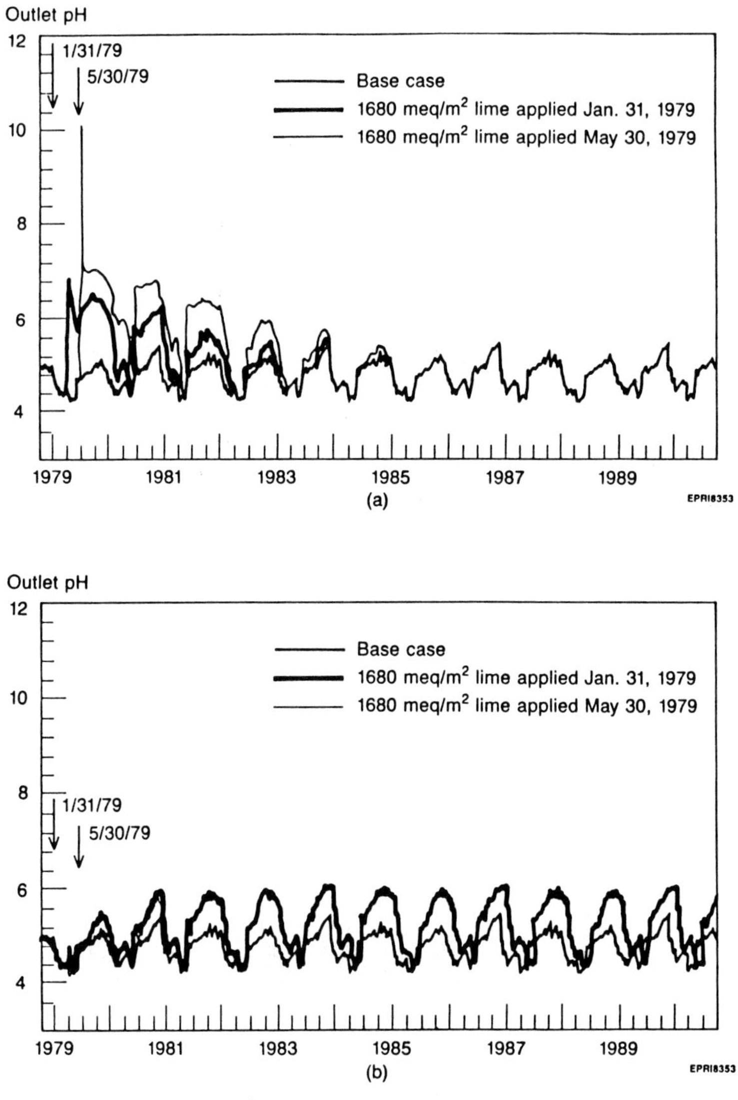

Figure 4: Simulated pH response to liming during different seasons: (a) lime applied directly to lake surface; (b) lime applied to adjacent land catchment (David and Goldstein, 1988; Figure 1).

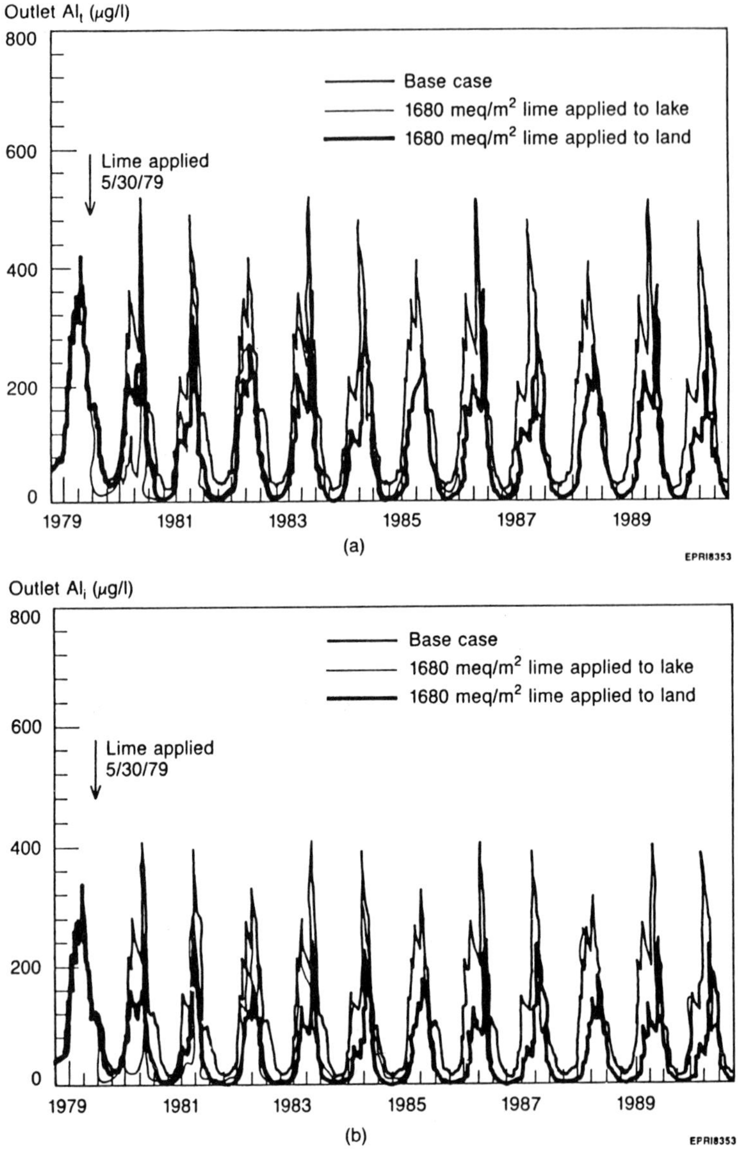

Figure 5: Simulated lake outlet Al responses to spring liming events: (a) total aluminum (Al_t) response; (b) inorganic aluminum (Al_i) response (David and Goldstein, 1988; Figure 9).

By varying the season of treatment Davis and Goldstein (1988) were able to simulate the effects of snowmelt hydrology on treatment timing. A comparison of pH responses (**Figures 4a and 4b**) indicated that a significant washout effect occurred following lime application to lake ice during winter, while the catchment treatment was relatively insensitive to the season of application.

In addition to pH responses, other important parameters such as aqueous Al behavior (**Figure 5**) were simulated. The results indicated that not only did catchment liming produce a more effective reduction in Al, but also that toxic inorganic Al dominated the total Al response to neutralization of both Woods Lake and catchment (Davis and Goldstein, 1988).

Finally, the response of several soil parameters simulated by the model, such as base saturation (**Figure 6**) and soil pH (**Figure 7**), were instrumental in developing the concluding hypothesis that enhancing soil base saturation through natural cation exchange processes may be a very effective technique for neutralizing watersheds and adjacent surface waters (Davis and Goldstein, 1988).

5. PRACTICAL CONSIDERATIONS IN USING SIMULATION MODELS

Results from model applications such as that discussed above must be interpreted within the proper context, because model and data uncertainties can be substantial. Models are useful for developing and testing hypotheses and integrating field data, but simulation results must be qualified and verified during field monitoring activities. Often, discrepancies between simulated and observed behavior provide opportunities to refine our understanding of the processes involved.

Inaccurate model calibrations or predictions may indicate inadequate representation of certain processes important in a watershed. They may also indicate the presence of unforeseen conditions in a catchment which might affect the efficacy of a given treatment plan. Once the culprit processes have been identified, a three-tiered research approach should be followed to maximize the effectiveness of model refinements:

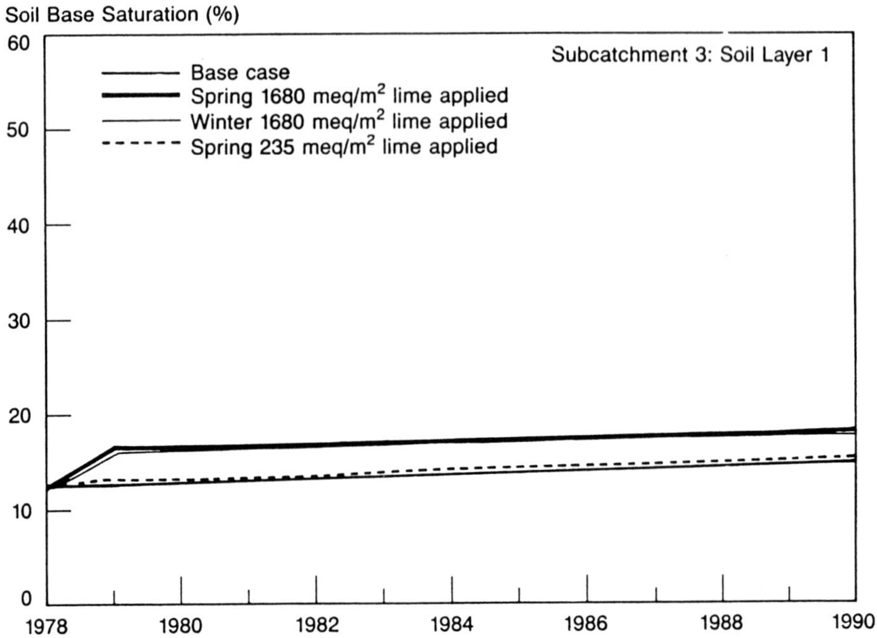

Figure 6: Soil base saturation in response to terrestrial liming: soil layer 1 of subcatchment 3 (David and Goldstein, 1988; Figure 5).

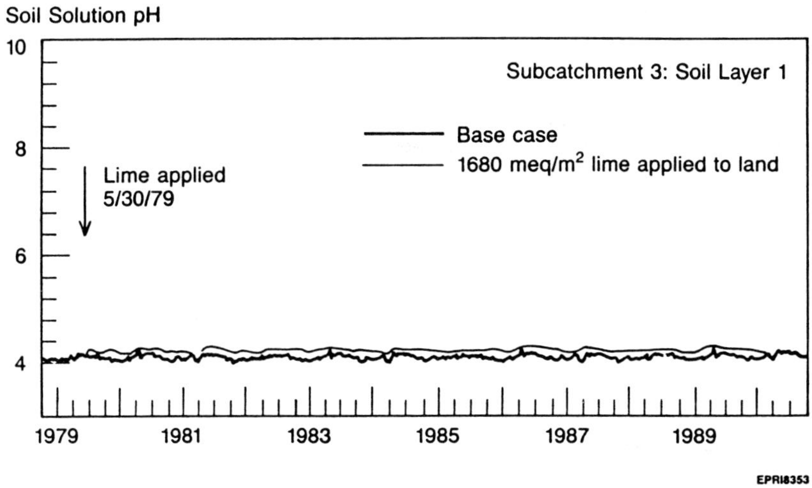

Figure 7: Soil solution pH response to Spring Terrestrial Liming: Only the hiogh dose (1680 meq m^{-2}) spring application (shown here) had a detectable response relative to the base case (Davis and Goldstein, 1988; Figure 8).

a) Literature Research and Expert Consultation: In order to formulate the process algorithms, it will be necessary to survey relevant literature and consult with experts in various fields. This will insure maximum utilization of past research results and minimize the duplication of effort. Every attempt should be made to adapt available algorithms to the model before original formulations from first principles are considered. New algorithms should be consistent with the basic model approach. For example, in the case of a model like ILWAS, all new representations should emphasize robust mechanistic processes rather than empirical analogs to insure compatibility with other model components.

b) Field Research: Once the algorithms have been revised, it may be necessary to parameterize and calibrate the model based on field data from the system being treated. Both historical data and baseline field data collected from experimental plots during the pre-treatment phase of the field monitoring effort should be used to achieve an accurate working version of the model prior to lime treatment.

c) Laboratory Research: It is conceivable, and perhaps likely, that initial literature and field research may not be complete for fully developing and parameterizing the algorithms necessary to properly simulate a key process. In such situations, laboratory work such as soil column experiments and plant dose-response tests may be required to supplement (a) and (b).

6. RECOMMENDED MODEL REFINEMENTS

As discussed above, in many respects the ILWAS model provides an appropriate vehicle for examining complex interactions associated with watershed acidification and neutralization. Nevertheless, Davis and Goldstein (1988) identified several areas where the current ILWAS model formulation may be insufficient for accurately simulating processes that may be important. Elements from other simulation models can be incorporated to overcome some of these deficiencies, but original model formulations may also be required.

Two major soil chemistry processes are either absent or not well formulated. Calcite dissolution in the upper soil horizons and lake sediments currently is not represented in ILWAS, but is probably important in controlling the speed and effectiveness of terrestrial liming. Several of the other liming models discussed above (Sverdrup and Warfvinge, 1985; Warfvinge and Sverdrup, 1984; 1988; DePinto et al., 1987) have well developed dissolution algorithms that may be incorporated within the framework of the ILWAS model.

Seasonally high flows associated with spring snowmelt contribute to significant pulses of acidity in Adirondack surface waters (Schofield et al., 1985). This phenomenon has been simulated successfully using ILWAS in both unlimed (Gherini et al., 1985) and limed systems (Davis and Goldstein, 1988). Moreover, the liming simulations suggest terrestrial liming effectively may buffer these seasonal acid pulses (Davis and Goldstein, 1988). However, the biogeochemical response of watersheds to small scale episodic hydrologic events (e.g., storms) is not well understood. These responses also may be important, but over far shorter time scales than have been simulated to date, time-step modifications will be necessary to further investigate this.

The current ILWAS model algorithm simulates several terrestrial vegetation processes: interception storage, evapotranspiration, foliar exudation, oxidation, leaf wash-off, litter fall and decay, nutrient uptake, CO_2 exchange, root respiration, and soil nitrification (Gherini et al., 1985). However, these processes are simulated by relatively constant rate parameters that are largely insensitive to biogeochmical feedback mechanisms associated with dynamic changes in the health of terrestrial biota (Davis and Goldstein, 1988).

Simulating the effects of liming on terrestrial biota is important for two major reasons. First, the efficacy of terrestrial liming as a feasible mitigation strategy may be determined not only by its ability to restore aquatic systems, but by its concurrent effects on forest health. The potential harmful (e.g., toxic) and/or beneficial (e.g., fertilization) effects on vegetation, fauna, and soil microbiology currently are not well understood. Second, terrestrial catchment biogeochemistry greatly influences adjacent surface water chemistry (Schofield et al., 1985). Soil microbes, plant roots and soil macrofauna all affect the nutrient balance and physical characteristics of the soil profile. Changes in these processes resulting from liming may have both direct and indirect impacts on soil and surface water quality that have not been simulated (Davis and Goldstein, 1988).

Finally, the current ILWAS model structure divides an entire watershed into several subcatchments. Each is treated as a homogeneous system in which soils, geology, hydrology and vegetation cover do not vary internally (Gherini et al., 1985). For most watershed processes, this level of aggregation is appropriate, but for some soil and vegetation classes (e.g., riparian and shoreline vegetation and wetlands) localized processes in hydrologically dominant portions of the catchment may disproportionately influence surface water chemistry. The ability to simulate potentially significant spatial heterogeneity within a subcatchment may be essential for interpreting field results in some cases.

7. SUMMARY

Mechanistic computer models can provide new insights into the complex processes involved in surface water acidification and neutralization. Applied within an appropriate context, computer models can be useful for interpreting field data and exploring natural processes. Simulation results can help in the design and implementation of treatment plans in the field. Likewise, field studies are essential for developing, verifying and refining computer models that are accurate predictive and analytical tools. Used in conjunction with ongoing field research efforts, mechanistic models will continue to facilitate a better understanding of the complex acidification process and potential mitigation strategies.

Living Lakes, Inc., EPRI, and others are currently developing a joint project to field test terrestrial liming at Woods Lake, providing an excellent opportunity to verify the Davis and Goldstein (1988) simulations and to draw on existing neutralization models to further refine ILWAS as a robust and comprehensive resource management tool.

EXPERIMENTAL LIMING OF WATERSHEDS: AN INTERNATIONAL COOPERATIVE EFFORT BETWEEN THE UNITED STATES AND WEST GERMANY

Robert W. Brocksen
Heinz W. Zoettl
Donald B. Porcella
Reinhard F. Huettl
Karl-Heinz Feger
Joe Wisniewski

ABSTRACT

An intergrated, international cooperative program has been initiated to manipulate two watersheds. This will include soil liming and/or fertilization of two watersheds, one in New York State and another in the Black Forest of West Germany. A description of the watersheds and apporach is documented in this paper.

1. BACKGROUND

For more than a decade, "new type" forest damages have been observed in the Black Forest of West Germany. In addition to increased needle losses, foliar discoloration symptoms typical of acute nutrient deficiencies are frequently observed in coniferous forests. Occasionally, similar symptoms also occur in beech trees. Accelerated soil acidification and induced nutritional disturbances have also been observed. Both man-made and natural influences are considered potential factors for causing these recent damages. To test this hypothesis, numerous small experimental liming/fertilization studies have been conducted. Results indicate that revitalization of declining trees and stands is possible by improving the chemical soil status and the nutrient supply via liming and/or fertilization (Zoettl and Huettl, 1987; Huettl, 1987; Huettl and Wisniewski, 1988). However, the findings from these experiments indicate that further conclusive results can only be obtained through studying complete ecosystems or through watershed manipulations.

Similar problems concerning lake acidification and forest decline have been observed in the Adirondacks and other areas in the northeastern U.S. Many extensive scientific

Water, Air, and Soil Pollution **41** (1988) 455–471.

investigations have been conducted in these regions. These investigations include research on aquatic and terrestrial systems, including the liming of lakes and streams (Porcella, 1989) and the fertilization of forest areas. It was hypothesized that the treatment of the soil via liming/fertilization should lead to improved water quality and forest vitality. The above efforts have reinforced this hypothesis. To integrate the aquatic and terrestrial components, the study of whole ecosystems via manipulations of watersheds.

From an historical viewpoint, forest liming and fertilization have a long tradition, particularly in the Black Forest. However, in previous applications, the main objective of liming was to increase forest productivity mainly by stimulating raw humus transformation. Therefore, little attention was given to the effects of this manipulation on the whole ecosystem including impacts on the bio-, pedo- and hydrosphere. However, recent investigations into older liming and fertilization practices in West Germany revealed that liming and fertilization might have a pronounced beneficial and sustained effect on forest vitality. Due to the lack of scientific design of these practices, no conclusion on causal mechanisms with respect to "new type" forest damages can be drawn.

In the U.S., liming has been practiced as a mitigation tool to reduce acidity in lake and stream waters for more than 30 years (Hasler et al., 1951). Due to ecological as well as technical considerations, it was suggested that liming forest watersheds might be a much more appropriate way of improving water quality over the long-term rather than liming lakes and streams directly. In addition, this approach would probably improve overall vitality of soils and terrestrial vegetation.

These two similar research approaches in the "old and new world" to manage forest ecosystems/watersheds have resulted in the initiation of an integrated, international cooperative effort. Research projects will focus on characterization of the principal processes and mechanisms related to the manipulation of the soil via liming and/or fertilization of a forested watershed.

2. EXPERIMENTAL DESIGN AND OBJECTIVES

The U.S. ecosystem/watershed experimental area is located in the State of New York, Herkimer County (Adirondacks), and is called the Woods Lake watershed (**Figure 1**). At its outlet, Woods Lake has a drainage area of 2.07 km^2. The surface area of the lake is 0.23 km^2 and its watershed area is 1.92 km^2. Approximately 30% of the watershed area is drained by numerous small intermittently flowing swales (**Figure** 2). The remaining 70% are drained by seven perennial or intermittent streams. The Woods Lake watershed has 0.04 km^2 of wetlands of which 0.03 km^2 are adjacent to the lake shore. The remaining 0.01 km^2 of wetlands are submerged by a beaver dam in the headwaters of the northwest inlet tributary. The geology of the Woods Lake watershed consists of pre-Cambrian age metamorphic bedrock and is comprised of charnockitic mangerite and pyroxene gneiss.

The bedrock of the Woods Lake watershed is not extensively fractured suggesting that groundwater storage in the bedrock and transport to the lake is minimal. In the watershed, numerous bedrock outcrops occur. Areas upgradient of these irregularities often remain saturated throughout the year, except for periods of extended dry weather. Surface and sub-surface run-off from these areas is commonly diverted to intermittent streams and springs that occur at breaks in the bedrock ridges. Less than 10% of the Woods Lake basin is covered by surficial deposits greater than 3 m deep. Seismic refraction work at Woods Lake indicates that thicker glacial deposits, up to 10 m deep, occur only along the northwest shore and in pockets along the northeast part of Woods Lake (**Figure 3**). As with most of the Adirondack region, aeolian silts mantle much of the Woods Lake watershed. These deposits are predominantly silts with some coarse and fine sands and usually average 30 to 35 centimeters in thickness. Typically, these are less permeable deposits than the underlying terrace and may impede infiltration. Fluvial sands, found only in underlying areas, make up only a minor part of the surficial deposits in the Woods Lake watershed. Table I reflects various Woods Lake watershed characteristics. Figure 4 reveals information about the location of hydrologic data collection at Woods Lake watershed.

The overall objective of the Woods Lake experimental watershed liming study is to determine the ecological effects of the addition of $CaCO_3$ and/or fertilizers to portions of the

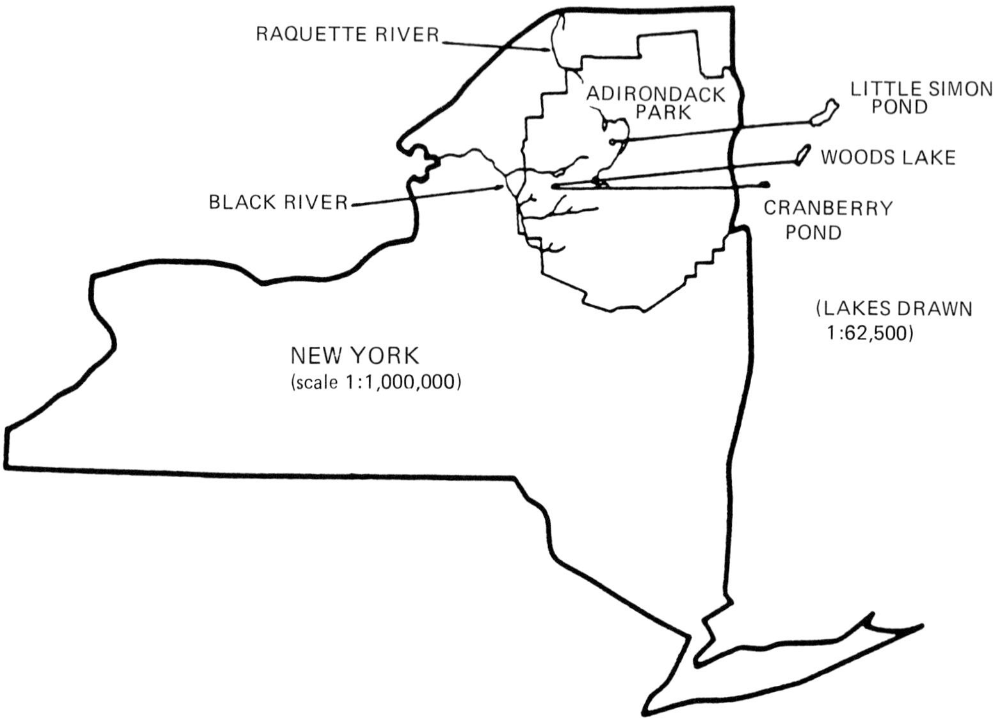

Figure 1: Outline map of New York State showing the boundary of the Adirondack State Park, the Black and Raquette Rivers, and the location of the three LAMP study lakes: Woods Lake as well as Cranberry Pond and Little Simon Pond.

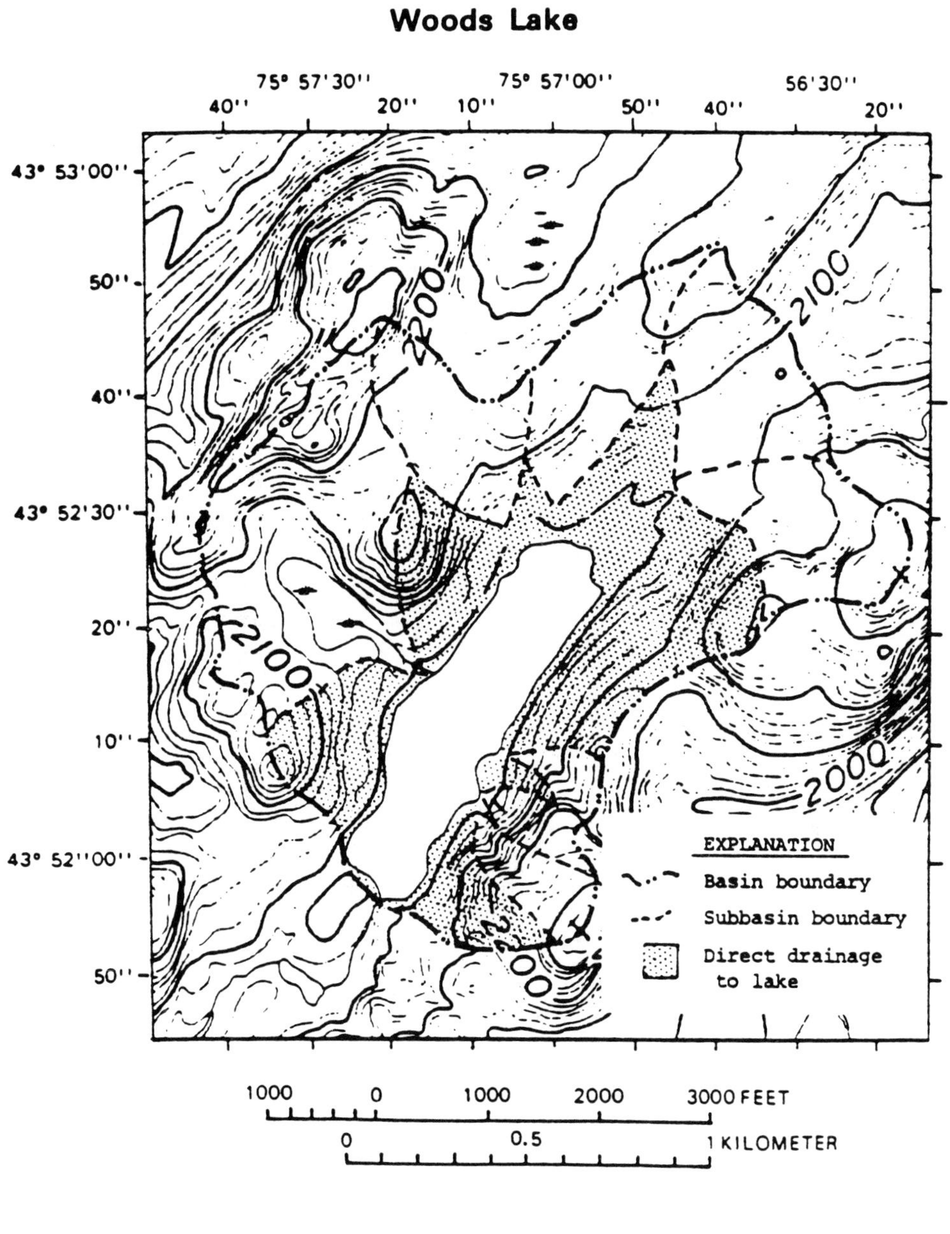

Figure 2: Drainage pattern of the Woods Lake watershed.

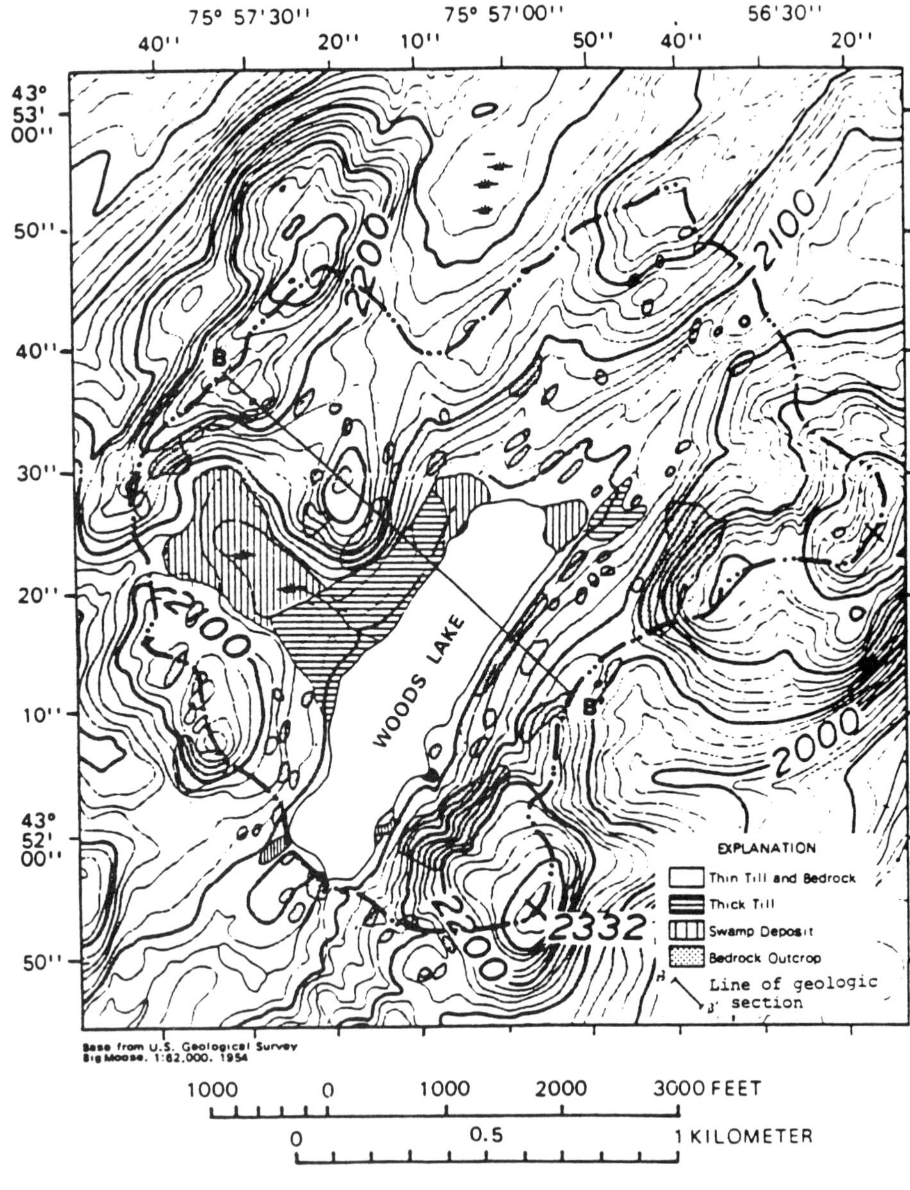

Figure 3: Surficial geology of the Woods Lake watershed.

Table I.

Woods Lake watershed characteristics.

Latitude	43°52'
Longitude	71°58'
Watershed Surface Area (Total)	2.07 km^2
Lake Surface Area	0.23 km^2
Terrestrial Watershed Area	1.92 km^2
Lake Volume	8.0 x 10^5 m3
Lake Surface Altitude (NVGD 1929)	606
Lake/Watershed Area Ratio	1:8.3
Maximum Relief (m)	122
Average Lakewater Residence Time	214 days
Maximum Lake Depth	12 m
Mean Lake Depth	3.5 m
Percent Forest Cover	96
Percent Open	4%
<u>Overburden</u>	
AverageDepth	2.2 m
Basin Area containing till deposits less than 10 ft thick.	9.2%
Wetlands and Swamp	2.3%

Table I. (Cont'd.)

Soil Horizon Data

Horizon	Thickness (cm)	Hydraulic Conductivity cms^{-1}	1:1 pH	1:50 pH 10^{-4} N $CaCl_2$
01	15	1.3×10^{-3}	2.95	3.42
02	13	1.3×10^{-3}	----	----
A2	4	1.0×10^{-4}	3.62	3.79
B2	50	3.0×10^{-3}	4.22	3.97
C	90	7.7×10^{-3}	4.92	4.97

Vegetation

Percent Confierous	10.7
Percent Maples	48.0
Percent Beech-Birch	41.3

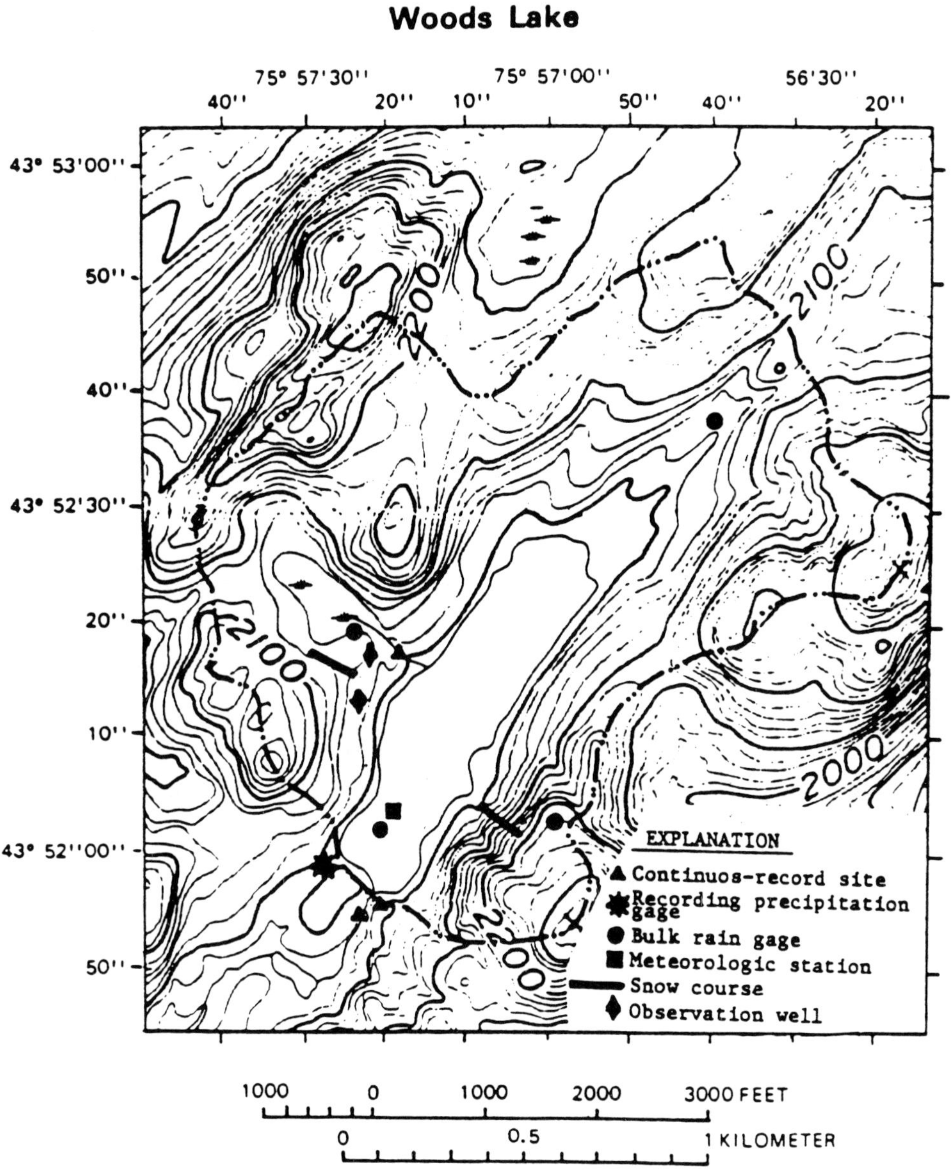

Figure 4: Location of hydrologic data collection sites at Woods Lake.

catchment area. The treatment will occur in fall 1989, after a year of pre-treatment study and plot experiments. The project is designed to examine four principal research hypotheses:

- Terrestrial application of $CaCO_3$ and/or fertilizers will increase base saturation of soil cation exchange sites and, thereby, effectively neutralize acidic inputs.
- In comparison with lake liming, a more gradual and longer duration change in lake water pH and acid neutralizing capacity will occur with terrestrial liming.
- Concentrations of Al and trace metals will decrease in lake water, increasing survival of aquatic biota.
- Episodic acidic pulses will be effectively neutralized over a longer duration both lake wide and in near-shore areas.

These principal research hypotheses will be tested together with an additional set of hypotheses aimed at identifying secondary effects of liming and/or fertilizing of terrestrial and wetland communities.

The West German research project will be conducted in the southern part of the Black Forest in southwestern Germany (State of Baden-Wuerttemberg) at the Schluchsee Experimental Watershed area, an area where well drained podsols and podsolic brown forest soils are the most common soil types. The bedrock consists of "Baerhalde" - granite which is extremely poor in Ca and Mg. The forest stand consists of 20 to 60 yr-old Norway spruce. Moderate tree damage occurs as tip-yellowing in older needle year-classes associated with a pronounced magnesium deficiency (Zoettl et al., 1987; Feger et al., 1988a,b).

Figures 5 and 6 show the general location in the Black Forest and the design of the Schluchsee Experimental Watershed area. The area is presently being investigated as part of the ARINUS-project (Zoettl et al., 1987; Feger et al., 1988a). Within this area, an additional watershed will be chosen to initiate the liming/fertilization (and control) study in spring 1989.

The major objective of the West German project is the investigation of the principal processes and causal mechanisms of "new type" forest damage on an ecosystem basis as following the manipulation of the soil by liming (fertilization) in a forest watershed.

The project is focused on similar hypotheses:

- Terrestrial application of limestone/fertilizers will decrease soil acidity and reduce the concentration of toxic metals.

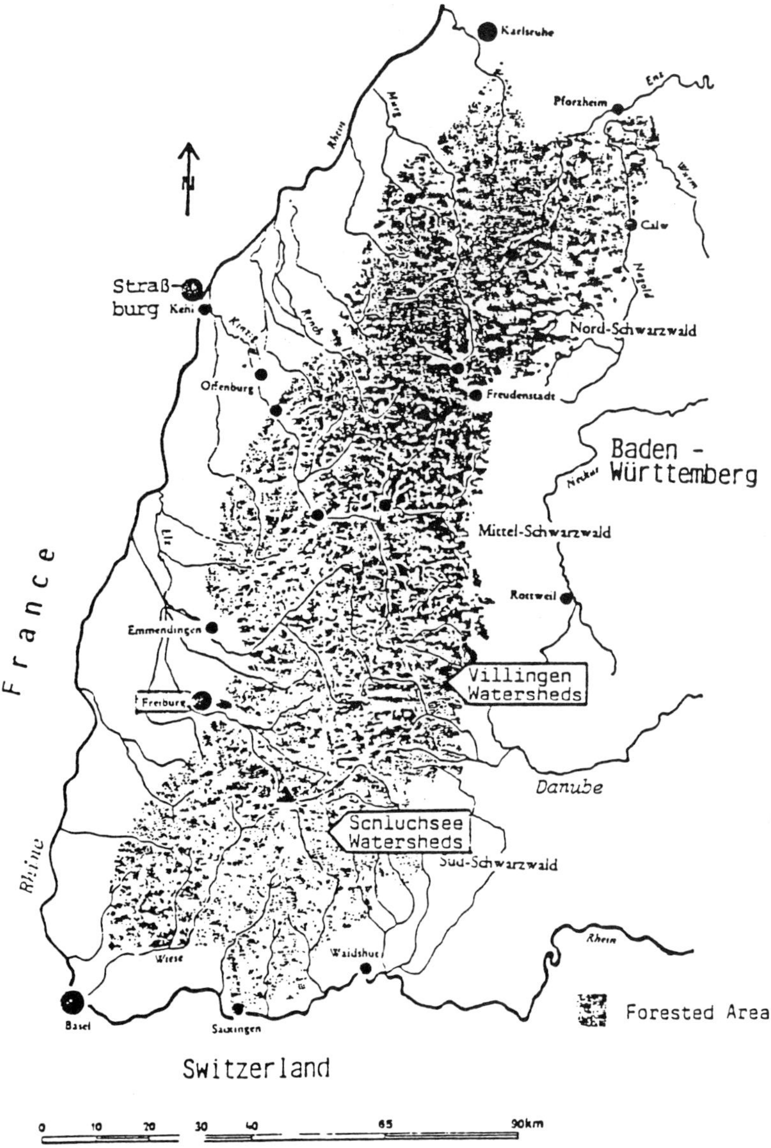

Figure 5: General location of the Schluchsee Experimental Watershed area in the Southern Black Forest (Federal Republic of Germany).

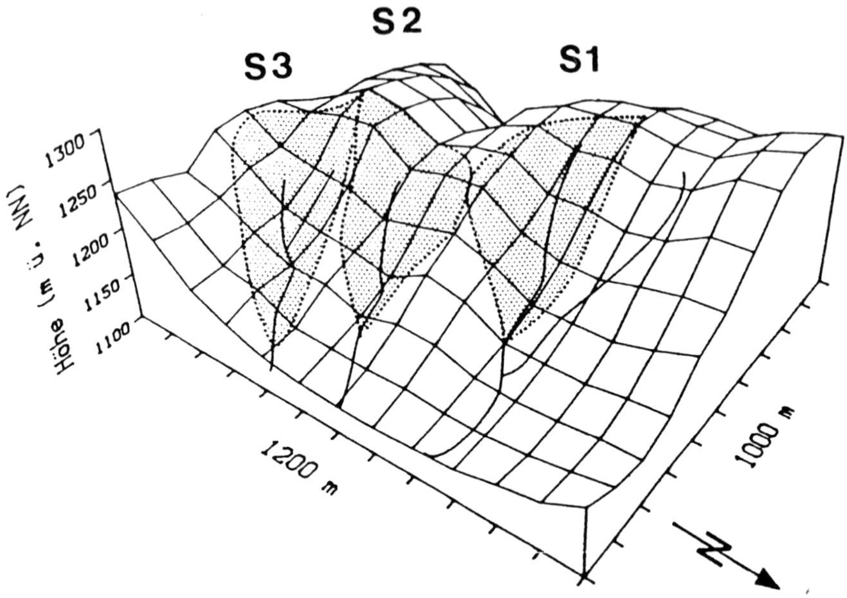

Figure 6: The Schluchsee Experimental Watershed area (Hohe m. u. NN = elevation above sea level in m.)

- Reduced soil acidity and toxic metals will lead to improvements in the overall health of terrestrial vegetation.
- Treatment will lead to improved quality of water in the watershed.
- The effects of treatment can be assessed and evaluated by quantifying nutrient and water fluxes in the forest ecosystem.

Figure 7 shows a simple conceptual ecosystem (watershed) model indicating the different compartments and fluxes that will be investigated and measured. Special investigations will be carried out to obtain detailed information e.g., about N mineralization in the soil organic layers, root and rhizosphere status, tree growth patterns, cambium conductivity, soil characteristics, and foliar histology.

3. A COOPERATIVE EFFORT

The Woods Lake and Schluchsee Watershed experimental studies have similar objectives, approaches and measurement techniques (**Table II**). Although the experimental design is similar, there are two significant differences in the projects: 1) purpose of treatment and 2) geographic factors that will make the results more interesting and allow a broader application of the results. The purpose of the West German Watershed study is to evaluate terrestrial treatment in terms of its ability to restore optimum function to managed forests. The effects of the treatment on aquatic ecosystems is but a secondary priority. The purpose of the U.S. study is directed at an aquatic ecosystem with an analysis of terrestrial effects of somewhat lesser importance. The different purposes will affect the quantity and composition of the treatment. Geographic and climatic differences will also affect the projects' results.

Integration of the two projects will come largely come from application of the Integrated Lake-Watershed Assessment Study (ILWAS) model to both systems. This model simulates all of the important cations and anions in watershed soil/water systems (Gherini et al., 1985; Davis and Goldstein, 1988), allowing project scientists to predict the effects in advance of treatment and to design the study protocols to make the important measurements. In addition, and perhaps most important, the experiments in the two dissimilar watersheds will provide the

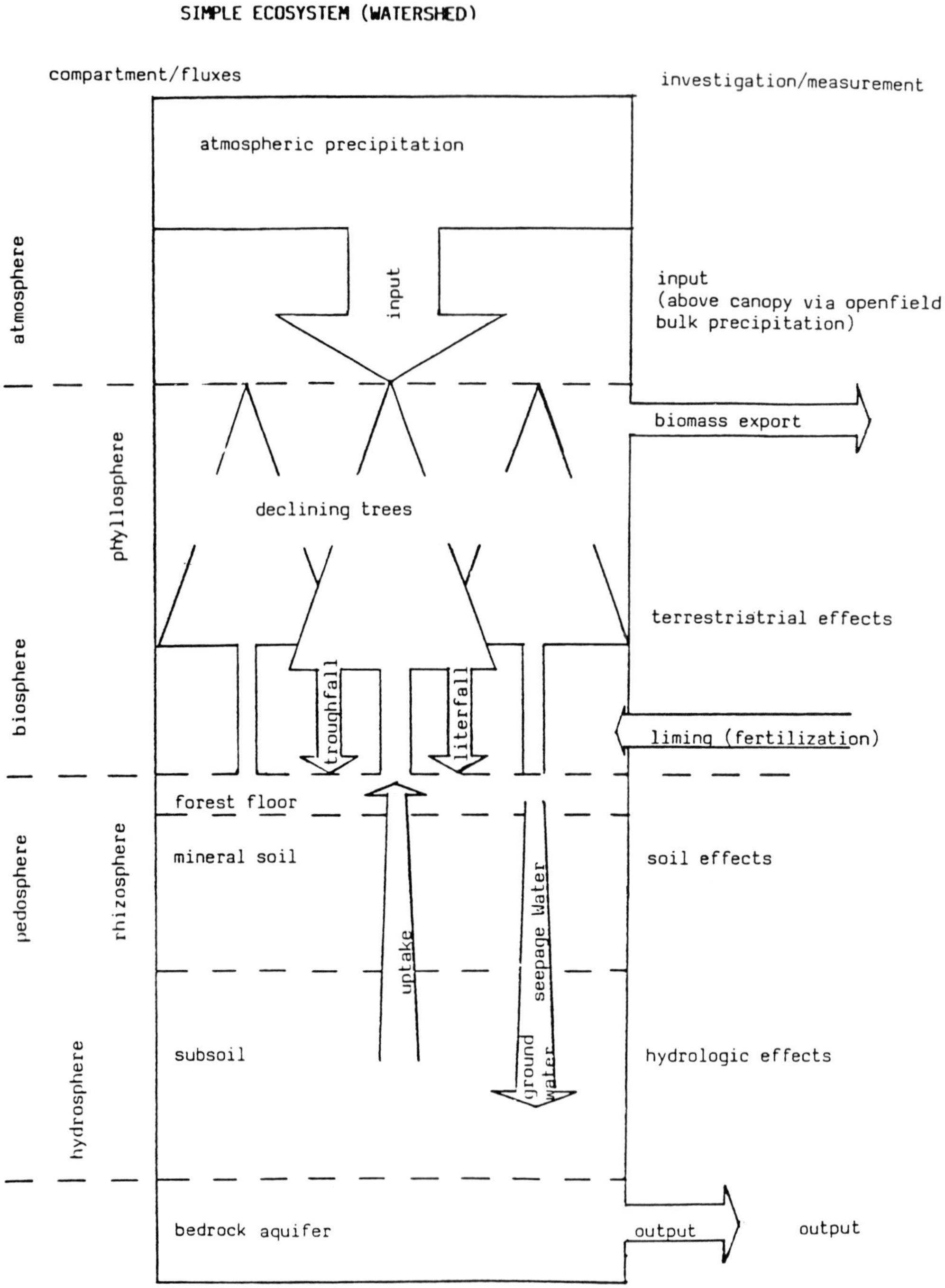

Figure 7: Simple ecosystem (watershed model).

Table II

Similarties and Differences between the Cooperative Pilot Projects in West Germany and the U.S.

Different (–)/Similar (+)	U.S.	W. Germany	Description U.S./West Germany
name	(-)	(-)	Woods Lake Watershed/Schluchsee Forest Experimental Watershed
geology	±	±	basepoor granite/basepoor glacial sands (derived from granite)
soils	+	+	podzols/acid brown earths
climate	+	+	sub-continental
precipitation	-	-	7700/7800 mm yr^{-1}
elevation	-	-	400 m/7200 m above sea level
forest	-	-	natural/unnatural
trees species	-	-	mixed/coniferous (Norway spruce)
age	+	+	20 to 60 yr
forest damages	(+)	(+)	foliar losses, moderate decline symptoms/Mg deficiency symptoms
atmospheric deposition	+	+	moderate S, H^+ and N input
ecosystem onset	+	+	
watershed (catchment) study	+	+	

TABLE II: (Cont'd).

Different (-)/Similar (+)	U.S.	W. Germany	Description U.S./West Germany
manipulation (liming study)	+	+	
terrestrial effects	+	+	forest vitality and nutrition
soil effects (mineral soil)	+	+	chemical and physical soil status
organic layer	+	+	incubation assays
soil microbiology	+	-	only U.S. (eventually also in W. Germany through U.S. support)
rhizosphere	-	+	only W. Germany (biomass, distribution, nutrient contents); eventually also in U.S.
aquatic effects	+	+	(water chemistry)
aquatic biota	+	-	only U.S.
output	+	+	
on-line data exchange system	+	+	provided by U.S.
QA/QC (quality assurance, quality control)	+	+	provided by U.S.

most rigorous check of the model to date. Previous validation studies of the ILWAS model have been conducted in about 30 lakes in the United States (Gherini et al., 1985); however, the experimental modification of the soil will provide a much more difficult test of the model.

Additional watershed studies such as the Loch Fleet liming project in the United Kingdom (Brown, 1988, this volume) and the ARINUS and SOLLING projects in West Germany will be assessed in light of the results obtained in this study.

ACKNOWLEDGMENT

The authors wish to acknowdedge the numerous West German and U.S. individuals who have contributed to the initiation and implementation of this project. These individuals are far too numerous to name, so we will not be specific here.

REFERENCES

Abrahamsen, H. and Matzow, D.: 1984, *Verh. Internat. Verein Limnol.* **22**: 1981.

Acid Magazine: 1988a, **6**; 2, National Environmental Protection Board, Box 1302, S-171 25 Solna, Sweden.

Acid Magazine: 1988b, **6**; 10, National Environmental Protection Board, Box 1302, S-171 25 Solna, Sweden.

Acid Magazine: 1987, **1**; 10, National Environmental Protection Board, Box 1302, S-171 25 Solna, Sweden.

Acre, R.G., and Boyd, C.E.: 1975, Effects of agricultural limestone on water chemistry, phytoplankton productivity, and fish production in soft water ponds, *Trans. Am. Fish. Soc.* **104**: 308.

Adams, T.B. and Brocksen, R.W.: 1988, Dose-relationships for the addition of limestone to lakes and ponds in the northeastern United States, *Water, Air, and Soil Poll.*, this volume.

Ahl, T.: 1985, *Water Quality Bulletin* **10**: 82.

Aldinger, E.: 1987, Elementgehalte im Bodfen und in Nadelin verschieden stark geschaedigter Fichten-Tannen-Bestaende auf Praxiskalkungsflaechen im Buntsandstein-Schwarzwald. Freiburger Bodenkdl. Abh., **19**: 266.

Alenas, I.: 1986, Kalkningsporjektet Härskogen, Vattenkemish och biologisk respons på kalkningsåtgärder i sju västsvenska sjöar. Swedish Water Air Pollution Institute, Gothenburg, L 86/201: 90 (in Swedish).

Almer, B., Dickson, W., Eskström, C., and Hörnström, E.: 1978, Sulfur pollution and aquatic ecosystem, 272, In: *Sulfur in the Environment II, Ecological Impact*, J.O. Nriago, Ed. John Wiley, New York.

Almer, B.: 1972, The effect of acidification on fish stocks in lakes on the west coast of Sweden. Inform. Inst. Freshw. Res., Drottningholm **12**: 47 (in Swedish with English summary.)

American Public Health Association: 1978, *Standard methods for the examination of water and wastewater*. American Public Health Assoc., Washington, D.C., 1193.

American Public Health Association, American Water Works Association, and Water Pollution Control Federation: 1981, *Standard Methods for the Examination of Water and Wastewater*. American Public Health Assoc., Washington, D.C., 1134.

Amrhein, C., Jurinak, J.J. and Moore, W.M.: 1985, *Soil Sci. Soc. of Amer. J.* **49**, 1393.

Anderson, L.G. and Settle, R.F.: 1979, *Benefit-Cost Analysis: A Practical Guide*, Lexington Books, Toronto.

Anderson, B.I., Alenas, I., and Hultberg, H.: 1984, Rep. Inst. Freshw. Res., Drottningholm **61**: 16.

Andersson, P., and Borg, H.: 1988, *Can. J. Fish. Aquat. Sci.*, in press.

Andersson, P. and Nyberg, P.: 1984 Drottningholm 61, 34.

Appelberg, M.: 1986, The crayfish Astacus in acid and neutralized environments, University of Uppsala, Department of Limnology, 150.

Arnold, D.E.: 1988, *Water, Air and Soil Poll.*, this volume.

Arnold, D. E., Bender, P. M., Hale, A. B., and Light, R. W.: 1981, 15 In: R. Singer, (ed.) *Effects of Acidic Precipitation on Benthos*. North American Benthological Society, Hamilton, N. Y., 154.

Arnold, D. E., Light, R. W., and Paul, E. A.: 1985, *Vulnerability of Selected Lakes and Streams in the Middle Atlantic Region to Acidification: a Regional Survey*, U. S. Fish and Wildlife Service, Biological Report 80(40.19), Washington, D.C., 133.

Baath, E., Berg, B., Lohm, U., Lundgren, B., Lundkivst, H., Rosswall, T., Soederstroem, B., and

Wiren, A.: 1980, *Pedabiologia*, **20**: 85.

Baker, J.P.: 1984, Effects of acidification on phytoplankton. Chapter 5 Pages 47 In: A.P. Altshuller and R.A. Linthurst (eds.), the Acidic Deposition Phenomenon and Its Effects. Critical Assessment Review Papers, Vol. II, Washington, D.C. EPA-600/8-830-016BF.

Baker, J.P.: 1982, Effects on fish of metals associated with acidification. In: *Acid Rain/Fisheries*, 165 R.E Johnson, ed. American Fisheries Society, Bethesda, MD.

Baker, J.P. and Schofield, C.L.: 1980, Aluminum toxicity to fish as related to acid precipitation and Adirondack water quality. In: Ecological Impact of Acid Precipitation, 292. D. Drablos and Tollan, A., eds. Oslo-As, Norway.

Barnes, R.B.: 1975, The determination of specific forms of aluminum in natural water. *Chem. Goal* **15**: 177.

Beamish, R.J.: 1976, *Water, Air, and Soil Poll.* **6**: 501.

Beese, F. and Waraghae, A.: 1985, AFZ **43**: 1164.

Beggs, G.L., Gunn, J.M., and Olver, C.H.: 1985, *Ont. Fish. Tech. Rep. Ser.* **17**: 14.

Bendall, B.E. and McNicol, D.K.: 1987, Fish predation, lake acidity and the composition of aquatic insect assemblages. *Hydrobiologia* **150**: 193.5

Bender, P. M.: 1978, *Studies on the Periphyton Communities of Two Infertile Mountain Streams with Differing pH*. M. S. Thesis, The Pennsylvania State University, University Park, 42.

Bengtsson, B., Dickson, W., and Nyberg, P: 1980, Liming acid lakes in Sweden, *Ambio* **9**: 34.

Bernes, C. and Thörnelöf, E.: 1988, Present and future acidification of Swedish lakes: Model calculations based on an extensive survey, In: *Environmental impact models to assess regional acidification*, IIASA/Reidel, 30.

Björklund, I., Borg, H., and Johansson, K.: 1984, *Ambio* **13**: 118.

Blake, L.: 1981, *New York Fish and Game J.* **28**: 208.

Booth, G.M., Hamilton, J.G., and Molot, L.A.: 1987, Final Technical Report, Experimental Lake Neutralization Studies in Ontario 1981-1986. Volume I and II.

Booth, G.M., Hamilton, J.G., and Molot, L.A.: 1986, Liming in Ontario: short-term biological and chemical changes, *Water, Air, and Soil Poll.* **31**: 709.

Booty, W.G.: 1983, *"Watershed acidification model and the soil acid neutralization capacity concept."* Ph.D. dissertation. McMaster University, Hamilton, Ontario, 194.

Borg, H.: 1986, *Water, Air and Soil Poll.* **30**: 1007.

Bosch, C., Pfannkuch, EW., Rehfuess, K.E., Runkel, K.H., Schramel, P. and Senser, M.: 1986, *Forstw. Cbl.* **105**: 218.

Boyd, C.E. and Cuenco, M.L.: 1980, Refinements of the lime requirement procedure for fish ponds. *Aquaculture* 21: 293.

Brakke, D.F., Landers, D.H., Linthurst, R.A., and Messer, J.: 1985, *Lake Line* **6**: 14.

Brettum, P. and Hindar, H., 1985. Effects of lime treatment on the biological system. 79 In: K. Baalsrud, A. Hindar, M. Johannessen and D. Matzow (eds.), *Lime Treatment Against Acid Water.* Norwegian Institute for Water Research, Oslo, Norway.

Britt, D.L. and Fraser, J.E.: 1983, Effectiveness and Uncertainties Associated with the Chemical Neutralization of Acidified Surface Waters. Proc. Int. Symp. on Lake Restoration, Protection, and Management. EPA 440/5-83-001, 96.

Brocksen, R.W.: 1988, *Water, Air and Soil Poll.*, this volume.

Brocksen, R.W., Filbin, G.F., Bonoff, M.B., Gremillion, P.T., Danehy, R., and Fraser, J.E.: 1987, The effect of limestone application on water quality in fifteen acidified New York and Massachusetts lakes. In: Proceedings of the 80th Annual Meeting of the Air Pollution Control Association, 1. June 21-26, 1987. New York, NY.

Brocksen, R., Zoettle, H., Porcella, D., Huettl, R., Feger, K-H, Wisniewski, J.: 1988, *Water, Air and Soil Poll.*, this volume.

Brown, D.J.A.: 1983, *Bull. Environ. Contam. Toxicol.* **30**: 582.

Brown, D.J.A.: 1982, *Water Air and Soil Pollut.* **16**: 343.

Brown, D.J.A., Howells, G.D., Dalziel, T.R.K. and Stewart, B.R.: 1988, *Water, Air and Soil Poll.*, this volume.

Brown, J.M. and Goodyear, C.D.: 1987, *Acid Precipitation Mitigation Program: Research methods and protocols.* U.S. Fish and Wildlife Service, National Ecology Center, Leetown, WV. NEC-87: 27.

Brown, D.J.A.: 1983, *J. Fish., Biol.* **18**, 31.

Brown, D.J.A: 1988, *Water, Air and Soil Poll.*, this volume.

Brown, D.J.A., Howells, G.D., Dalziel, T.R.K., and Stewart, B.R.: 1988, *Water, Air and Soil Poll.*, this volume.

Buckler, D.R., Mehrle, P.M., Cleveland, L., and Dwyer, J.: 1987, *Water, Air, Soil Poll.* **35**: 97.

Bukaveckas, P.A.: 1989, Effects of calcite treatment on primary producers in acidified Adirondack lakes. I. Response by phytoplankton communities. *Can. J. Fish. Aquat. Sci.* 46, in press.

Bukaveckas, P.A.: 1988, Effects of calcite treatment on primary producers in acidified Adirondack lakes. II. Response of macrophyte communities. *Lake Reserv. Manage.* **4**:107.

Bukaveckas, P.A. and Driscoll, C.T.: 1988, Changes in transparency associated with altered lake acidity: results from experimental neutralizations of acidified Adirondack lakes, in press.

Campbell, S., Bartoshesky, J., Heimbuch, D., Janicki, A., and Petrimoulx, H.: 1987, Relationships between acid deposition, watershed characteristics, and stream chemistry in Maryland's Coastal Plain. Prepared for the Maryland Department of Natural Resources, Power Plant Research Program, Annapolis, MD., Report No. AD-87-8.

Casselman, J.M., and Harvey, H.H.: 1973, *Prog. Fish Cult.* **35**: 218.

Casselman, J.M.: 1987, *'Determination of age and growth of fishes'*, In: A.H. Weatherley and H.S. Gill (eds.), *The Biology of Fish Growth*, Academic Press, London, 209.

Central Electricity Generating Board (CEGB): 1988, Loch Fleet News No. 10. Status report of the Loch Fleet Project, Central Electricity Research Laboratories, 38.

Chen, C.W., Gherini, S.A., Hudson, R.J.M., and Dean, J.D.: 1983, *The Integrated Lake-Watershed Acidification Study, Volume 1: Model Principles and Application Procedures.* Electric Power Research Institute RP1109-5. EA-3221.

Charles, D.F., 1982: Limnological studies on 50 Adirondack Mountain (NY) lakes. Ph.D. Dissertation, Indiana University, Bloomington, IN.

Clawson, M.: 1959, Methods of measuring the demand for and value of outdoor recreation. In: *Resources for the Future*, Washington, D.C. **10**:36.

Colquhoun, J., et al.: 1984, *"Acidity Status Update of Lakes and Streams in New York State*, New York Department of Environmental conservation.

Correll, D.L., Miklas, J.J., Hines, A.H, and Schafer, J.J.: 1987, *Water, Air, and Soil Poll.* **35**: 63.

Correll, D.L., Goff, N.M., and Peterjohn, W.T.: 1984, Ion balances between precipitation inputs and Rhode River watershed discharges. In: (ed.), O.P. Bricker, *Geologic Aspects of Acid Deposition*, 7: 77. *Acid Precipitation Series*, 7. Stoneham, MA.

Cowling, E. B.: 1982, *Environ. Sci. Technol.* **16**: 110A.

Crawshaw, D.H., Diamond, D., Prigg, R.F., Cragg Hine D., Robinson, J.F.: 1987, Acidification of surface waters in Cumbria and South Pennines. North West Water Authority. Final Report to Department of Environment (Contract PECD 7/10/89), NWWA, Dawson House, Warrington, WA5 3LW, UK.

Dahl, K.: 1927, *Salmon and Trout Magazine*, **46**: 35.

Davis, J.E., and Goldstein, R.A.: 1988, "Simulated Response of an Acidic Adirondack Lake-Watershed to Various Liming Mitigation Strategies." *Water Res. Research* **24**: 525.

Degerman, E.: 1987, Humic lakes - A literature survey with emphasis on fish and acidification. Environmental Protection Board, Rep. 3415, 72 (In Swedish with English summary).

Denniger, W.: 1988, *Holz- und and Zentralbl.* **22**: 322.

DePinto, J.V., Scheffe, R.D., Booty, W.G., and Young, T.C.: 1989, *Can. J. Fish. Aquat. Sci.*, **46**, in press.

DePinto, J.V., Scheffe, R.D., Young, T.C., Booty, W.G., and Rhea, J.R.: 1987, Use of Acid Lake Reacidification Model (ALaRM) to Assess Impact of Bottom Sediments on Calcium Carbonate Treated Lakes. Proc. North Amer. Lake Management Soc., Portland, OR (Nov., 1986), in press.

Derome, J., Kukkola, M., and Maelkoenen, E.: 1986, Forest liming on mineral soils - results of Finnish experiments. National Swedish Environmental Protection Board, report 3084; 107.

Dickson, W.: 1985, Liming in Sweden, *Liming Acidic Waters: Environmental and Policy Concerns*, Center for Environmental Information, Inc., Albany, NY, 29.

Dillon, P.J., Reid, R.A., and Girard, R.: 1986, *Water, Soil, and Poll.* **31**: 59.

Dillon, P.J., Yan, N.D., Scheider, W.A., and Conroy, N.: 1979, Acidic lakes in Ontario, Canada: characterization, extent, and responses to base and nutrient additions, *Arch. Hydrobiol. Beih. Ergebn. Limnol.* **13**: 317.

DN: 1987, Handlingsplan for kalking i Norge (Plan for liming in Norway). Directorate for Nature Management, 32 (in Norwegian).

Dozier, B.J. and Richerson, P.J.: 1975, An improved membrane filter method for the enumeration of phytoplankton. *Verh. Internat. Verein. Limnol.* **19**: 1524.

Drever, J.I.: 1982, *The Geochemistry of Natural Waters*, Prentice-Hall, Inc., Englewood, NJ, 388.

Driscoll, C.T., Fordham, G.F., Ayling, W.A., and Oliver, L.M.: 1989a, *Can. J. Fish. Aquat. Sci.*, **46**, in press.

Driscoll, C.T., Ayling, W.A., Fordham, G.F., and Oliver, L.M.: 1989b, *Can. J. Fish. Aquat. Sci.*, **46**, in press.

Driscoll, C.T., Baker, J.P., Bisogni, J.J.Jr., and Schofield, C.: 1980, *Nature* **284**: 161.

Driscoll, C.T. and Newton, R.M.: 1985, Chemical characteristics of acid-sensitive lakes in the Adirondack region of New York. *Environ. Sci. Technol.* **19**: 1018.

Driscoll, C.T.: 1984, A procedure for the fractionation of aqueous aluminum in dilute acidic water. *Int. J. Environ. Anal. Chem.* **16**: 267.

Driscoll, C.T. and Schafran, G.C., 1984.

Dutkowsky, D. and Menz, F.C.: 1988, "A Cost Function for Neutralizing Acidic Adirondack Surface Waters." *J. Envir. Eco. Manage.*, in press.

Ecological Analysts: 1983, Survey of water quality and flow in Maryland coastal zone streams: a data report. Prepared for the Maryland Department of Natural Resources, Power Plant Research Program, Annapolis, MD. J-86-1.

Edman, G., Fleisher, S., Fritz, Ö., and Stribe L.: 1988, Högvadsån 1978-1986, Försurad, kalkad, pånyttfodd, Falkenbergs kommun och Länsstyrelsen i Hallands län, 181 (in Swedish with English summary).

Ekström, C. and Hörnström E.: 1986, Acidification and Liming Effects on Phyto- and Zooplankton in some Swedish West Coast Lakes, National Environmental Protection Board, PM 1864, 108.

Emler, P.W., Jr.: 1988, Living Lakes, Inc., personal correspondence.

Engblom, E. and Lingdell, P-E.: 1983, Bottenfaunans anvandarhet some pH-indikator, National Environmental Protection Board, PM 1741, 181 (in Swedish with English abstract).

Engblom, E. and Lingdell, P-E.: 1985, Hur påverkar kalkdoserare bottenfauna?, National Environmental Protection Board, PM 1994, 81. (in Swedish with English summary).

Eriksson, F., Hörnström, E., Mossberg, P. and Nyberg, P.: 1983, Ecological effects of lime treatment of acidified lakes and rivers in Sweden. *Hydrobiologia* **101**: 145.

Eriksson, F., Hörnström, E., Mossberg, P. and Nyberg, P.: 1982, Ecological effects of lime treatment of acidified lakes and rivers. National Swedish Board of Fisheries, Institute for Freshwater Research. Drottingholm, Sweden.

Eriksson, M.O.G., and Tengelin, B.: 1987, *Hydrobiol.* **146**: 187.

Eriksson, M.O.G., L. Henrikson, B.-I. Nilsson, G. Nyman, H.C. Oscarson, A.E. Stenson and Larsson, K.: 1980, Predator-prey relations important for the biotic changes in acidified lakes. *Ambio* 9: 248.

Evans, D.O., Henderson, B.A., Bax, N.J., Marshall, T.R., Oglesby, R.T., and Christie, W.J.: 1987, *Can. J. Fish. Aquat. Sci.* 44 (Suppl. 2): 448.

Evans, R.A: 1989, *Can. J. Fish. Aquat. Sci.*, **46**, in press.

Evers, F.-H.: 1986a, Forstudeungung und Nitratbelastung von Sicker-, Oberflaechen- und Quellwasser. IMA-Querschnittsseminar "Restabilisierungsmab- nahmen - Duengung". Karlsruhe, West Germany, 15/16 April, 1986. Proceedings 57.

Evers, F.-H.: 1986b, Abschliebende Bewertun - weiterer Forschungsbedarf, IMA-Querschnittsseminar "Restabilisierungsmabnahmen - Duengung". Karlsruhe, West Germany, 15/16 april 1986. Proceedings, 98.

Farley, D.A. and Werritty, A.: 1988, *Journal of Hydrology*, in press.

Feger, K.H., Brahmer, G. and Zoettl, H.W.: 1988b, Project Arinus: Effects of restabilization measures (fertilization) and atmospheric depositions on the cycling of nitrogen and sulfur in the eco- and hydrosphere of Black Forest sites. Paper presented at COST 612 Workshop: Effects of land use in catchments on the acidity and ecology of natural surface waters' organized by the Commission of the European Communities, University of Wales, Institute of Science & Technology, Cardiff/Wales, 11 - 13 April, 1988.

Feger, K.H., Zoettl, H.W. and Brahmer, G.: 1988a, Project ARINUS: II. Einrichtung der Messstellin and Vorlaufphase - KFK-PEF-Berichte **35** 27.

Fiskeristyrelens: 1982, *Räd och riktlinjer för kalkning av sjöar och vattendraf.* Technical Report, 1982. Maddelande frän Fiskeristyrelsen nr. 1.

Fiskeristyrelsen och Statens Naturvårdsverk.: 1981, Kalkning av sjöar och vattendrag. Inform. Inst. Freshw. Res., Drottningholm **4**: 182, (in Swedish with English summary).

Flick, W.A., Schofield, C.L., and Webster, D.A.: 1982, Remedial Action for Intensive Maintenance of Fish Stocks in Acidified Waters, 287. In: R.E. Johnson (ed.) Acid Rain/ Fisheries, Amer. Fish Soc., Bethesda, MD.

Fordham, G.F. and Driscoll, C.T.: 1989, Short-term changes in the acid/base chemistry of two acidic lakes following calcium carbonate treatment. *Can. J. Fish. Aquat. Sci.* **46**, in press.

Fraser, J.E. and Britt, D.L.: 1982, *Liming of acidified waters: A review of methods and effects on aquatic ecosystems.* U.S. Fish and Wildlife Service, Eastern Energy and Land Use Team, FWS/OBS-80/40.13.

Fraser, J.E., Saunders, W.P., von Ottingen, S.L., Britt, D.L., and Sverdrup, H.U.: 1985, Ecological effects of base addition to lake ecosystems. Internatioinal Science and Technology Inc., 161.

Fraser, J.E. and Gremillion, P.T.: 1988, "Selecting an Appropriate Level of Technology for Lake Liming: Logistic and Cost Considerations." International Science and Technology, Inc.

Fraser, J., Britt, D., Kinsman, J., DePinto, J.V., Rogers, P., Sverdrup, H. and Warfvinge, P.: 1985, *APMP Guidance Manual - Volume II: APMP liming materials and methods.* Technical Report, U.S. Fish and Wildlife Service, Eastern Energy and Land Use Team.

Fraser, J. D., Hinckley, R., Burt, R. and Severn, R.R.: 1982, Feasibility study to utilize liming as a technique to mitigate surface water acidification. EPRI EA 2362, Project 1109-14 Interim Report. Electric Power Research Institute, Palo Alto, CA.

Frost, S., Huni, A., and Kersjhaw, W.E.: 1971, *Can. J. Zool.* **49**: 167.

Funke, W.: 1986, Tiergesellschaften im Oekosystem "Fichtenforst" (Protozoa, Metazoa-Inverteprata) - Indikatoionen von Veraenderungen in Waldoekosystemen. Forschungsbericht KfK - PEF, 9.

Funke, W. and Jans, W.: 1986, Der Einflub von Kalk-, Mineralduengergaben und kuenstlicher Meregnung auf die Bodenfauna in Fichtenforsten. IMA-Querschnittsseminar "Restabilisierungsmabnahmen - Dunengung." Karlsruhe, West Germnay, 15/16 April 1986. Proceedings 50.

Gherini, S.A., Chen, C.W., Mok, L., Goldstein, R.A., Hudson, R.J.M., and Davis, G.F.: 1985, *Water, Air and Soil Pollut.* **26**: 425.

Gloss, S.P., and Schofield, C.L.: 1988, *Water, Air and Soil Poll.*, this volume.

Gloss, S.P., Schofield, C.L., and Sherman, R.E.: 1988, An evaluation of New York State lake liming data and the application of models from Scadinanvian lakes to Adirondack lakes. U.S. Fish and Wildlife Service, Progress Rept. NERC-88/04. **41**.

Gloss, S.P., Schofield, C.L., Spateholts, R.L., and Polonski, B.A.: 1989, Survival, growth, reproduction and diet of brook trout *(Salvelinus fontinalis)* in acidic lakes after liming. *Can. J. Fish. Aquat. Sci.* **46**, in press.

Gloss, S.P., Schofield, C.L., and Spateholts, R.L.: 1987, Conditions for reestablishment of brook trout *(Salvelinus fontinalis)* populations in acidic lakes following base addition. *Lake Reserv. Manage.* **3**: 412.

Goddard, C.I., Loftus, D.H., MacLean, J.A., Olver, C.H., and Shuter, B.J.: 1987, *Can. J. Fish. Aquat. Sci.* **44** (Suppl. 2): 239.

Godfrey, P.J., Rudy III, A. and Zajicek, O.T.: 1985, The Massachusetts Acid Rain Monitoring Project. A.R.M. Phase I: 240. Phase II: 18. Water Resources Research Center, University of Massachusetts, Amherst, MA.

Goldstein, R.A., Chen, C.W., and Gherini, S.A.: 1985, *Water, Air and Soil Poll.*, **26**: 327.

Goulden, P.D. and Afgahn, B.K.: 1970, An automated method for determining mercury in water. Technicon, ADV. Autoanal **2**: 317.

Grande, M., Andersen, S., and Sevaldrud, I.: 1980, Stocking of brook trout (*Salvelinus fontinalis*) in acidic lakes 1975-1978. SNSF-project IR 68/80, 88 (in Norwegian).

Grande, M., Muniz, I.P., and Andersen, S.: 1978, *Verh. Int. Verein Limnol.* 20, 2076.

Greening, H., Janicki, A., and Saunders, W.P.: 1987, An evaluation of stream liming effects on water quality and yellow perch spawning in Maryland Coastal Plain streams. 1986 Final Report. Prepared for the Maryland Department of Natural Resources, Tidewater Administration, Annapolis, MD.

Gross, G.M.: 1971, Carbon determination, In: Carver, R.E. (ed.) Procedures in Sedimentary Petrology, *Wiley Interscience* 573.

Gunn J.M. and Keller W.: 1984, *Fisheries* **9**: 19.

Gunn, J.M., McMurtry, M.J., Bowlby, J.N., Casselman, J.M., and Liimatainen, V.A.: 1987, *Trans. Am. Fish. Soc.* **116**: 618.

Gussone, H.A.: 1987, *Forést- und Holzw.* **6**: 158.

Haines, T.A. and Akielaszek, J.: 1983, *A regional survey of chemistry of headwater lakes and streams in New England: Vulnerability of acidification.* U.S. Fish and Wildlife Service, Eastern Energy and Land Use Team, FWS/OBS-80/40.15.

Haines, T.A., Pauwels, S.J., and Jagoe, C.H.: 1986, *Predicting and evaluating the effects of acidic precipitation on water chemistry and endemic fish populations in the northeastern United States.* U.S. Fish and Wildlife Service, Eastern Energy and Land Use Team, Biol. Rep. 80(40.23).

Haines, T.A.: 1981, "Acid Precipitation and its Consequences for Aquatic Ecosystems: A Review." *Trans. Am. Fish. Soc.* **1106**: 669.

Hale, A. B.: 1978, *A Comparative Study of the Insect Communities in Two Woodland Streams, with Notes on the Growth and Fecundity of Two Species of Leuctra (Plecoptera)*, M. S. Thesis, The Pennsylvania State University, University Park, 50.

Hall, R.J., Likens, G.E., Fiance, S.B., and Hendrey, G.R.: 1980, *Ecology* **61**: 976.

Hall, L.W., Jr.: 1987, *Water, Air and Soil Poll.* **35**: 87.

Hall, L.W., Jr., Pickney, A.E., and Horseman, L.O.: 1985, *Trans. Am. Fish. Soc.* **114**: 861.

Hamilton, J.G., Gunn, J.M., Booth, G.M., Beggs, G.L., Rietveld, H.J., and Nunro, J.R.: 1988, *Can. J. Fish. Aquat. Sci.*, in review.

Hartmann, F.K. and Jahn, G.: 1959, Ueber die Wirking von Kalkungen auf Waldboeden verschiedenen Naehrstoff- und Basengehaltes in soziologischoekologischer Betrachtung. Der Wald braucht Kalk, 3. Auflage, 30.

Hasler, A.D., Brynildson, O.M., and Helm, W.T.: 1951, Improving conditions for fish in brown-water bog lakes by alkalization, *J. Wildl. Mgmt.* **15**: 347.

Hasselrot, B. and Hultberg, H.: 1984, Liming of acidified lakes and streams and its consequences for aquatic ecosystems. *Fisheries* **9**: 4.

Haung, I., Kottke, I., and Oberwinkler, F.: 1986, Vergleichende Untersuchungen der Mykorrihizaen in verschiedenen Bodentiefen von Geduengten und ungeduengten Standorten. IMA-Querschnittsseminar "Restabilierungsmabnahmen - Duengung". Karlsruhe, West Germany, 15/16 April, 1986. Proceedings, 78.

Hendrey, G.R.: 1987, *Water, Air, and Soil Poll.* **35**: 1.

Hendrey, G.: 1980, Effects of acidity on primary productivity in lakes: phytoplankton. In: D. Shriner et al. (eds.) Atmospheric Sulfur Deposition - Environmental Impact and Health Effects. Ann Arbor Science, Ann Arbor, MI.

Henrikson, L., Nyman, H.G., Oscarson, H.G. and Stenson, J.A.E.: 1984, Changes in the zooplankton community after lime treatment of acidified lake. *Verh. Internat. Verein. Limnol.* **22**: 3008.

Herricks, E.E., and Cairns, J., Jr.: 1977, The effects of lime neutralization of acid mine drainage on stream ecology, Proc. 32nd Indust. Waste Conf. Purdue Univ., 477.

Hicks, J.R.: 1956, A Revision of Demand Theory. Oxford University Press.

Hildebrand, E.: 1988, *Forst-u. Holz* **43**: 51.

Hillman, D.C., Potter, J.F., and Simon, S.J.: 1986, *National Surface Water Survey, Eastern Lake Survey (phase I - synoptic chemistry) analytical methods manual.* EPA 600/4-86-009. Office of Research and Development, Environmental Monitoring Systems Laboratory, Las Vegas, Nev.

Hillman, D.C., Pia, S.H., and Simon, S.J.: 1986, Analytical methods manual for the National Surface Water Survey Stream Survey (Middle Atlantic Phase I, Southeast Screening and Middle Atlantic Episode Pilot). Prepared for the Environmental Monitoring Systems Laboratory, Office of Research and Development, U.S. EPA, Las Vegas, NV. Final Report.

Hindar A.: 1984, pH-trend and lime dissolution in liming of small lakes in Gjerstad, Aust-Agder. Report 14-1984. The Norwegian Liming Project, 69 (in Norwegian).

Hindar, A., Gunnerod, T.B. and Baalsrud, K.: 1985, *Händbok i kalking av surt van.* Technical Report, ISBN 82-90404-32-8.

Hindar, A. and Kleiven, E., Haraldstad, Ø, Raddum, G.G., Wright, R.F., Fjellheim, A. and Johannessen, M.: 1987, Liming in Norway 1984-1986. DN-report, 2-1987, 100 (in Norwegian).

Hindar, A. and J.P. Nilssen: 1984, Annual Report, Gjerstad 1982/1984. Kalkingsprosjektet Rapport 21. Norwegian Institute for Water Research, Oslo, Norway.

Hindar, A. and Rosseland, B.: 1988, *Water, Air and Soil Poll.*, this volume.

Hornstrom, E. and Ekstrom, C.: 1985, Liming effects on plankton in Swedish west coast lakes. Draft Report, Statens Naturvardsverk, Solna, Sweden.

Howells, G.D. (ed.): 1986, The Loch Fleet Project: A Report on the Pre-Intervention Phase (1) 1984-1986, Publ. CEGB, SSEB, NSHEB, BC.

Howells, G.D. and Brown, D.J.A.: 1985, *Water, Air and Soil Poll.* **31**: 817.

Howells, G.D. and Dalziel, T.R.K. (eds.): 1988, The Loch Fleet Project: A Report on the Intervention Phase (2) 1986-1987, Publ. CEGB, SSEB, NSHEB, BC.

Huettl, R.F.: 1987, "Neuartige" Waldschaeden, Ernaehrungestoerunge und Duengung. AFZ, **42**: 289.

Huettl, R.F. and Wisniewski, J.: 1987, Fertilization as a tool to mitigate forst decline associated with nutrient deficiencies. *Water, Air and Soil Poll.*, **33**: 265.

Huettl, R.F.: 1988, Diagnostic fertilization and amelioration of forest decline, in press.

Huttl, R.F.: 1985, 'Neuartige' Waldschaeden und Naehrelementversorgung von Fichtenbestaenden (Picea abies Karst.) in Suedwestdeutschland. Freiburger Bodenkdl. Abh., 16, 195.

Huttl, R.F.: 1987, *AFZ* **42**: 289.

Huettl, R.F.: 1988, "New type" forest damages in Central Europe. Report of World Resources Institute, Washington, D.C., U.S.A., in press.

Huettl, R.F. and Zoettl, H.W.: 1986, Diagnostische Duengungsversuche in geschaedigten Nadelbaumbestaenden Suedwestdeutschlands. IMA-Quereschnitts-seminar 'Restabilierungsmabnahmen - Duengung', Karlsruhe, FRG, 15/16 April 1986, 3.

Hultberg, H. and Anderson, I.B.: 1982, Liming of acidified lakes induced long-term changes. *Water, Air and Soil Poll.* **18**: 311.

Hunn, J.B., Cleveland, L., and Buckler, D.R.: 1985, Additional calcium reduces effects of low pH and aluminum on developing brook trout eggs and fry. Research Information Bulletin. U.S. Fish and Wildlife Service, Columbia National Fisheries Research Laboratory, Columbia, MO.

Hutchinson, T.C., and Havas, M.: 1986, *Water, Air and Soil Poll.* **28**: 319.

Håkansson, L.: 1986a, Liming - mercury. Project plan 1985-1989., National Environment Protection Board, Rep. 3097, 29. (in Swedish).

Håkansson, L., 1986b, Liming - mercury. Working plan. National Environmental Protection Board, Rep. 3099, 187 (in Swedish).

Håkansson, L., Nilsson, A., and Andersson T.: 1988, *Environ. Poll.* **49**: 145.

Imboden, D.M. and Schwarzenbach, R.P.: 1985, "Spatial and Temporal Distribution of Chemical Substances in Lakes: Modeling Concepts." In: W. Stumm (ed.), *Chemical Processes in Lakes.* John Wiley & Sons, New York.

Impact Assessment, Interim Report, U.S. - Canada Memorandum of Intent on Transboundary Air Pollution, Work Group No. 1, February 1981.

International Liming Workshop.: 1982, *The liming of acidified waters: Issues and research needs.* U.S. Fish and Wildlife Service, Eastern Energy and Land Use Team, FWS/OBS-80/40.14.
Interagency Task Force on Acid Precipitation (ITFAP): 1982, *National Acid Precipitation Assessment Plan.* ITFAP, Washington, D.C.

Isermann, K.: 1987, *AFZ* **42**: 997.

Ivanhnenko, T.: 1988, *Water, Air and Soil Poll.*, this volume.

Jackson, S.T. and Charles, D. F.: 1988, Aquatic macrophytes in Adirondack (NY) lakes: Patterns of species composition in relation to environment. *Can. J. Bot.*, in press.

Janicki, A., Saunders, W.P., and Ross, E.: 1986, A retrospective analysis of the frequency and magnitude of pH events in some Atlantic Coast striped bass spawning grounds. Prepared for the U.S. Fish and Wildlife Service.

Janicki, A. and Greening, H.: 1988, An evaluation of stream liming effects on water quality and anadromous fish spawning in Maryland Coastal Plain streams: 1987 results. Prepared for Living Lakes, Inc., Washington, D.C., and the Maryland Department of Natural Resources, Power Plan Research Program, Annapolis, MD. Final report.

Janicki, A. and Greening, H.: 1987, A summary of Maryland stream chemistry data. Prepared for the Maryland Department of Natural Resources, Power Plant Research Program, Annapolis, MD. Report No. AD-87-11.

Janicki, A., and Cummins, R.: 1983, An analysis of survey data on the chemistry of twenty-three streams in the Chesapeake Bay watershed: Some implications of the impact of acid deposition. Prepared for the Maryland Department of Natural Resources, Power Plant Research Program, Annapolis, MD. Report No. PPSP-MP-51.

Jeffrey, S.W. and Humphrey, G.F.: 1975, New spectrophotometric equations for determining chlorophylls a, b, c_2 in higher plants, algae and natural phytoplankton. *Biochem. Physiol. Pflanzen* **167**: 191.

Johannessen, M., and Hindar, A.: 1987, Mitigation studies, In: Acidification and Water Pathways. Proceedings of an International Symposium. Bolkesjo, Norway (ISBN-82-554-0485-6), 325.

Johnston, A.E., Goulding, K.W.T. and Poulton, P.R.: 1986, *Soil Use and Management* **2**: 3.

Kalkingsprosjektet: 1985, Håndbok i kalking av surt vann (Handbook on Liming of Acid Water), 48.

Kaupenjohann, M.: 1988, Chemischer Bodenzustand und Naehrelementversorgung immissionsbelasteter Fichtenbestaende in NO-Bayern (in preparation).

Keller, W., Pitblado, J.R., and Conroy, N.I.: 1986, *Water, Air and Soil Poll.* **29**: 285.

Kerr, J.J.: 1983, "A Methodology for Estimating Recreational Fishing Benefits of Reduced Acid Deposition" Prepared for: The James C. Penney Foundation, Inc. January.

Klauda, R.J. and Palmer, R.E.: 1986, Laboratory and field bioassay experiments on blueback herring from Maryland Coastal Plain streams. Prepared for the Maryland Department of Natural Resources, Power Plant Research Program, Annapolis, MD. Report No. PPSP-AD-15.

Knapp, C.M., Saunders, W.P., Heimbuch, D.G., Greening, H.S., and Filbin, G.J.: 1988, Maryland Synoptic Stream Chemistry Survey report: Estimating the number and distribution of streams affected by or at risk from acidification. Prepared for the Maryland Department of Natural Resources, Power Plant Research Program, Annapolis, MD Report No. AD-88-2.

Kratz, K., Williams, M., and Cooper, S.D.: 1988, *Bull. North Amer. Benthol. Soc.* **5**: 71.

Kretser, W.A. and Colquhoun, J.R.: 1984, "Treatment of New York's Adirondack Lakes by Liming" *Fisheries* **9**: 1.

Kreutzer, K.: 1987, Field experiments on the influence of acid irrigation and liming in Norway spruce ecosystems. In: Perry, R., Harrision, R.M., Bewll, J.N.B. and Lester J.N., 1987: Acid Rain: scientific and technical advances, 641.

Kreutzer, K.: 1988, Stickstoff-Haushalt und Nitrataustrag in: Koch, G.: 1988, *AFZ*, **43**: 66.

Kreutzer, K. and Zelles, L.: 1986, *Forstw. Cbl.* **105**: 314.

Kwiatkowski, R. and J. Roff, 1976. Effects of acidity on the phytoplankton and primary productivity of selected northern Ontario lakes. *Can. J. Bot.* **54**: 2546.

L'Abée-Lund, J.H. and Kleiven, E.: 1987, The fish in Tjønnstrond, Telemark after catchment liming. DN-report, 3-1987, 24 (in Norwegian).

Lahmersdorf, N.: 1985, *AFZ* **41**: 1155.

Lazarek, S., 1986. Responses of the Lobelia-epiphyte complex to liming of an acidified lake. *Aq. Bot.* **25**: 73.

Lean, D.R.S. and Burnison, B.K.: 1979, An evaluation of errors in the ^{14}C method of primary production measurement. *Limnol. Oceanogr.* **24**: 917.

Leivestad, H., Muniz, I.P., and Rosseland, B.O.: 1980, Acid stress in trout from a dilute mountain stream. In: Drabløs, D. and Tollan, A. (ed.), Ecological impact of acid precipitation, 318, SNSP-project.

Leivestad, H., Hendry, G., Muniz, I.P., and Sneivik, E.: 1976, Effects of acid precipitation on freshwater organisms, 87, In: Braekke, F. (Ed): Impact of acid precipitation on forest and freshwater ecosystems in Norway, S.N.S.F.-Project, FR 6/76, Oslo-As.

Lessmark, O. and Thörnelöf E.: 1986, Liming in Sweden, *Water, Air and Soil Poll.* **31**: 809.

Lessmark, O., Degerman, E. Johlander, A., and Sjölander, E.: 1986, Effects of liming on fish immediately downstream of lime dosers. Inform. Inst. Freshw. Res., Drottningholm **5**: 31 (in Swedish with English summary).

Lessmark, O.: 1987, Ground liming as a method of mitigating acidification of surface waters. Inform. Inst. Freshw. Res., Drottningholm **9**: 38 (in Swedish with English summary).

Likens, G.E., Barmann, F.H., Pierce, R.S., Eaton, J.S. and Johnson, N.M.: 1977, *Bio-Geo-Chemistry of a forested ecosystem*, Springer-Verlag, New York, Heldelberg, Berlin.

Likens, G. E., Bormann, F. H., and Johnson, N. M.: 1972, *Environment* **14**: 33.

Liming Project: 1987, Liming of acid water, Baalsrud, K. (ed.), The Norwegian Liming Project. Final report, 147.

Linthurst, R.A., Landers, D.H., Eilers, J.M., Brakke, D.F., Overton, W.S., Meier, E.P. and Crowe, R.E.: 1986, Characteristics of Lakes in the Eastern United States, Volume I. Populations descriptions and physico-chemical relationships. EPA/600/4-86/007a, US-EPA, Washington, D.C., 136.

Little, I.M.D.: 1957, A Critique of Welfare Economic. Oxford University Press, London.

Living Lakes, Inc. Annual Report 1986.

Living Lakes, Inc.: 1987a, Field operations program implementation manual for aquatic liming and fish restoration demonstration program, Washington, D.C.

Living Lakes, Inc.: 1987b, *Living Lakes Data Base Users Guide*, Living Lakes, Inc., Washington, D.C.

Living Lakes, Inc.: 1987c, *Field Operations Program Implementation Manual for Aquatic Liming and Fish Restoration Demonstration Program*, Living Lakes, Inc., Washington, DC.

Living Lakes, Inc.: 1987d, *Audit Operating Plan for the Independent Quality Assurance and Quality Control Auditing Program*, Living Lakes, Inc., Washington, D.C.

Lovell, H. L.: 1971, *Calculating Alkali Requirements and Costs*, Mine Drainage Research Laboratory, The Pennsylvania State University, University Park, 2.

Lynch, J. A., Corbett, E. S., and Rishel, G. B.: 1985, *Atmospheric Deposition: Spatial and Temporal Variation in Pennsylvania 1984*, Report LW8505, Institute for Research on Land and Water Resources, The Pennsylvania State University, University Park, 97.

Magnuson, J.J., Baker, J.P., and Rahel, E.J.: 1984, *Phil. Trans. Royal Soc. London B* **205**: 255.

Maine Department of Inland Fisheries and Wildlife. "Planning for maine's Inland Fish and Wildlife: Species Assessments and Strategic Plans" 1986-1991. *Fisheries*, 2: 1.

Maitland, P.S., Lyle, A.A. and Campbell, R.N.B.: 1987, Acidification and fish in Scottish Lochs, Institute of Terrestrial Ecology (NERC) Report (ISBN-1-870393-04-X).

Marcus, M.D., Parkhurst, B.R., Baker, J.P., Creager, C.S., Fannin, T.E., Ingersoll, C.G., Mount, D.R., and Rahel, F.J.: 1986, *An Evaluation and Compilation of the Reported Effects of Acidification on Aquatic Biota. Volume 1: Compiled Data*, EPRI EA-4825, Electric Power Research Institute, Palo Alto, CA, 197.

Martin, N.V.: 1952, *Trans. Am. Fish. Soc.* **81**: 111.

Matuszek, J.E.: 1987, 'An assessment of the current impact and potential risks of acid deposition on smallmouth bass populations in Ontario', Ontario Ministry of Natural Resources, Ont. Fish. Acid. Rep. Serv. 87-09, Toronto, Ontario, 22.

Matzner, E. 1985: Auswirkungen von Duengung und Kalkung auf den Elementumsatz und die Elementverteilung in Zwei Waldoekosystemen im Solling. AFZ, **41**: 1143.

Matzow, D., Rosseland, B.O., and Skoghjeim, O.K.: 1985, Effects of Lime treatment on fish In: Baalsrud, K. (ed.). Liming of acid water. Final report. The Norwegian Liming Project, 147.

Mayer, R.: 1981, Natuerliche und anthropogene Komponenten des Schwermetall-Haushaltes von Waldoekosystemen. Goettinger Bodenkundl. Ber. **70**: 292.

McKensie, G.W.: 1979, "Consumer Surplus Without Apology: A Comment." American Economic Review 69:3, 465.

McKinley, K.: 1985, A Technical Report on the Lake Acidification Mitigation Project. Report 1397-03-85-CR. General Research Corporation, McLean, VA.

Mehrle, P., Finger, S., Buckler, D., and Ludke, L.: 1985, Striped bass decline in Chesapeake Bay may be related to acidic environments. Research Information Bulletin No. 85-21. U.S. Fish and Wildlife Service, Columbia National Fisheries Research Laboratory, Columbia, MO.

Menz, F.C. and Driscoll, C.T.: 1983, "An Estimate of the Costs of Liming to Neutralize Acidic Adirondack Surface Waters." *Water Resources Research*, **19**: 5.

Merritt, R.W. and Cummins, K.W.: 1984, *An Introduction to the Aquatic Insects of North America*, 2nd ed., Kendall- Hunt, Dubuque, IA, 722.

Meyer, P., et al.: 1983, "An Updating Analysis of Differential Sport Fishing Values for Columbia River Salmon and Steelhead." May.

Middleton, A.C. and Rovers, F.A.: 1976, Average pH, *J. Water Poll. Cont. Fed.*, **48**: 395.

Mierle, G., Clark, K., and France, R.: 1986, *Water, Air, and Soil Poll.* **31**: 593.

Mills, K.H., Chalanchuk, S.M., Mohr, L.C., and Davies, I.J.: 1987, *Can. J. Fish. Aquat. Sci.* **44** (Suppl. 1): 114.

Minnesota Department of Natural Resources: 1985, "Indexing Minnesota Fish Lakes Relative to Potential Susceptibility to Acidic Deposition" Division of Fish and Wildlife. Special Publication No. 138. August.

Minshall, G.W. and Minshall, J.N.: 1978, *Archiv fuer Hydrobiologie* **83**: 324.

Minshall, G.W. and Kuehne, R.A.: 1969, *Archiv fue Hydrobiologie* **66**: 169.

Molnia, B.F.: 1979, *Journal of Sed. Pet.* **44**: 589.

Molot, L.A., Hamilton, J.G., and Booth, G.M.: 1986, *Water Res.* **20**: 757.

Monitor, 1985, Sura och forsurade vatten, Naturvardsverket informerar, 180, National Environmental Protection Board, Box 1302, S-171 25 Solna, Sweden, (in Swedish).

Morgan, E.L., Eagleson, K.W., Waver, T.P., and Isom, B.G.: 1986, Linking automated biomonitoring to remote computer platforms with satellite data retrieval in acidified streams, *Impact of Acid Rain and Deposition of Aquatic Biological Systems*, ASTM STP 928, Amer. Soc. Testing and Materials, Philadelphia, 84.

Muniz, I.P. and Leivestad, H.: 1980, Acidification-effects on freshwater fish. In: Drablos, D. and Tollan, A. (ed.), Ecological impact of acid precipitation, 84, SNSF-project.

National Atmospheric Deposition Program: 1987, *NADP/NTN Annual Data Summary: Precipitation Chemistry in the United States, 1986*, Natural Resource Ecology Laboratory, Fort Collins, CO., 363.

National Environmental Protection Board: 1985, Action Program against Air Pollution and Acidification, 77.

Nesse, J.C.: 1948, Development and status of pond fertilization in central Europe, *Trans. Am. Fish. Soc.* **76**: 335.

Nilsson, N.A. and Johansson A.: 1985, A comparative study of the benthos of limed and non-limed streams; Effects of different liming methods. Inform. Inst. Freshw. Res. Drottningholm, **11**: 56, (in Swedish with English summary).

Nilsson, N.-A.: 1972, *J. Fish. Res. Bd.*, Canada **29**: 693.

Norusis, M.J.: 1986a, *SPSS/PC+*, SPSS, Inc., Chicago, IL.

Norusis, M.J.: 1986b, *Advanced Statistics SPSS/PC+*, SPSS, Inc., Chicago, IL.

Nyberg, P.: 1984, *Phil. Trans. R. Soc., Lond. B* **305**: 549.

Nyberg, P.: 1987, Reclamation of acidified Arctic char (*Salvelinus alphinus* (L)) lakes in Sweden by means of liming, *Verh. Internat. Verein. Limnol.*, x.

Nyberg, P., Appelberg, M., and Degerman, E.: 1986a, *Water, Air and Soil Poll.* **31**: 669.

Nyberg, P., Degerman, E., Esktröm, C., and Hörnström E.: 1986b, Acid-sensitive Arctic char (*Salvelinus alpinus*) lakes in southern and central Sweden, Inform. Inst. Freshw. Res., Drottningholm **6**, 240 (in Sweden with English summary).

Nyberg, P. and Thörnelöf: 1988, *Water, Air and Soil Poll.*, this volume.

O'Dell, J., Gabor, J., and Dintaman, R.: 1975, Survey of the anadromous fish spawning areas. Prepared for the Maryland Department of Natural Resources in cooperation with the United States Department of Commerce.

Oden, S.: 1976, *Water, Air, and Soil Poll.* **6**: 137.

Oden, S.: 1968, *The Acidification of Air and Precipitation and its Consequences on the Natural Environment*, Swedish Nat. Sci. Res. Council, Ecology Committee. Bull. 1. Swedish National Research Council, Stockholm (in Swedish).

Omernik, J. and Griffith, G.: 1986, *Environ. Manage.* **10**: 829.

OMOE (Ontario Ministry of the Environment): 1982, 'Sudbury environmental study. Studies of lakes and watersheds near Sudbury Ontario: final limnological report', Ontario Ministry of the Environment Rep. SES 009/782, Toronto, Ontario.

Peters, N.E.: 1984, Hydrologic Analysis of Woods and Panther Lake Basins in the West-Central Adirondack Mountains. In: The Integrated Lake-Watershed Acidification Study, Vol. 4: Summary of Major Results. EA-3221, Vol. 4, RP 1109-5. Electric Power Research Institute, Palo Alto, CA 222.

Paterson, K.: 1987, Loch Fleet Project: Water Sampling and Analysis Procedures. SSEB Report No. BRT/87/71.

Plummer, N.L., Wigley, T.M.L. and Parkhurst, D.L.: 1978, *Am. J. Sci.* **278**: 179.

Porcella, D.B.: 1987, Overview of approach and liming of lakes for the Lake Acidification Mitigation Project (LAMP), *Lake and Reservoir Management* **3**: 401.

Porcella. D.B.: 1988, *Water, Air and Soil Poll.*, this volume.

Porcella, D.B.: 1989, Mitigation of acidic conditions in lakes: An overview of an ecosystem perturbation experiment. *Can. J. Fish. Aquat. Sci.* **46**, in press.

Qvenild, T.: 1986, Stocking of brook trout in Norway. Fisk og Fiskestell 7, 1 (in Norwegian).

Raddum, G.G., Hagenlund, G., and Halvorsen G.A.: 1984, Effects of lime treatment on the benthos of Lake Sondre Boksjo, Rep. Inst. Freshw. Res., Drottningholm **61**: 167.

Raddum, G.G. and Fjellheim, A.: 1987, 77, In: Witters, H. and Vanderborght, (eds.), *Ecophysiology of acid stress in aquatic organisms*. Belgian Royal Zool. Soc., 287.

Raddum, G.G.: 1979, Effects of low pH on insect larvae. SNSF-project, IR 45/79, 58 (in Norwegian).

Raddum, G.G.: 1980, Comparison of benthic invertebrates in lakes with different acidity. In: Tollan, A. and Drabløs, D. (eds): Ecological impact of acid precipitation, 330, SNSF0- project.

Raddum, G.G. and Fjellheim, A.: 1984, *Verh. Int. Verein Limnol.* **22**: 1973.

Raddum, G.G., Brettum, P., Matzow, D., Nilssen, J.P., Shov, A., Svealv, T., and Wright, R.F.: 1986, *Water, Air and Soil Poll.* **31**: 721.

Reading, H.G. ed.: 1986, *Sedimentary Environments and Facies*, Blackwell Scientific Publications, Boston 34.

Reger, D.B.: 1931, Randolph County Report, West Virginia Geological Survey, Wheeling News Litho. Co., 150.

Renberg, I, Hellberg, T. and Nilsson, M.: 1985, Effects of acidification on diatom communities as revealed by analyses of lake sediments. In: F. Andersson and B. Olsson, *Lake Gärdsjön-An acid lake and its catchment*, 219, *Ecological Bulletins*, **37**.

RGAR: 1987, Acid Deposition in the United Kingdon 1981-1985. Report on the UK Review Group of Acid Rain, Warren Spring Laboratory, Stevenage, UK.

Riely, P.L. and Rockland, D.B.: 1988, "A Review of Fisheries Economic Valuation Methods." Prepared for: The USDA Forest Service Wildlife and Fisheries, Washington D.C. February.

Ritter, T., Kottke, I., and Oberwinkler, F.: 1986: Vergleich der Vitalitaet von Mykorrhizaen unterschiedlich geduengter Flaechen mittels FDA-Vitalflourogromierung. IMA-Duerschnittsseminar "Revitalisierungsmabnahmen - Duengung". Karlsruhe, West Germany, 15/16 April, 1986. Proceedings 83.

Rockland, D.B.: 1987, "Economic Concepts Related to the Comparison of Relative Values of Timber and Fishery Resources." Prepared for: Society of American Foresters, Minneapolis, Minnesota. October.

Rockland, D.B.: 1985, "The Economic Benefits of a Fishery Resource: A Practical Guide" Technical Report I. SFI Economics Program.

Rockland, D.B. and Oatis, P.H.: 1987, "A Preliminary Economic Assessment of Liming Acidified Lakes and Ponds in Massachusetts."

Rodenkirchen, H.: 1986, *Forstw. Cbl.* **105**: 338.

Rorison, I.H.: 1980, The Effects of Soil Acidity on Nutrient Availability and Plant Response. Effects of Acid Precipitation on Terrestrial Ecosystems. Plenum, N.Y.

Rosseland, B.O., Skogheim, O.K., and Sevaldrud, I.H.: 1985, *Water, Air and Soil Poll.*, **30**: 65.

Rosseland, B and Hindar, A.: 1988, Liming lakes, rivers and catchments in Norway, this volume.

Rosseland, B.O., Sevaldrud, I. H., Svalastog, D., and Muniz, J.P.: 1980, Studies of freshwater fish populations-effects of acidification on reproduction, population structure, growth and food selection. In: Drabløs, D. and Tollan, A., (eds.): Ecological impact of acid preciptiation, 336, SNSF-project.

Rosseland, B.O. and Skogheim, O.K.: 1984a, *Fisheries* **9**: 10.

Rosseland, B.O. and Skogheim, O.K.: 1984b, Rep. Inst. Freshw. Res. Drottningholm **61**: 186.

Rosseland, B.O., Skogheim, O.K., Abrahamsen, H., and Matzow, D.: 1986a, *Can. J. Fish. Aquat. Sci.* **43**: 1888.

Rosseland, B.O., Skogheim, O.K., Kroglund, F., and Hoell, E.: 1986b, *Water, Air and Soil Poll.* **30**: 751.

Ruffin, E.: 1832, An Essay on Calcareous Manure. Revised 1852.

Ryckman, J.R. and Lockwood, R.N.: 1985, "On-Site Creel Surveys in Michigan 1975-82. Michigan Department of Natural Resources. Fisheries Division. Fisheries Research Report No. 1922. December.

Samuelson, P.A.: 1947, Foundations of Economic Analysis. Cambridge.

Saunders, W.P., Britt, D.L., Kinsman, J.D., DePinto, J., Rodgers, P., Effler, S., Sverdrup, H., and Warfvinge, P.: 1985, APMP Guidance Manual Volume I: APMP Research Requirements. U.S. Fish and Wildlife Service, Eastern Energy and Land Use Team, Biol. Report 80 (40.24), 207.

Saunders, W.P., Britt, D.L., Kinsman, J.D., DePinto, J., Rodgers, P., Effler, S., Sverdrup, H., and Warfvinge, P.: 1985, APMP Guidance Manual - Volume II: APMP Research Requirements. U.S. Fish and Wildlife Service, Eastern Energy and Land Used Team, Biol. Rep. 80(40.25), 197.

Schaffner, W.R., 1989. The effects of neutralization and addition of brook trout (Salvelinus fontinalis) on the limnetic zooplankton communities of two acidic lakes. *Can. J. Fish. Aquat. Sci.* **46**, in press.

Schafran, G.C. and Driscoll, C.T.: 1987, *Env. Sci. and Tech.* **21**: 988.

Scheffe, R.D., DePinto, J.V., Bilz, K.L., and West, K.A.: 1986, *Water, Air, and Soil Poll.*, **31**: 857.

Schierl, R. and Rodenkirchen, H.: 1987, Effects of artificial acid rain on Al, Cd, Mn and Zn in soil solutions and forest vegetation. In: Perry, R., Harrison, R.M., Bell, J.N.B. and Lester, J.N., 1987: Acid Rain: scientific and technical advances. 663.

Schindler, D.W.: 1988, *Science* **239**: 149.

Schneider, J.C.: 1986, "'Acid Rain': Significant to Michigan lakes and their Fisheries." *The Michigan Academician.* **18**: 1.

Schneider, J.C.: 1978, "Predicting the standing crop of fish in Michigan lakes" Michigan Department of Natural Resources. Fisheries Division. Fisheries Research Report No. 1860. October.

Schofield, C.L., Galloway, J.N., and Hendrey, G.R.: 1985, *Water, Air and Soil Poll.*, **26**: 403.

Schofield, C.L., Gloss, S.P., Plonski, B., and Spateholts, R.: 1989, *Can. J. Fish. Aquat. Sci.*, **46**: in press.

Schofield, C.L., et al.: 1987, "Production and growth efficiency of brook trout *(Salvelinus fontinalis)* in relation to food supply and water quality changes in two Adirondack mountain (New York) lakes following liming." *Can. J. Fish. Aquat. Sci.*, in press.

Schofield, C.L, Gloss, S.P. and Jospehson, D.: 1986, "Extensive evaluation of lake liming, restocking strategies, and fish population response in acidic lakes following neutralization by liming." U.S. Fish and Wildlife Service, Eastern Energy and Land Use Team. Interim Progress Report. NEC-86/18. 117.

Schofield, C.L. and Driscoll, C.T.: 1987, *Biogeochemistry* **3**: 63.

Schofield, C.L.: 1976, Acid precipitation: effects on fish. *Ambio* **5**: 228.

Schreiber, R.K., Goodyear, C.D., and Hartman, W.A.: 1988, Challenges and progress in the federal/state liming research program, *Lake Line* **8**: 8.

Schreiber, R.K. and Rago, P.J.: 1984, *Fisheries* **9**: 31.

Schreiber, R.K. and Britt, D.L.: 1987, *Fisheries* **12**: 2.

Schreiber, K.: 1988, *Water, Air and Soil poll.*, this volume.

Segdwick, S.D.: 1973 *Trout Farming Handbook*: Seeley, Service and Co., London, 14.

Seibt, G. and Reemtsma, J.B.: 1977, Ertragskundliche und bodenkundliche Ergebnisse langfristiger Kalkduengungsversuche im nord- und westdeutschen Bergland - weitere Kalkungsversuche. Schriftenr. Forstl. Fak. der Univ. Goettingen und NFV **50**: 89.

Seibt, G. and Wittich, W.: 1977, Eritragskundliche und bodenkundliche Ergebnisse langfristiger Duengungsversuche im nord- und westdeutschen Bergland. Der aelteste Kalkduengungsversuch in Neuenheerse und bodenkundliche Grundlagen. Schriftenr. Forstl. Fak. der Univ. Goettingen und NFV **50**: 1.

Sevaldrud, I.H., Muniz, I.P., and Kalvenes, S.: 1980, Loss of fish population in southern Norway. Dynamics and magnitude of the problem. In: Drabløs, D. and Tollan, A. (ed), Ecological impact of acid precipitation, 350, SNSF-project.

Sevaldrud, I.H. and Skogheim, O.K.: 1986, *Water, Air and Soil Poll.*, 30: 381.

Shepard, M.: 1986, "Restoring life to acidified lakes." EPRI Journal. April/May.

Shuter, B.J., MacLean, J.A., Fry, F.E.J., and Regier, H.A.: 1980, *Trans. Amer. Fish. Soc.* **109**: 1.

Simmons, K.R. and Halliwell, D.B.: 1986, Ecological and chemical responses of Whetstone Brook during continuous treatment with calcium carbonate. U.S. Fish and Wildlife Service, Nat. Ecol. Res. Center-Leetown, Kearneysville, WV. Draft Annual Rep.

Skogheim, O.K. and Rosseland, B.O.: 1984, Rep. Inst. Freshw. Res. Drottningholm **61**: 186.

SNV: 1981, Liming of lakes and rivers, Inf. Søtvattenslab. Drottningholm. Report 4/1981, 201 (in Swedish).

Sorg, C. and Loomis, J.: 1986, "Economic value of Idaho sports fisheries with an update on valuation techniques." NAJFM **6**: 494.

Speziale, B.J., Schreiner, S.P., Giammatteo, P.A. and Schindler, J.E.: 1984, Comparison of DMF, DMSO and acetone for extraction of phytoplankton chlorophyll. *Can. J. Fish. Aquat. Sci.* **41**: 1519.

Statens Energiverk and Statens Naturvårdsverk, 1987, Rapport 3412, 44 (in Swedish).

Staubitz, W.W, and Zarriello, P.: 1989, Basin and hydrologic characteristics of two acidic Adirondack lakes. *Can. J. Fish. Aquat. Sci.* **46**, in press.

Stokes, P.: 1986, Ecological effects of acidification on primary producers in aquatic ecosystems. *Water, Air and Soil Poll.* **30**: 421.

Stoner, J.H.: 1987, Llyn Brianne - Acid waters project. First Technical Summary Report, Welsh Water Authority, Llanelli, Dyfed SA15 4EL.

Sunde, S.E.: 1926, Surt vann dreper laks- og ørretyngel (Acid water kills fry of salmon and trout). Norsk Jæg.- og Fisk. foren. Tidsskr., **2**: 1 (in Norwegian).

Sutcliffe, D.W. and Carrick, T.W.: 1973, *Freshw. Biol.* **3**: 437.

Sverdrup, H.U.: 1983, *Chemica Scripta* **22**: 12.

Sverdrup, H.U.: 1984, *Chemica Scripta* **22**: 8.

Sverdrup, H.U.: 1985, Calcite Dissolution Kinetics and Lake Neutralization. Doctoral Dissertation. Eng. CODEN: LUTKDH/TKKT/1002/1-169/1985.

Sverdrup, H.U.: 1986, *Water, Air and Soil Poll.* **31**: 689.

Sverdrup, H.U.: 1986, *Water, Air, and Soil Poll.* **31**: 827.

Sverdrup. H.U. and Bjerle, I.: 1983, *Vatten* **39**: 41.

Sverdrup, H.U., Rasmussen, R. and Bjerle, I.: 1984, *Chemica Scripta* **24**: 53.

Sverdrup, H.U. and Warfvinge, P.: 1984, *Vatten* **40**: 340.

Sverdrup, H.U. and Warfvinge, P.: 1985, *Water Resources Research* **21**: 1373.

Sverdrup, H.U., Warfvinge, P. and Bjerle, I.: 1983, *Vatten* **40**: 164.

Sverdrup, H.U., Warfvinge, P. and Bjerle, I.: 1985a, Teori og beregningsexempel for kalkningen av en sur sjo (Theory and calculation example for lake liming). Technical Report, Dept. of Chemical Engineering II, Lund Institute of Technology.

Sverdrup, H.U., Warfvinge, P. and Fraser, J.: 1985b, The dissolution efficiencies of different stream liming methods. In: Hans C. Martin, *International Conference on Acid Precipitation*, Muskoka, Canada, Part 2, 827, Riedel Publishing Company.

Sverdrup, H.U., Warfvinge, P. and Molot, L.: 1985c, A Scandinavian model used to predict the reacidification of limed lakes in Nova Scotia and Ontario, Canada. In: Hans C. Martin, *International Conference on Acid Precipitation*, Muskoka, Canada, Part 2, 689, Riedel Publishing Company.

Tideström, H.: 1984, Kalkning av rinnande vatten; Utvärdering av kalkdoseringsä nlaggningar. National Environmental Protection Board, Rep. 1873, 128.

Turnpeny, A.W.H., Dempsey, .H., Davis, M.H., and Fleming, J.M.: 1988, *J. Fish Biol.* **32**: 101.

U.S. Environmental Protection Agency: 1983, *Methods for chemical analysis of water and wastes*. EPA-600/4-79-020. Office of Research and Development, Environmental Monitoring and Support Laboratory, Cincinnati, Ohio.

Ulrich, B., 1986, *Forstw. Cbl.* **105**: 421.

Ulrich, B. and Keuffel: 1970, *Forstarchiv* **41**: 25.

Umweltbundesamt: 1986, *Daten zur Unwelt.* 1986/87. 550.

USEPA 1983 (Revised) Methods for Chemical Analysis of Water and Wastes, Method 200.0, Atomic Absorption Methods. EPA 600/4-79. 020 Cincinnati.

Villella, R. and Schreiber, R.K.: 1986, *Lake Line* **6**: 30.

Violette, D.M.: 1985, "A model to estimate the economic impacts on recreational fishing in the Adirondacks from current levels of acidification." Presented at the AERE workshop on Recreation Demand Modeling, May.

VMND (Vereinigte-Wirtschafts-Dienste): 1987, Waldschaeden in den Niderlanden und Deutschland am staerksten. VWD - Landwirtschaft und Eraehrung **165**: 30.

Vollenweider, R.A.: 1969, A manual on methods of measuring primary production in aquatic environments. Blackwell Scientific Publications, Oxford and Edinburgh 213.

Waller, R.M. and Ayer, G.R.: 1975, Water Resources of the Black River Basin, New York. NYS Dept. Environ. Cons. Basin Planning Rpt., BRB-1, 205.

Warfvinge, P. and Sverdrup, H.: 1984, "Soil Liming and Runoff Acidification Mitigation" In: *Lake and Reservoir Management*, EPA 440/5-84-001. USEPA, Wash. D.C. 389.

Warfvinge, P. and Sverdrup, H.: 1988, *Water Resources Research* **24**: 701.

Warfvinge, P. and Sverdrup, H.: 1988a, Watershed liming, Proc. 7th Annual Conference of North American Lake Management Society, Orlando, Florida.

Warfvinge, P. and Sverdrup, H.: 1988b, Soil liming as a measure to mitigate acid runoff, *Water Resource Research,* in press.
41.5

Warfvinge, P., Sverdrup, H. and Bjerle, I.: 1984, Dissolution of calcite powder sinking in a column of acidic water in relation to a theoretical lake liming model. *Chemica Scrita* **24**: 67.

Warfvinge, P., Sverdrup, H., Norrström, A.C. and Jacks, G.: 1987, A model for dissolution of limestone in soils and neutralization of soil systems. In: *Acidification and water pathways*, 137, Norwegian National Committee for Hydrology, ISBN 82-554-0486-4.

Waters, T.F. and Ball, R.C.: 1957, Lime application to a soft-water, unproductive lake in northern Michigan, *J. Wildl. Mgmt.* **21**: 385.

Watt, W.D.: 1986, The case for liming some Nova Scotia salmon rivers, *Water, Air, and Soil Poll.* **31**: 775.

Weaver, T.: 1988, Tennessee APMP Annual Report 1987, Laurel Branch. U.S. Fish and Wildlife Service, Nat. Ecol. Res. Center-Leetown, Kearneysville, WV. Draft Annual Rep.

Wenzel, W. and Ulrich B.: 1988, *Forst u. Holz* **43**: 12.

West Virginia Department of Natural Resources: 1986, Performance report: acid water studies, project number F-24-R-5, 9.

Wetzel, R.G.: 1983, *Limnology.* 2nd Edition. W.B. Saunders and Company. New York, 767.

Wiederholm, T.: 1984, 508, In: Resh, V.H. and Rosenberg, D.M. (eds.), *The Ecology of Aquatic Insects*. Praeger, The Netherlands, 623.

Wiener, J.G. and Eilers, J.M.: 1987, *Lake and Reservoir Management* **3**: 365.

Williams, T.G. and Turpin, D.H.: 1987, Photosynthetic kinetics determine the outcome of competition for dissolved inorganic carbon by freshwater microalgae: implications for acidified lakes. *Oecologia* **73**: 307.

Willis, R.D.: 1976, "Consumer Surplus Without Apology," American Economic Review, **66**: 589.

Wilson, E.J., Roberts, D.J. and Skeffington, R.A.: 1988, The effect of lime application to the Loch Fleet catchment and on the soil chemical properties - an initial report. CEGB Report No. TPRD/L/3283/R88.

Wright, R.F.: 1982, Changes in the chemistry of Lake Hovvatn, Norway, following liming and reacidification. Acid Rain Research, NIVA. Report 6/1986, 68 (in Norwegian with English summary).

Wright, R.F.: 1984, Liming of Lake Hovvatn. Water quality prior to and following liming. Liming Project, Report 3-1982, 63 (in Norwegian).

Wright, R.F.: 1985, *Can. J. Fish Aquat. Sci.* **42**: 1103.

Wright, R.F. and Johannssen, H.: 1980, Input-output budgets of major ions at gauged catchments in Norway. In: Drablos, D. and Tollan, A. (Eds.), Proc, Int. conf. Ecol. Impact Acid Precip, Sandefjor, SNFC Project, Oslo-As, Norway, 440.

Wright, D., Lawrenz, R., Popp, W. and Dank, M.: 1988, *A cooperative study of the ecological response of Thrush Lake, Cook County, Minnesota, to mitigative/protective base addition.* U.S. Fish and Wildlife Service, Nat. Ecol. Res. Center-Leetown, Kearneysville, WV. Draft Annual Rep.

Wright, R.F., Lotse, E. and Semb., A: 1988, *Nature* **334**: 670.

Wright, R.F. and Skogheim, O.K.: 1983, *Vatten* **39**: 301.

Yan, N.D. and Dillion, P.J.: 1984, Experimental neutralization of lakes near Sudbury, Ontario. In: J. Nriagu (ed.) Environmental Impacts of Smelters. John Wiley and Sons, Toronto, Canada.

Yan, N.D., and Miller, G.E.: 1984, 'Effects of deposition of acid and metals on chemistry and biology of lakes near Sudbury, Ontario.' In: J.O. Nriagu (ed.), *Environmental Impact of Smelters*, Advances in environmental Science and Technology, **15**, John Wiley and Sons, New York, 243.

Yan, N.D., Scheider, W.A., and Dillon, P.J.: 1977, 'Chemical and biological changes in Nelson Lake, Ontario following experimental elevation of lake pH', Proc. 12th Can. Symp.: 1977, *Water Pollution Reserch* Canada. **213**.

Yan, N., 1979. Phytoplankton, community of an acidified, heavy metal contaminated lake near Sudbury, Ontario: 1973-1977. *Water, Air and Soil Poll.* **11**: 43.

Young, T.C., DePinto, J.V., Rhea, J.R. and Scheffe, R.D.: 1989, Calcite dose determination, treatment efficiency and sediment dose response to whole lake neutralization. *Can. J. Fish. Aquat. Sci.* **46**, in press.

Zoettl, H.W., Feger, K.H. and Brahmer, G.: 1987, Projekt Arinus: I. Zielsetzung und Auisgangslage. KFK-PEF-Berichte, **12**: 269.

Zoettl, H.W. and Huettl, R.F.: 1986, Nutrient supply and forest decline in southwest-Germany. *Water, Air, and Soil Pollut.*, **31**: 449.

Zoettl, H.W.: 1987, *Forstw. Cbl.* **106**: 105.

Zurbuch, P.: 1984, *Fisheries* **9**: 42.

Zurbuch, P., Menedez, R. and Woodrum, J.E.: 1988, *West Virginia Dogway Fork Project, Cooperative Acid Precipitation Mitigation Program.* U.S. Fish and Wildlife Service, Nat. Ecol. Res. Center-Leetown, Kearneysville, WV. Draft Annual Rep.

Zwezschwitz, V.E.: 1985, *Forstw. Cbl.* **104**: 205.

INDEX OF AUTHORS

Adams, T.B.
Living Lakes, Inc.
1000 Vermont Avenue, N.W.
Suite 900
Washington, D.C. 20005
U.S.A.; 137

Arnold, D.E.
U.S. Fish and Wildlife Service
Pennsylvania Cooperative Fish and Wildlife Research Unit
7 Ferguson Building
University Park, PA 16802
U.S.A.; 385

Booth, G.M.
BAR Environmental
Nicholas Beaver Park
Brock Road, R.R. #3
Guelph, Ontario N1H 6H9
Canada; 75

Brocksen, R.W.
Living Lakes, Inc.
1000 Vermont Avenue, N.W.
Suite 900
Washington, D.C. 20005
U.S.A.; 85, 137, 455

Brown, D.J.A.
Central Electricity Research Laboratories
Kelvin Avenue
Leatherhead
Surrey KT22 7SE
United Kingdom; 25, 409

Bukaveckas, P.A.
Department of Biology
Indiana University
Bloomington, IN 47405
U.S.A.; 223

Casselman, J.M.
Ontario Ministry of Natural Resources
Lake Ontario Fisheries Station, R.R. 4
Picton, Ontario P3E 5P9
Canada; 113

Davis, J.E.
Environmental Sciences Consultant
2311 Carlmont Drive, Suite #2
Belmont, CA 94002
U.S.A.; 435

Dalziel, T.R.K.
Central Electricity Research Laboratories
Kelvin Avenue
Leatherhead
Surrey KT22 7SE
United Kingdom; 25, 417

Dickson, A.
Central Electricity Research Laboratories
Kelvin Avenue
Leatherhead
Surrey KT22 7SE
United Kingdom; 417

Dodge, D.P.
Fish Community and Habitat Section
Ontario Ministry of Natural Resources
99 Wellesley Street West
Toronto, Ontario M7A 1W3
Canada; 75

Emler, Jr., P.W.
Living Lakes, Inc.
1000 Vermont Avenue, N.W.
Suite 900
Washington, D.C. 20005
U.S.A.; 85

Feger, K-H.
Institute of Soil Science
asnd Forest Nutrition
University of Freiburg
Bertoldstrasse 17
D-7800 Freiburg i. Br.
West Germany; 455

Gloss, S.P.
Water Research Center
University of Wyoming
Laramie, WY 82071
U.S.A.; 241

Greening, H.S.
International Science &
Technology, Inc.
11260 Roger Bacon Drive
Reston, VA 22090
U.S.A.; 359

Gunn, J.M.
Fisheries Branch
Ontario Ministry of Natural
Resources
Queen's Park
Toronto, Ontario M7A 1W3
Canada; 113

Hindar, A.
Norwegian Institute for
Water Research
Regional Office - Sorlandet
Televeien 1, N-4890 Grimstad
Norway; 17, 165

Howells, G.D.
Central Electricity Research
Laboratories
Kelvin Avenue
Leatherhead
Surrey KT22 7SE
United Kingdom; 25

Huettl, R.F.
Institute of Soil Science
and Forest Nutrition
University of Freiburg
Bertoldstrasse 17
D-7800 Freiburg i. Br.
West Germany; 95, 455

Ivahnenko, T.I.
Department of Geology and Geography
West Virginia University
425 White Hall
Morgantown, WV 26506
U.S.A.; 311

Janicki, A.
International Science & Technology,
Inc.
11260 Roger Bacon Drive
Reston, VA 22090
U.S.A.; 359

Keller, W.
Ontario Ministry of the Environment
199 Larch Street
Sudbury, Ontario P3E 5P9
Canada; 75, 113

Marcus, M.D.
Western Aquatics, Inc.
203 Grand Avenue
P.O. Box 546
Laramie, WY 82070
U.S.A.; 279

McMurtry, M.J.
Ontario Ministry of Natural Resources
Queen's Park
Toronto, Ontario M7A 1W3
Canada; 113

Nyberg, P.
Tage Erlander Laboratory
Box 51
S-684 01 Munkfors
Sweden; 3

Porcella, D.B.
Electric Power Research Institute
3412 Hillview Avenue
Palo Alto, CA 94304
U.S.A.; 43, 455

Powell, M.J.
Fish & Wildlife
Ontario Ministry of Natural Resources
Sudbury, Ontario P3E 5P9
Canada; 113

Proctor, M.V.
Central Electricity Research Laboratories
Kelvin Avenue
Leatherhead
Surrey KT22 7SE
United Kingdom; 417

Rauch, H.W.
Department of Geology and Geography
West Virginia University
425 White Hall
Morgantown, WV 26506
U.S.A.; 331

Renton, J.J.
Department of Geology and Geography
West Virginia University
425 White Hall
Morgantown, WV 26506
U.S.A.; 331

Richman, L.A.
Ontario Ministry of the Environment
135 St. Clair Avenue West
Suite 100
Toronto, Ontario P3E 5P9
Canada; 75

Riely, P.L.
Sport Fishing Institute
1010 Massachusetts Avenue, N.W.
Washington, D.C. 20001
U.S.A.; 293

Rosseland, O.
Norwegian Institute for Water Research
P.O. Box 33 - Blindern,
N-0313 Oslo 3
Norway; 17, 165

Rockland, D.B.
Sport Fishing Institute
1010 Massachusetts Avenue, N.W.
Washington, D.C. 20001
U.S.A.; 293

Schofield, C.L.
Department of Natural Resources
Cornell University
Ithaca, NY 14853
U.S.A.; 241

Schreiber, R.K.
National Ecology Research Center
U.S. Fish and Wildlife Service
Box 705
Kearneysville, WV 25430
U.S.A.; 53

Sherman, R.E.
Department of Natural Resources
Cornell University
Ithaca, NY 14853
U.S.A.; 241

Skinner, W.D.
The Pennsylvania State University
Pennsylvania Cooperative Fish and Wildlife Research Unit
7 Ferguson Building
University Park, PA 16802
U.S.A.; 385

Spotts, D.E.
Pennsylvania Fish Commission
450 Robinson Lane
Bellefonte, PA 16802
U.S.A.; 385

Stewart, B.R.
Central Electricity Research Laboratories
Kelvin Avenue
Leatherhead
Surrey KT22 7SE
United Kingdom; 25

Sverdrup, H.
Department of Chemical Engineering
Lund Institute of Technology
Box 124
Chemical Center, S-221 00 Lund
Sweden; 189

Thørneløf, E.
National Environmental Protection Board
S-171 24 Solna
Sweden; 3

Tomassini, F.D.
Ontario Ministry of the Environment
P.O. Box 213
125 Resources Road
Rexdale, Ontario M9M 5L1
Canada; 75

Warfvinge, P.
Department of Chemical Engineering
Lund Institute of Technology
Box 124
Chemical Center, S-221 00 Lund
Sweden; 189

Wisniewski, J.
Wisniewski and Associates, Inc.
6862 McLean Province Circle
Falls Church, VA 22043
U.S.A.; 455

Zoettl, H.W.
Institute of Soil Science and Forest Nutrition
University of Freiburg
Bertoldstrasse 17
D-7800 Freiburg i. Br.
West Germany; 455

SUBJECT INDEX

Al
  toxic metal, 3, 17, 25, 43, 53, 75, 137, 165, 189, 279, 359, 409
ALaRM model, 43, 435
Alkalinity measurements, 3
Anadromous fish, 359
Atmospheric pollutants
  HNO_3, 95
  hydrocarbons, 3
  NH_3, 3
  NH_x, 95
  NO_x, 3, 95
  O_3, 95
  S, 3
  SO_2, 95

Bass, 75, 113
Beetles, water, 25
Benefit-cost analysis, 293

Ca
  effects on sphagnum, 25
Ca concentration, 3, 17, 25, 43, 53, 75, 95, 137, 165, 241, 279, 331, 359, 417
Canada
  liming program, 75, 113
Chlorophyll levels, 223
Cl concentration, 137
Color measurements, 3, 75
Conductivity measurements, 3, 75, 137
Cray fish, 3

DeAcid model, 75, 137, 359, 435

F concentration, 53
Fe
  toxic metal, 3
Fertilizer
  forest treatment, 95, 455
Fish, 3, 17, 25, 43, 53, 75, 85, 113 165, 189, 223, 279, 359, 385, 409
Fisheries, 137
Forest
  nutritional disturbance, 95, 455
Forest liming
  Living Lakes, 455
  Loch Fleet, 25
  West Germany, 95, 455

Groundwater
  nitrate, 95

Hg concentration, 3
HNO_3
  atmospheric pollutant, 95
Hydrocarbons
  atmospheric pollutant, 3

ILWAS model, 435
Insects, 43
Invertebrates, 165, 385

K concentration, 95

Lake liming
  Canada, 75, 113
  LAMP, 43, 223
  Living Lakes, 85, 137, 279, 359, 455
  Norway, 17, 165
  Sweden, 3, 189
  US, 53, 85, 137, 223, 241, 455
  West Germany, 95, 455
LAMP
  liming program, 43, 223
Liming
  benefit cost analysis, 293
Liming methods, 189, 385
Liming programs
  APMP, 53
  Canada, 75, 113
  LAMP, 43, 223

Living Lakes, 85, 137, 279, 359 455
Loch Fleet, 25, 409, 417
NAPAP, 53
Norway, 17, 165
Sweden, 3, 189
US, 53, 85, 137, 223, 241, 331, 359, 385, 455
West Germany, 95, 455

Living Lakes
liming programs, 85, 137, 279, 359, 455

Loch Fleet
liming programs, 25, 409, 417

Maine
liming costs, 293

Mayfly nymphs, 25

Metal
Al, 3, 17, 25, 43, 53, 75, 137, 165, 189, 279, 359, 409
Ca, 3, 17, 25, 43, 53, 75, 95, 137, 165, 241, 279, 331, 359, 417
F, 53
Fe, 3
Mg, 3, 95
Mn, 3, 43

Mg concentration, 3, 95

Michigan
liming costs, 293

Minnesota
liming costs, 293

Mn
toxic metal, 3, 43

Model
ALaRM, 43, 435
DeAcid, 75, 137, 359, 435
ILWAS, 435, 455
SLIM, 435
Sverdrup, 189, 241

N, total
concentration, 53, 137, 279

New Hampshire
liming costs, 293

New York
liming costs, 293

NH_3
atmospheric pollutant, 3

NH_x
atmospheric pollutant, 95

Nitrate
groundwater, 95

NO_x
atmospheric pollutants, 3, 95, 189, 279

Norway
liming program, 17, 165

O_3
atmospheric pollutant, 95

Ontario
liming program, 75, 113

P, total
concentration, 53, 95, 137

Pennsylvania
liming costs, 293

Perch, 17, 75, 113

Phytoplankton, 43, 75, 223

River liming
Norway, 17, 165
Sweden, 3

S
atmospheric pollutant, 3, 189, 223

Salmon, 3, 53, 165, 189

Sampling program, 3, 53, 75, 85, 113, 137, 165, 223, 279, 331, 359, 385

Scotland
liming program, 25, 409, 417

Sediment
lake, 25
stream, 331

Simulation models, 435

SLIM model, 435

SO_2
atmospheric pollutant, 95
Soil
leaching, 95
liming, 189, 455
Soil chemistry, 25, 417, 455
Sphagnum
Ca effects, 25
Stream chemistry, 417
Stream liming
Loch Fleet, 25, 409, 417
Scotland, 25, 409, 417
Sweden, 3, 189
US, 53, 75, 331, 359, 385
Sucker, 53, 223
Sweden
liming programs, 3, 189
Sverdrup model, 189, 241

Trout, 3, 17, 25, 53, 75, 113, 165, 223, 331, 385, 417

US
liming program, 53, 85, 137, 223, 241, 331, 359, 385, 455

Watershed liming
Norway, 165
Scotland, 25, 409, 417
simulation models, 435
US, 455
West Germany, 455
Water quality
effects of liming, 279
parameters, 53
West Germany
liming program, 95, 455
West Virginia
liming costs, 293

Zn
toxic metal, 43, 279
Zoobenthos, 75
Zooplankton, 43, 53, 75

ANNOUNCEMENT

RESTORATION OF AQUATIC AND TERRESTRIAL SYSTEMS

Proceedings of a Special Water Quality Session Dealing with the Restoration of Acidified Waters in conjunction with the Annual Meeting of the North American Fisheries Society held in Toronto, Ontario, Canada, 12–15 September 1988.

Guest Editors

ROBERT W. BROCKSEN
Living Lakes, Inc., Washington, D.C., U.S.A.

and

JOE WISNIEWSKI
Wisniewski and Associates, Inc., Virginia, U.S.A.

Please note that a hardbound edition of this special issue of *Water, Air, and Soil Pollution,* Vol. 41, Nos. 1–4 (September/October 1988), is available from the publishers.

ISBN 0–7923–0111–0 Prices: Dfl. 200,–/$99.00/£64.00